T0191951

Sustainable Textiles: Production, Processing, Manufacturing & Chemistry

Series Editor

Subramanian Senthilkannan Muthu, Head of Sustainability, SgT and API, Kowloon, Hong Kong

This series aims to address all issues related to sustainability through the lifecycles of textiles from manufacturing to consumer behavior through sustainable disposal. Potential topics include but are not limited to: Environmental Footprints of Textile manufacturing; Environmental Life Cycle Assessment of Textile production; Environmental impact models of Textiles and Clothing Supply Chain; Clothing Supply Chain Sustainability; Carbon, energy and water footprints of textile products and in the clothing manufacturing chain; Functional life and reusability of textile products; Biodegradable textile products and the assessment of biodegradability; Waste management in textile industry; Pollution abatement in textile sector; Recycled textile materials and the evaluation of recycling; Consumer behavior in Sustainable Textiles; Eco-design in Clothing & Apparels; Sustainable polymers & fibers in Textiles; Sustainable waste water treatments in Textile manufacturing; Sustainable Textile Chemicals in Textile manufacturing. Innovative fibres, processes, methods and technologies for Sustainable textiles; Development of sustainable, eco-friendly textile products and processes; Environmental standards for textile industry; Modelling of environmental impacts of textile products; Green Chemistry, clean technology and their applications to textiles and clothing sector; Eco-production of Apparels, Energy and Water Efficient textiles. Sustainable Smart textiles & polymers, Sustainable Nano fibers and Textiles; Sustainable Innovations in Textile Chemistry & Manufacturing; Circular Economy, Advances in Sustainable Textiles Manufacturing; Sustainable Luxury & Craftsmanship; Zero Waste Textiles.

More information about this series at https://link.springer.com/bookseries/16490

Ali Khadir · Subramanian Senthilkannan Muthu
Editors

Biological Approaches in Dye-Containing Wastewater

Volume 1

Editors
Ali Khadir
Western University
London Ontario, ON, Canada

Subramanian Senthilkannan Muthu
SgT Group and API
Hong Kong, Kowloon, Hong Kong

ISSN 2662-7108 ISSN 2662-7116 (electronic)
Sustainable Textiles: Production, Processing, Manufacturing & Chemistry
ISBN 978-981-19-0547-6 ISBN 978-981-19-0545-2 (eBook)
https://doi.org/10.1007/978-981-19-0545-2

This Springer imprint is published by the registered company Springer Nature Singapore Pte Ltd.
The registered company address is: 152 Beach Road, #21-01/04 Gateway East, Singapore 189721, Singapore

Contents

About the Editors

Ali Khadir is an environmental engineer and a member of the Young Researcher and Elite Club, Islamic Azad University of Shahre Rey Branch, Tehran, Iran. He has published several articles and book chapters in reputed international publishers, including Elsevier, Springer, Taylor & Francis, and Wiley. His articles have been published in journals with IF of greater than 4, including Journal of Environmental Chemical Engineering and International Journal of Biological Macromolecules. He also has been the reviewer of journals and international conferences. His research interests center on emerging pollutants, dyes, and pharmaceuticals in aquatic media, advanced water, and wastewater remediation techniques and technology.

Dr. Subramanian Senthilkannan Muthu currently works for SgT Group as Head of Sustainability, and is based out of Hong Kong. He earned his Ph.D. from The Hong Kong Polytechnic University, and is a renowned expert in the areas of Environmental Sustainability in Textiles & Clothing Supply Chain, Product Life Cycle Assessment (LCA), and Product Carbon Footprint Assessment (PCF) in various industrial sectors. He has five years of industrial experience in textile manufacturing, research and development and textile testing and over a decade of experience in life cycle assessment (LCA), carbon and ecological footprints assessment of various consumer products. He has published more than 100 research publications, written numerous book chapters, and authored/edited over 100 books in the areas of Carbon Footprint, Recycling, Environmental Assessment, and Environmental Sustainability.

Overview of Biological Technologies for Azo Dye Removal

L. P. Silva Júnior, I. R. M. Câmara, A. B. S. da Silva, F. M. Amaral, F. Motteran, B. S. Fernandes, and S. Gavazza

Abstract The textile industry segment has been continuously expanded and is one of the most expressive consumers of chemical products, dyes, and water. As a result, the sector is responsible for generating a large amount of industrial wastewater. These effluents are highly toxic and potentially carcinogenic and mutagenic. Azo dyes are widely used in the textile industry; thus, they are commonly found in its wastewater. This chapter approaches the main processes of azo dyes removal and biodegradation, highlighting the most used and effective reactor configurations. While anaerobic reactors are most efficient for combined color and organic matter removal, aerobic reactors are required to mineralize aromatic amines, byproducts of azo dye degradation. Therefore, combination of anaerobic and aerobic reactors is desired. The following topics are also covered: the influence of operational parameters; effectiveness of applying one- or two-stages (aerobic and anaerobic) processes; and metabolism, stoichiometry, and byproduct formation. Sequencing Batch Reactors (SBR) are the most used one-stage system, showing good removal efficiencies. Upflow Anaerobic Sludge Blanket (UASB) and Expanded Granular Sludge Blanket (EGSB) reactors were effectively used as anaerobic reactors, while different biofilm aerobic reactors showed excellent performance on removing aromatic amines.

Keywords Textile wastewater · Effluent treatment · Recalcitrant effluent · Aromatic amines · Environmental contamination · Biodegradation · Anaerobic–aerobic degradation · Reactor configuration · Two-stage systems · Metabolism of microorganisms

1 Introduction

In 2020, the global textile market was estimated at US$1000 billion, with expected growth given in Compound Annual Growth Rate (CAGR) of 4.4% from 2021 to

L. P. Silva Júnior · I. R. M. Câmara · A. B. S. da Silva · F. M. Amaral · F. Motteran · B. S. Fernandes · S. Gavazza (✉)
Department of Civil Engineering, Federal University of Pernambuco (UFPE), Acadêmico Hélio Ramos Avenue, s/n, Recife, PE 50,740–530, Brazil
e-mail: savia.gavazza@ufpe.br

A. Khadir et al. (eds.), *Biological Approaches in Dye-Containing Wastewater*, Sustainable Textiles: Production, Processing, Manufacturing & Chemistry,
https://doi.org/10.1007/978-981-19-0545-2_1

2028. Pacific Asia alone accounted for over 47% of the global revenue in 2020 [48]. China is the world leader in the production and export of textile and clothing raw materials, followed by the United States, which is considered the main producer and exporter of raw cotton, in addition to being the largest importer of raw textiles and clothing [83]. India is considered the third-largest textile manufacturing industry [83]. The largest textile producer countries in the world, by region, are the United States in North America, Brazil in Central and South America, Germany in Europe, Saudi Arabia in the Middle East & Africa, and China in the Asia Pacific [48].

The textile segment is one of the most expressive consumers of chemical products, dyes, and water. As a result, this sector is responsible for discharging a large amount of highly toxic and potentially carcinogenic and mutagenic wastewater [11, 47]. This wastewater is characterized by high contents of organic matter, suspended solids, chemical oxygen demand (COD), salinity, and frequently, sulfate. Low biochemical oxygen demand (BOD) and COD ratio, together with intense color, complement the composition and give so much challenge to treat this wastewater [29, 121].

The most used dyes in the industries, among the organic dyes, are of the azo type. Characterized by the chromophore azo bond (-N = N-) in the molecular structure, they present toxic characteristics. In addition, their biodegradation leads to the formation of aromatic amines, which are often equally or more toxic to the environment than the dye itself [2, 5, 90].

In order to treat the wastewater from the textile industry, physicochemical processes have been frequently studied and applied [84, 98, 99, 115]. However, these technologies present high energy demand and maintenance cost and may not mineralize the toxic compounds, which makes them less viable for low-income regions. These treatment methods also generate large amounts of non-inert chemical sludge, which increases the cost of treatment and disposal and makes the process often not feasible for small producers and small-scale industries [60].

In this context, biological processes are interesting alternatives because they are low-cost and ecologically friendly technologies, removing color and reducing toxicity, frequently achieving compound biomineralization [121]. Among the existing technologies, the following stand out: anaerobic processes [4, 18, 39], aerobic processes [43, 51], and more recently, since the 90 s, the combination of anaerobic and aerobic processes [9, 121] has been studied applying one-stage [7, 8, 16] and two-stage systems [1, 12], 36; [58, 87, 94, 112]. These solutions become interesting, since the biological degradation process of azo dyes starts in an anaerobic environment for the reductive cleavage of azo bonds, leading to color removal. Nonetheless, the byproducts from azo dye degradation are often hazardous recalcitrant compounds that are better degraded in an aerobic environment [17].

Finally, this chapter elucidates the main technologies for the biological treatment of effluents from the textile industry, combining aspects of engineering, chemistry, and biological process.

2 Anaerobic Reactors

2.1 *Color Removal of Azo Dye Textile Effluent in Anaerobic Reactors*

The mineralization of azo dyes begins with the reductive cleavage of the azo bond, which occurs more easily under anaerobic conditions due to the low redox potential (≤ -50 mV) [11, 79]. This reaction occurs with the aid of the enzyme azoreductase [29]. In this process, breaking the double bond between nitrogen atoms involves the transfer of 4 electrons, as shown in Fig. 1.

In the reductive cleavage process, the azo dye acts as the electron acceptor, due to the electron withdrawing capacity of the azo bond. The presence of an electron donor is required and reducing equivalents, biologically formed by the fermentation of easily degrading organic matter, commonly perform this role [2]. The addition of co-substrates is encouraged in order to increase the amount of reducing equivalents available in the environment [28, 29, 103]. Nevertheless, their use in excessive concentrations can cause unwanted effects on the system, such as decreased color removal efficiency [22].

The use of redox mediators can improve the dye degradation process [29, 102]. Since in anaerobic mixed cultures, the production of reducing equivalents is not a limiting factor, the electron transfer to the azo bond can become the drawback. In this context, redox mediators are chemical compounds that have low molecular weight and act as accelerators of azo bond cleavage. First, the mediators are enzymatically reduced using the available organic matter as the electron donor. Then, there is a chemical transfer of the captured electrons to the azo dye, resulting in the recovery of the mediator in its oxidized form [19, 20].

Given the electrophilic characteristic of the azo bond ($-N = N-$), that is, its high electron affinity, azo dyes are better degraded under oxygen-free conditions [93, 111]. This happens because oxygen in its free form (O_2) or in its combined form competes with the dye, as an acceptor, for the electrons of the organic matter. As a result, organic matter removal may be high, but color removal may be low [2]. The presence of color in textile effluents is not only an aesthetic problem but also an

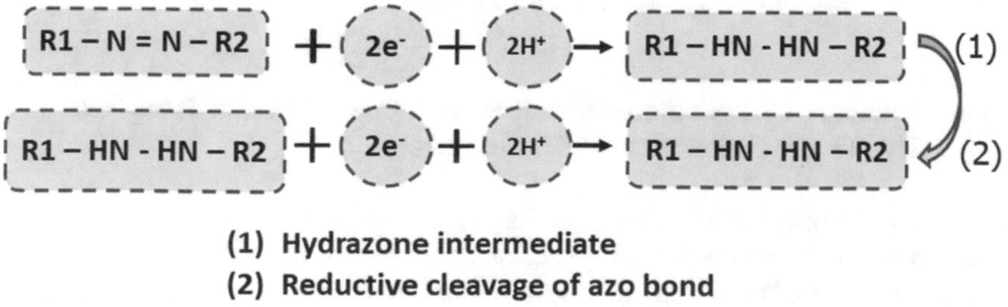

R1 – N = N – R2 + 2e⁻ + 2H⁺ → R1 – HN - HN – R2 (1)

R1 – HN - HN – R2 + 2e⁻ + 2H⁺ → R1 – HN - HN – R2 (2)

(1) Hydrazone intermediate

(2) Reductive cleavage of azo bond

Fig. 1 Electron requirements for reactions steps of reductive cleavage of azo bonds

environmental one, with negative impacts on water bodies, inhibiting photosynthesis in water. Thus, high color removal efficiency is desirable and the anaerobic reactors are the most used for this purpose.

2.2 Configuration of the Anaerobic Treatment Systems

Anaerobic reactors are used under different configurations, such as anaerobic filter, packed-bed, anaerobic rotating disc, fluidized bed reactor (FBR), structured bed reactor (SBR), UASB (Upflow Anaerobic Sludge Blanket) reactor and expanded granular sludge bed (EGSB) reactor [121]. This chapter will mainly focus on the performance of UASB reactors for textile wastewater treatment, since it is the most used configuration with this purpose. Figure 2 shows a didactic schematic of the most used anaerobic reactors.

The UASB reactor is a widely used anaerobic reactor configuration for color and organic matter removal in textile wastewater [11, 18, 20, 28, 103]. Initially designed for treating industrial effluents at mesophilic temperatures (between 20 and 45 °C), it is also used for treating domestic effluents [40]. In a UASB reactor, the effluent passes through a dense blanket of anaerobic sludge, in an upward flow, reaching a three-phase separation unit (solid/liquid/gas) [71]. Due to the high biomass concentration in the reactor, it tolerates high organic loads and is widely used in the treatment of complex effluents [29]. When treating highly toxic compounds, such as azo dyes, these reactors. can reach high treatment efficiency, ensuring stability [2, 20, 28, 103].

As a result of low sludge production, the generation of energy in the form of biogas, and its low costs in comparison to aerobic treatments, the UASB reactor is considered a sustainable technology. Furthermore, it is considerably versatile and goes from small to large scales [70]. Mechanical agitation is not used in UASB reactors to avoid disaggregation and granule shear. Additionally, biogas production or effluent recycling can promote contact between biomass and effluent.

The study on textile effluent treatment in anaerobic reactors has advanced during the 1980s and 1990s [9]. Nonetheless, isolating these azo dye-degrading cultures remains a challenging task. Still, studies with anaerobic reactors inoculated with mixed cultures have been widely used as a single treatment for textile effluents during the 1990s and 2000s [20, 22, 28, 103, 122].

2.2.1 Influence of Operational Conditions on Color Removal from Azo Dye Textile Effluent by UASB Reactors

Modifying the operational settings of the reactor can optimize color removal efficiency and should be chosen with caution to achieve a more robust treatment of complex effluents. When high color removal efficiencies are the target, the optimal

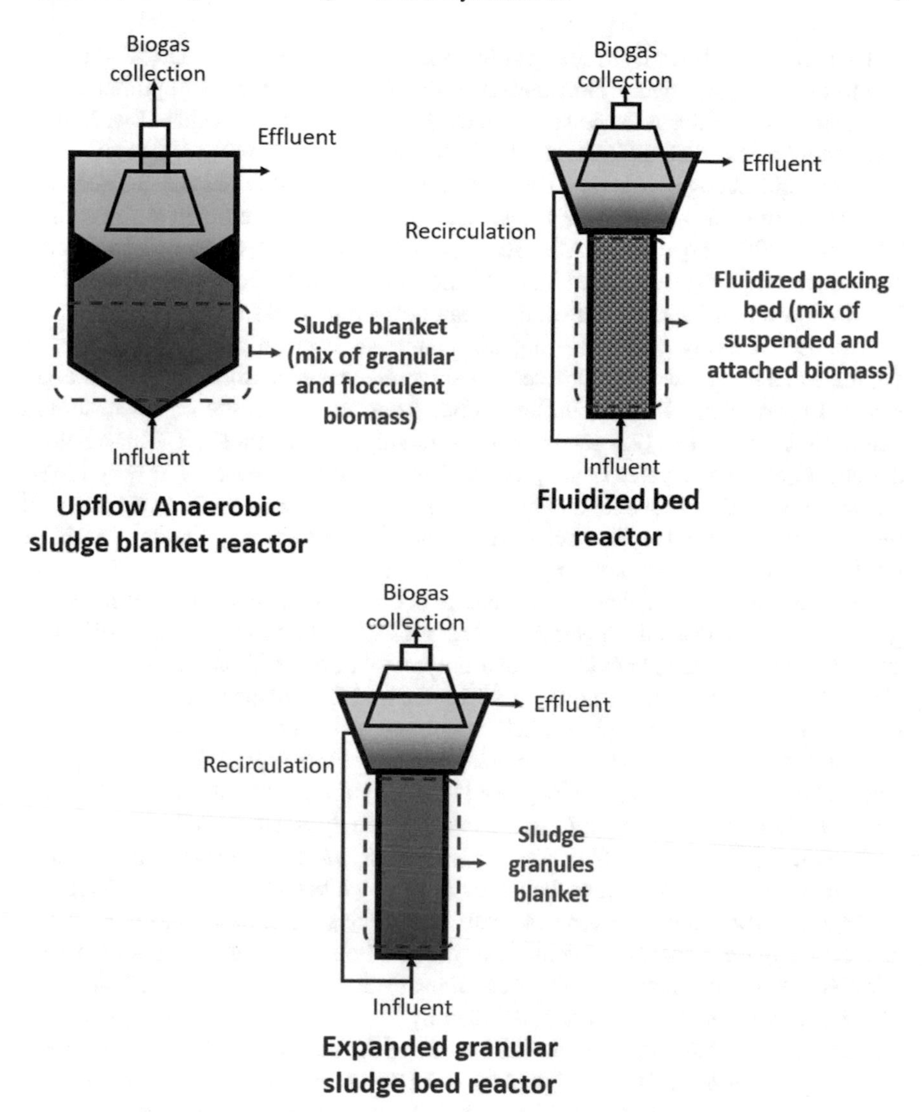

Fig. 2 Common anaerobic reactor configurations used for degradation of azo dyes

conditions necessary for dye cleavage must be considered, such as hydraulic retention time (HRT), the choice of co-substrate, the use of a redox mediator, process temperature, initial dye concentration, and others [121].

The higher the hydraulic retention time (HRT), the longer the biomass will be in contact with the effluent. When HRT is increased, the organic load applied to the reactor decreases, and the overall removal efficiency is expected to increase. Razo-Flores et al. [103] assayed 160 mL UASB reactors with 75 mg/L (0.25 mM*) of the azo dye azodisalicylate, using glucose as co-substrate. An increase in HRT from 8 to 16 h and to 26 h led dye removal efficiencies to increase from 63.4% to 69.1% and to 88.9%, respectively. Nonetheless, a decrease in HRT below optimal values can lead to a significant decrease in removal rates, but when combined with other parameters it can yield different results. In the study of [20], using 100 mg/L of the azo dye Acid Orange 7 (AO7) (0.29 mM*) in a 1.3 L UASB reactor, the decrease in HRT from 6 to 4 h and to 2 h decreased color removal efficiency from 98 to 90% and to 70%, respectively. After adding 3 μM (1.1 mg/L*) of the redox mediator Anthraquinone-2,6-disulfonate (AQDS), color removal was above 90% for all HRTs. The configurations used by both authors can be seen in Table 1.

The use of easily-degradable organic matter as electron donor is indicated to facilitate azo bond cleavage. Different co-substrates provide different effluent color removal rates. Table 1 shows different configurations of the use of co-substrates. Chinwetkitvanich et al. [22] used tapioca starch as a co-substrate in UASB reactors, at 200 and 500 mg/L, when treating real effluents. Color removals of 58% and 57%, respectively, of the effluent containing blue dye, and 52% and 56%, respectively, of the effluent with red dye were achieved, suggesting these concentrations should be used for the decolorization of 150 color SU (space units).

Donlon et al. [28] inoculated two separate 160 mL UASB reactors with anaerobic granular sludge and used either glucose (1.3 g COD/L) or a mixture of volatile fatty acids (VFAs)—acetate, propionate, and butyrate—(1.5 g COD/L) as co-substrate. The authors reported the removal of 95.1% and 91.2%, respectively, of 0.35 mM (101 mg/L*) and 0.18 mM (52 mg/L*) of influent Mordant Orange 1 (MO1). When no co-substrates were received, the reactor failed after 50 days. The use of 10 g/L of dextrin (19.8 mM*) and 5 g/L of peptone (72 mM*) as co-substrate in 500 mL serum bottles by [11] treating 300 mg/L of the reactive azo dyes Black 5 (0.30 mM*), Red 2 (0.49 mM*), Red 120 (0.22 mM*), Yellow 3 (0.51 mM*), Yellow 15 (0.47 mM*) and Yellow 17 (0.44 mM*), resulted in color removal between 77.8% and 97.1%.

Redox mediators can be used as catalysts for the reductive cleavage process of azo dyes. Figure 3 shows a schematic of the azo bond cleavage process combined with the use redox mediators. Their use improved the decolorization efficiency of a synthetic effluent in the study performed by Dos Santos et al. [30]. The authors reported the speed up in the decolorization rate of 0.3 mM of azo dyes (Reactive Red 2 (RR2) (185 mg/L*), AO7 (105 mg/L*), or MY10 (110 mg/L*), by up to eightfold compared to the mediator-free bottles in an EGSB reactor, with all three redox mediators (RM) used, Anthraquinone-2,6-disulfonate (AQDS), anthraquinone-2-sulfonate (AQS) and riboflavin (vitamin B2). Cervantes et al. [20] also reported a decolorization increase from 70% to above 95% of 100 mg/L of azo AO7 (0.29 mM*), with different doses of ADQS, 3, 10 and 30 μM (1.1, 3.7 and 11.0 mg/L*) in a UASB reactor, even at low HRTs (2 h). Similar results were reported by [4], treating a synthetic effluent with 0.06 mM (65 mg/L*) of the azo dye Direct Black 22 (DB22) in a UASB reactor, using 0.012 mM of Lawsone (2 mg/L*) and Riboflavin (45 mg/L*) as redox mediators. The use of Lawsone led to 90% of effluent decolorization and 87% of Riboflavin decolorization. However, when tested in real textile wastewater, the authors reported color removal of only 23% with Lawsone (when used with Sucrose as electron donor) and

Table 1 Important operational parameters of different anaerobic systems used for degradation of

Reactor type	Azo dye and concentration[a]	COD (mg/L)	HRT (hours)	Electron donors	Redox mediators	Color removal (%)	COD removal (%)	Reference
160 mL serum flasks	MO1 (0.18–0.35 mM)	910–1420	0.31–0.34 (days)	Acetate, propionate and butyrate, or glucose	–	91.8–95.1	81.8–86.6	Donlon et al. [28]
160 mL glass UASB bioreactors	Azodisalicyate (25–75 mg/L)	1000–3000	8, 16, and 26	Acetate, propionate and butyrate or glucose	–	72.8–98.8	84.1–94.8	Razo-Flores et al. [103]
500 mL serum flasks	RBk 5; RY 15; RR 2; RR 120; RY 3; RY 17 (300 mg/L)	–	Batch assays	dextrin, peptone	–	77.8–97.1	–	Beydilli et al. [11]
120 mL serum flasks	AO 7; AR 266; AY 137; AY 159; BR 23; DBk 19; DBk 22; DBl 53; DBl 71; DR 79; DR 81; DY 4; DY 12; DY 50; MO 1; MY 10; RB 5; RO 14; RO 16; RR 2; RR 4; RY 2 (0.3 mM (100–300 mg/L))	–	Batch assays	Acetate, propionate and butyrate	–	73–100	–	Van der Zee et al. [122]

(continued)

Table 1 (continued)

Reactor type	Azo dye and concentration[a]	COD (mg/L)	HRT (hours)	Electron donors	Redox mediators	Color removal (%)	COD removal (%)	Reference
1.3L lab-scale UASB	AO7 (100 mg/L)	–	2, 4, and 6	Acetate, propionate and butyrate	anthraquinone 2,6-disulfonate	>90	79–86	Cervantes et al. [20]
EGSB[d]	RR 2, AO 7, MY 10 (0.3 mM)	–	6	Acetate, propionate and butyrate mixed with glucose	Anthraquinone-2,6-disulfonate (AQDS), anthraquinone-2-sulfonate (AQS), riboflavin (vitamin B2), and cyanocobalamin (vitamin B12)	–	–	Dos Santos et al. [31]
100 mL glass serum flasks	DB 22 (0.06 mM)	1412–1599	Batch assays	Sucrose and ethanol	Lawsone, Riboflavin	83–93	24–69	Amorim et al. [4]
100 mL glass serum flasks	Dye mixture (317–450 mg Pt–Co/L)	1351–1530	Batch assays	Sucrose and ethanol	Lawsone, Riboflavin	23–38	26–65	Amorim et al. [4]
1.1L UASB	RR2 (40–400 mg/L)	> 1000	12	Glucose	Humic substances	67–98	36–73	Cervantes et al. [18]
3.5L UASBR[e]	DB 22 (0.06 mM)	1078	24	Ethanol	-	68 ± 5	77.7 ± 9.2	Florencio et al. (2021)

[a] Abreviations - First letter: A = Acid; B = Basic; C = Ciba; D = Direct; M = Mordant; R = Reactive. Second letter: O = Orange; Y = yellow; R = Red; Bk = Black; Bl = Blue; N = Navy
[b] WW = Wastewater
[c] WWTP = Wastewater Treatment Plant
[d] EGSB = Expanded Granular Sludge Bed
[e] UASBR = Upflow Anaerobic Structured-Bed Reactor

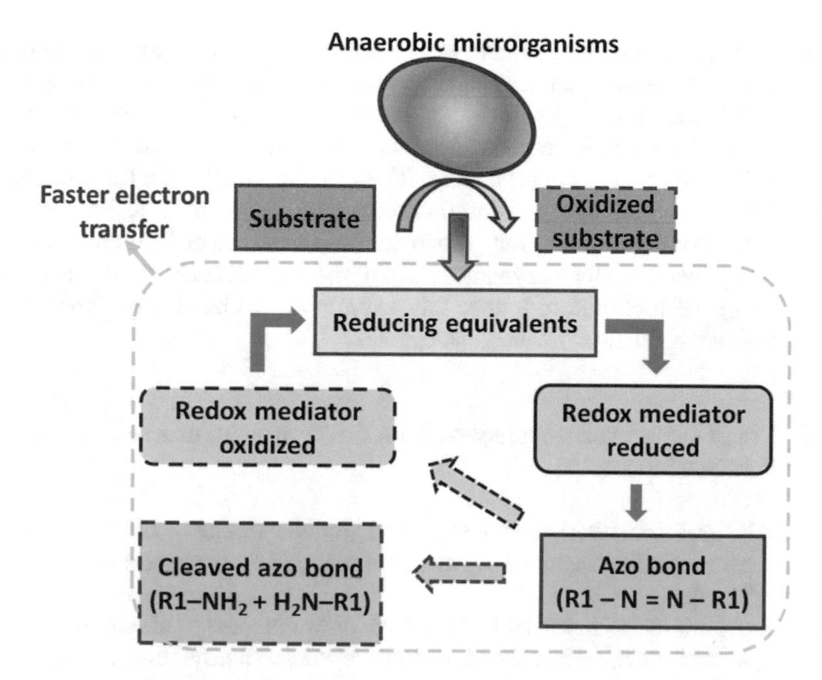

Fig. 3 Flowchart of the anaerobic reductive cleavage of azo bonds process aided by redox mediators

26% with Riboflavin (with ethanol as electron donor). The operational parameters of the reactors used by the authors are referred to in Table 1.

Process temperature influences the metabolism of the bacterial community, since it is directly related to the kinetics of the enzymatic reactions. Although most of the reactors are operated at a mesophilic temperature range, Dos Santos et al. [30] reported faster decolorization of the azo dyes RR2, AO7, and MY10 under thermophilic conditions (55 °C) compared to mesophilic conditions (30 °C), using EGSB reactor and AQDS as redox mediator. Furthermore, no lag phase was observed for the thermophilic conditions, suggesting this temperature range as advantageous over the mesophilic, for the tested conditions.

The initial concentration of the dye also has a great influence on its removal efficiency. This is mainly due to the toxicity characteristics of the dyes and the ability of microorganisms to overcome the toxicity, making some cultures more tolerant than others. High initial concentrations of azo dyes can inhibit community growth, leading to a decline in reactor efficiency. Beydilli et al. [11] analyzed the degradation of azo dye RR2 at initial concentrations of 50, 300, 500, 1000, and 2000 mg/L (0.08, 0.49, 0.81, 1.63, and 3.25 mM*) in 500 mL UASB reactors and verified that at the concentrations of 50 and 300 mg/L, complete color removal was achieved in the first 1 and 122 h of incubation, respectively, without inhibitory effects. Nevertheless, color removal was not completed within 400 h for the other concentrations, and there was an inhibition in the methanogenesis process. The different initial dye concentrations used by the authors are shown in Table 1.

Some characteristics of the azo dye itself must also be observed aiming to improve color removal efficiency. Van der Zee et al. [122] used 0.3 mM (100–300 mg/L) of 20 different kinds of azo dyes, including mono, di, tri, and tetra azo dyes (which have 1, 2, 3 or 4 azo bonds, respectively). The authors demonstrated that the rate of dye decolorization was not dependent on its molecular weight. On the other hand, chemical groups on the dye structure do influence color removal. Kulla et al. [69] demonstrated that the sulfonic group in azo dye molecules can decrease removal efficiency: (a) by hindering the enzymatic attack of the azoreductase, (b) by destroying the ability of the molecule to induce this enzyme or, (c) by its electronegativity, "shielding" the molecule from enzymatic attack.

2.2.2 Use of Other Anaerobic Systems on the Treatment of Azo Dye Textile Effluents

Fluidized bed reactor (FBR) and the expanded granular sludge bed (EGSB) reactor, originating from UASB reactors, have also shown good performance on decolorizing azo dye effluents.

The configuration of fluidized bed reactors allows the use of alternative support materials, such as sand, activated carbon, pumice stone, kinasite, tire scraping, glass beads, among others. Thus, the biomass adheres to the support material and is distributed in the reactor, increasing the contact surface between wastewater and microorganisms [10, 62, 85, 86, 92]. Maintaining the stability of this type of reactor in practice is more complicated than the UASB reactors, because it is necessary to control aggregate formation in the biofilm. This is performed adjusting the upflow speed of the liquid, size, and density of the dispersed particles of the influent wastewater, aiming to avoid the disaggregation of the biofilm formed in the support media [70]. Despite that, Haroun et al. (2009) reported color (65%) and soluble organic matter (98%) removal from a real textile effluent using FBR with activated carbon as support material, 0.6 g/L of glucose (3.3 μM*) as co-substrate, and 12 h HRT. Cirik et al. [24], in a study using a synthetic effluent of the azo dye Remazol Brilliant Violet 5R at concentrations from 100 to 200 mg/L (0.14–0.27 mM*), reported the improvement in FBR removal efficiencies. Using activated carbon as support medium color and organic matter removal, using glucose and ethanol as co-substrate, ranged from 60 and 76% to 75% and 99%, respectively.

In the Expanded Granular Sludge Bed (EGSB), granular biomass is used as inoculum. The granular bed has an expanded section, allowing greater contact between the influent wastewater and the microbial biomass. This configuration allows higher upflow speed than FBR, which can be boosted by effluent recirculation, improving the removal efficiency of toxic and recalcitrant compounds [70]. Dos Santos et al. [32] used EGSB reactor for the treatment of 1.35 g/L of the azo dye RR2 (2.19 mM) and reported 91% of color removal and 62% of organic matter removal. This shows that configurations of non-conventional or hybrid anaerobic reactors can be interesting alternatives in the treatment of. wastewater containing azo dyes, since they provide excellent results in both organic matter and color removal.

2.3 Byproducts of Azo Dye Reductive Cleavage and the Challenges for the Anaerobic Treatment

It is well established in literature, as demonstrated in the previous sections, that the cleavage of azo dyes requires an environment with low redox potential. Thus, anaerobic reactors are the most suitable for that job. Nonetheless, despite removing color from the effluent, the reductive cleavage of azo dyes results in compounds with at least one radical with an amino group (NH_2) [90, 116]. These byproducts are colorless compounds that are not metabolized in an anaerobic environment, due to the electronic stability of the benzene ring, which makes bond breakage hard and less energy-yielding for microbial cells in these environments [38, 42]. It is necessary to remove aromatic amines from the effluent, since they are even more toxic than the dye that produced them [2, 36, 37].

Kulla et al. [69] studied the azo dye Orange I, demonstrating that the sulfonic group present in the structure influences dye mineralization. The *Pseudomonas* strains K22 and KF46 used in the anaerobic degradation of the dye conducted the reductive cleavage of Orange I, partially removing the color from the synthetic effluent. However, one of the reductive cleavage byproducts, sulphanilic acid, was not metabolized, accumulating brownish-colored substances instead of releasing CO_2. The authors discuss that the non-utilization of the metabolite by microbial strains may occur because of the low cell permeability of the sulfonated molecules and their high electronegativity, which hinders enzymatic attack. Additionally, the increase in the remaining intracellular sulphanilic acid may have caused the destruction in the ability of the molecule to induce azoreductase, causing a decline in the cell ability to degrade the dyes.

The poor performance of anaerobic environments in degrading aromatic amines was also reported by Pereira et al. [96], who studied two simple aromatic amines, aniline and sulphanilic acid, in UASB reactors in the presence of nitrate and nitrite. Aniline was consumed with nitrate, but in a combination of nitrate and nitrite, aromatic amines ended up chemically reacting and forming other aromatic compounds. The amines formed had higher molecular weights and were more difficult to degrade and hence, more recalcitrant. Although these results are promising and similar studies have been reported on the anaerobic degradation of aromatic amines [80, 117], biological processes using oxygen as electron acceptor are still energetically favorable.

Currently, studies performed in anaerobic reactors focus more on increasing color removal efficiency along with the removal of azo dye byproducts and nutrients (nitrogen, phosphorus, carbon, sulfur), and on the generation of energy by the use of reactors combined with electrodes [3, 18, 35, 39, 44, 64]. Nonetheless, most of the reductive cleavage byproducts do not exempt aerobic treatment, since they are persistent in environments with low redox potential [59, 96]. Further studies using aerobic reactors for the complete mineralization of azo dyes have been conducted, which is discussed in the next session.

3 Aerobic Reactors

3.1 Color Removal of Azo Dye Textile Effluent in Aerobic Reactors

Azo dyes are not removed by conventional aerobic treatment systems nor rapidly degraded, since oxygen is the preferred electron acceptor [53, 61, 116]. Nevertheless, there have been indications that aerobic decolorization can occur in the presence of certain enzymes. Dye degradation under aerobic conditions is catalyzed mainly by azoreductase enzymes, but it also occurs in the presence of Nicotinamide Adenine Dinucleotide Hydrogen-Dichlorophenolindophenol reductase (NADH-DCIP reductase), malachite green reductase (MG reductase), and oxidative enzymes like lignin peroxidase and laccase [110]. Studies indicate that the action of some aerobic microorganisms combined with the enzyme azoreductase improves the kinetics of the azo dye degradation process [72]. In addition, some microorganisms that can aerobically decolorize azo dyes by the catalysis of aerobic or oxygen-insensitive azoreductases have been isolated [63, 82, 89]. Factors such as dye concentration, enzyme concentration, temperature, and intermediate complex formation rate also influence and contribute to the reduction of azo dyes [110].

It has also been observed that aerobic processes for the treatment of effluents that contain azo dyes are ineffective in most systems containing textile industry wastewater, although few microorganisms can partially or completely degrade dye molecules. However, the products formed during the anaerobic process, aromatic amines, can be demineralized or decomposed by the aerobic process. The aerobic treatment is essential for the degradation of the byproducts of the anaerobic degradation, and the removal of the toxicity generated by the production of aromatic amines.

3.2 Configuration of Aerobic Treatment Systems and Strategies to Enhance Color Removal

Several studies using bacteria, fungi, and algae capable of reducing azo dyes have been reported [27, 46, 105]. Bacteria are widely used for decolorizing azo dyes due to their high activity, extensive distribution, and strong adaptability [29, 95]. Nevertheless, the decolorization intermediates, such as aromatic amines, can inhibit the activity of a large number of bacteria [101]. On the other hand, fungi have a strong ability to degrade complex organic compounds by producing extracellular ligninolytic enzymes, including laccase, manganese peroxidase and lignin peroxidase, drawing attention to fungi in recent years [46, 101]. So far, some species, such as *Pleurotus ostreatus*, *Pichia* sp., *Penicillium* sp., and *Candida tropicalis*, have been confirmed to discolor azo dyes by adsorption or degradation [6, 55, 101].

Buitrón et al. [15] evaluated the aerobic degradation of the azo dye Acid Red 151 (AR151) by a sequencing batch biofilter packed with a porous volcanic rock, reaching up to 99% of color removal from an initial concentration of 50 mg/L (0.11 mM*) of AR151. Coughlin et al. [26] isolated two bacterial strains capable of reducing the azo bond of dyes AO7 and AO8, at 50 mg/L (0.14 mM*). One of the strains used both dyes as its only source of carbon, nitrogen, and energy, while the other could only reduce the azo bond when co-substrates (0.05% glucose and 0.2% NH_4SO_4) were present. Some azo dyes can be degraded by aerobic bacteria in the presence of suitable co-substrates [103]. The aerobic degradation of a simple azo compound (Orange Carboxy II) by *Flavobacterium* sp. was reported by [68].

Tan et al. [118] evaluated the aerobic decolorization and degradation of azo dyes both by suspended growing cells and immobilized cells of a newly isolated yeast strain, LH-F. The effects of different parameters on the decolorization of Acid Red B by growing cells of strain LH-F1 were investigated, including initial dye concentration (50–300 mg/L), concentrations of glucose (2–14 g/L), and ammonium sulfate (0.5–3.0 g/L), inoculation size (2%–10%, v/v), temperature (20–40 C), and pH (3–9). More than 90% of all six azo dyes were decolorized by strain LH-F1, and more than 98% of Acid Red B was removed within 10 h when the initial dye concentration was 50 mg/L.

The activated sludge system is the most promising aerobic system for treating wastewater containing azo dyes because it can be operated for a long period without too many concerns about the elimination of specific microbial strains from the treatment systems and frequent inoculation [74]. Aerobic granular sludge has attracted great attention for wastewater treatment containing azo dyes, since its compact structure and bead size allow them to resist the shock of toxic compounds and harsh environmental conditions. The more diverse the microbial population, the greater will be the removal of pollutants. Aerobic granules can be easily formed with azo dyes at low concentrations, such as 50 mg/L^{-1}, mixed with other biodegradable carbon sources, such as glucose, starch or ethanol [74]. Mata et al. [78] reported better granulation results with the addition of 20 mg/L of the azo dye AR14 to a synthetic textile effluent, using starch as carbon source, and organic matter concentration around 1000 mg/L^{-1}, using a batch reactor system for 62 days. Ibrahim et al. [52] developed granules capable of treating real textile wastewater from a mixture of sterilized activated sludge and bacterial species isolated from textile sludge.

Furthermore, during aerobic decolorization processes, the breakdown of compounds could be further degraded by monooxygenase and dioxygenase catalysis, which would induce the incorporation of O_2 into the aromatic ring of organic compounds before ring fission, as cited by [107]. Some azo dyes could be decolorized or even mineralized by certain microorganisms under aerobic conditions. Therefore, compared to the conventional two-stage system, aerobic processes with selected microbial strains become simple and economical alternatives for the treatment of textile effluents containing azo dyes. The conventional process of activated sludge treatment is one of the important aerobic treatment methods for azo dye degradation. In addition, it is often an effective and highly economical method that reduces organic pollutants present in several wastewaters. Nevertheless, the activated sludge

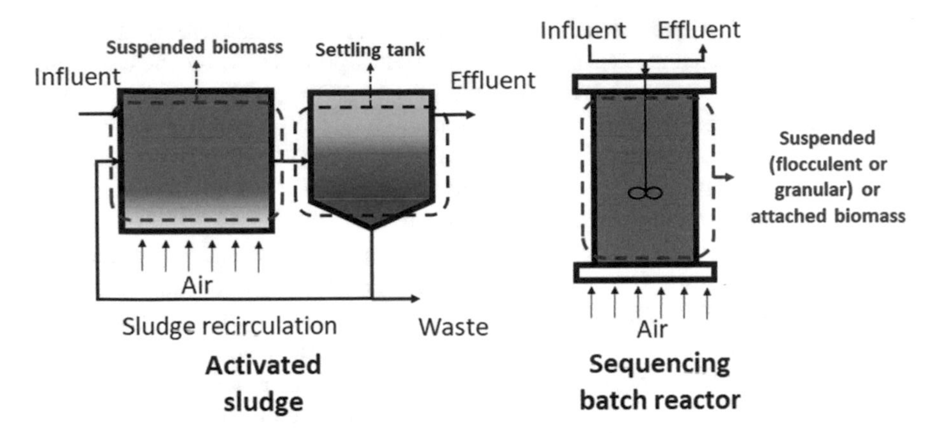

Fig. 4 Schematic of the main reactor configurations used for aerobic degradation of azo dyes

system used for aerobic azo dye treatment proved to be an ineffective process in most cases, since the degradation rate of azo bonds is significantly lower in the presence of aerobic microorganisms compared to the anaerobic process. Figure 4 shows a schematic of the activated sludge and sequencing batch reactors.

Different microbial cultures were isolated aiming at the discoloration and degradation of dyes under different conditions, such as anaerobic, microaerobic, aerobic, or alternating anaerobic and aerobic conditions. The consortia have proven to be more efficient in treating textile effluents containing azo dyes [81, 123]. Table 2 shows some types of aerobic effluent treatment systems containing azo dyes.

3.3 Removal of Azo Dye Textile Effluent Byproducts in Aerobic Reactors

The selection of the best option of biological treatment for the bioremediation of a specific type of industrial effluent is a difficult task, given the complex composition of this effluent. An efficient combination for the removal of azo dyes is the use of systems or processes in two or more stages, and the choice depends on effluent composition, dye characteristics, cost, toxicity of the degradation products, and future use of the treated effluent [113].

Despite the ineffectiveness of aerobic reactors in azo dye treatment, its use is studied as an anaerobic additional post-treatment for the mineralization of their byproducts. The most energetically favorable path for the biodegradation of aromatic amines is by introducing hydroxyl groups into the aromatic ring by oxygenase-like enzymes, destabilizing the ring's electronic shield and thus enabling its rupture [5, 33, 42]. Hence, the addition of aerobic post-treatment or the combination of anaerobic and aerobic environments to mineralize them to CO_2 and water is required [59, 121].

Table 2 Important parameters and efficiencies of aerobic systems used for degradation of azo dyes

Reactor type	Azo dye and concentration	COD (mg/L)	HRT (hours)	Electron donors	Biomass type	Color removal (%)	COD removal (%)	Reference
250 mL shaking flasks	Acid Brilliant Red GR (20–140 mg/L)	2–14 g/L of sucrose	24	Sucrose	–	97	–	[119]
250 mL shaking flasks	Acid Red B (50–300 mg/L)	–	8	–	Suspended growing and immobilized cells	97	–	[118]
Aerated SBB[a]	Acid Red 151 (25–50 mg/L)	–	Each cycle of the SBB consisted of four periods controlled by a timer: fill and draw (3 and 13 min, respectively), reaction (variable) and settle (30 min)	–	Immobilized	73	60	Buitron et al. (2004)
Aerobic granules	Acid Red 14 and Reactive Blue 19 (20–50 mg/L)	–	3.8 (228 min)	–	Aerobic granules	50–80	–	[74]
Aerobic	Indigo dye	–	96 (4 days)	–	Immobilized	97	97	61

[a]SBB = Sequencing Batch Biofilter
[b]WWTP = Wastewater Treatment Plant

The combination of an anaerobic process followed by an aerobic one for the treatment of textile effluents has already been highlighted in the early 1990s. A study by [50] with the azo dye Mordant Yellow (MY) 3, using mixed culture, switched the researchers' focus from treatments with a single anaerobic or aerobic unit to studying the combination of both environments. The authors verified that the complete mineralization of MY3 happened with the alternation between these processes. Thus, studies using aerobic-anaerobic packed bed reactors, aerobic-anaerobic fluidized bed reactors, and aerobic-anaerobic sequential batch or continuous-flow reactors have been better developed, and their systems were improved for the mineralization of azo dyes [9]. Anaerobic-aerobic combination has produced good results in studies conducted with synthetic and real effluents [1, 2, 16, 36, 41, 106].

Nonetheless, this combination results in the use of two treatment units, substantial energy demands, and trained personnel. Furthermore, aromatic amines are produced as intermediate compounds in the anaerobic degradation of azo dyes and are known to have toxicity equal to or greater than that of azo dyes, and many of these amines are carcinogenic [29, 97, 100]. It is noteworthy that the aerobic stage must be reached because it is necessary for the toxicity arising from azo dyes to be reduced.

The aromatic amines produced by the reduction of the azo bond are known to be mutagenic and carcinogenic and may have a higher degree of toxicity than the dye that originated them [13, 17, 21, 57]. Several studies on azo dye reduction indicate that most of the aromatic amines, produced in an anaerobic environment, are capable to be removed in an aerobic environment [1, 2, 16, 97, 121], thus confirming that two-stage systems are an excellent alternative for the complete removal of both the azo dyes and the byproducts generated in the treatment of wastewater containing these compounds.

4 Anaerobic-Aerobic Reactors

Due to the need for sequential anaerobic and aerobic environments for the mineralization of azo dyes, in a process similar to the one shown in Fig. 5, many studies have evaluated the efficiency and optimization of treating systems using a single reactor or sequential reactors under anaerobic and aerobic conditions. The application of these environments can be performed by one- or two-stage systems. In one-stage systems, the anaerobic and aerobic environments occur in the same compartment by alternating anaerobic and aerobic periods or using technologies that provide an oxygen concentration gradient for the microbial community, such as aerobic granules or biofilm reactors. In two-stage systems, firstly an anaerobic reactor is used, where microorganisms perform the reductive cleavage of azo bonds (first stage), followed by an aerobic reactor where the azo dye byproducts formed in the first stage are mineralized by aerobic or facultative organisms. In these systems, the anaerobic and aerobic biomasses are separated by two or more reactors.

Two-stage systems benefit from the separation of anaerobic and aerobic environments to provide specialized microbial communities necessary to perform the

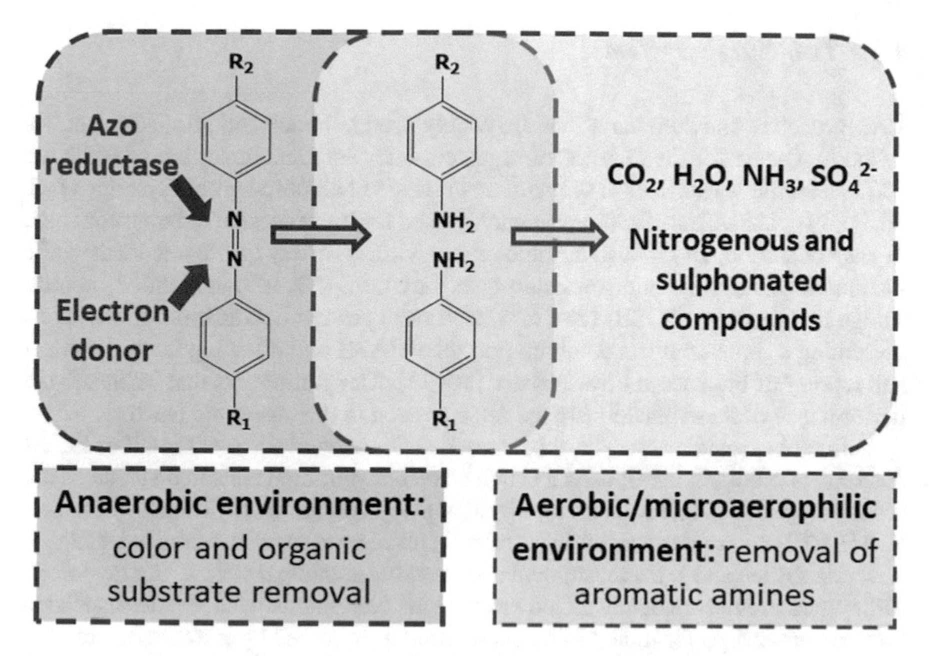

Fig. 5 Mineralization of azo dyes in sequential anaerobic-aerobic process

required reactions at each stage, without the anaerobic organisms being harmed by the presence of oxygen and the aerobic organisms, by the absence of oxygen. Nonetheless, some aromatic amines are not mineralized during the sequential anaerobic-aerobic stages, due to the need for other reactions that only occur in anoxic or microaerated environments [67, 73, 76, 122].

One-stage systems utilize the metabolic synergy of mixed microbial communities (presence of anaerobic, facultative, and aerobic organisms) to perform the metabolic reactions necessary for the mineralization of azo dyes. The variation of redox potential caused by the change between the anaerobic and aerobic environments in the same compartment might favor the degradation of recalcitrant aromatic amines, allowing specific reactions to happen, and consequently, the mineralization of aromatic amines that would not be degraded otherwise, in conventional two-stage anaerobic-aerobic reactors [124]. However, alternating between anaerobic and aerobic environments might harm the development of strict anaerobic or aerobic organisms, reducing the efficiency of one-stage systems [12, 76]. More recently, the possibility of using microaeration (continuous or intermittent) as the aerobic period has been observed [16, 79, 91]. Microaeration can reduce energy costs and minimize the harm to strict anaerobic organisms because of exposure to oxygen.

4.1 Two-Stage Systems

UASB reactors and their variations are widely used in bench and pilot-scale studies as the first stage for the removal of organic matter and decolorization of synthetic and real textile wastewater in sequential two-stage anaerobic-aerobic systems [1, 2, 36, 90, 114, 125]. This reactor is commonly used for its advantage of being compact, of easy operation, and robust, capable of receiving a variety of organic loads while maintaining stable operation even during its variations [23, 109]. Expanded granular sludge bed reactors (EGSB) have also been used for the degradation of azo dyes, presenting a more compact configuration than UASB and allowing intense contact and mixing of biomass and wastewater [30–32]. The parameters that influence the performance of these reactors are the same as seen in the anaerobic reactors.

Different aerobic reactors have been used as the second stage of UASB or EGSB reactors. O'Neill et al. [90] used a bench-scale conventional activated sludge as the second stage of a UASB reactor employed for the degradation of 150 mg/L of Procion Red H-E7B, an azo dye present in a synthetic textile wastewater. Activated sludge is well established and is the most popular system for municipal wastewater treatment. The system uses a completely mixed aerated tank continuously fed, where suspended microorganisms perform the biological treatment, followed by a settler that returns a part of the sludge to the aerated tank. In this study, the UASB reactor, which was operated with HRT of 24 h, was responsible for 60–61% of organic matter removal, while the total system removed from 66 to 88%. In general, it is commonly found that the anaerobic reactor is responsible for most of the organic matter removal. This behavior is favorable for biogas production, which is ideal for industrial applications. The system had maximum color removal of 77% while being fed with azo dye and 3.8 g/L of starch. The authors reported that increasing the ratio between starch and dye improved color removal in the system. In the aerobic reactor, microbial growth was limited by the low availability of carbon sources, since the growth of aerobic organisms has a much faster substrate consumption kinetics compared to anaerobic organisms.

Tan et al. [120] used a combination of EGSB and activated sludge to degrade synthetic wastewater containing the azo dye Mordant Yellow 10 and methanol as the electron donor. The EGSB reactor was operated with HRT varying from 16.8 to 29.3 h, which resulted in color removal higher than 97% even with azo dye concentration ranging from 57 to 192 mg/L. The first stage removed 100% of the methanol used, in all tested conditions. The authors reported minimum degradation of 88% of the aromatic amine 5-aminosalicylic acid in the aerobic reactor, but it was necessary to increase HRT in the aerobic reactor from 8.9 to 79.2 h to promote the development of organisms capable of degrading sulfanilic acid, which is another aromatic amine produced from the reductive cleavage of the azo bonds from Mordant Yellow 10. The system showed great color and organic matter removal; however, in order to degrade the aromatic amines produced, it was necessary to greatly increase the HRT of the aerobic system.

Amaral et al. [2] used a Submerged Aerated Biofilter—SAB as the second stage of a UASB in a pilot-scale system for the degradation of real textile wastewater from a medium-sized industry. Expanded clay spheres were used as a fixed support material inside the SAB. The biofilters allow the growth and maintenance of slow-growing organisms, such as those necessary for aromatic amine mineralization [34]. The authors reported that varying organic load from 1.84 to 2.42 kg of organic matter/$m^3.d^{-1}$ (HRT varying from 8 to 12) in the UASB reactors did not cause significant variation in the efficiency of the system. Nevertheless, the variation of wastewater composition highly influenced the decolorization efficiency of the system. SAB, which was operated with 6 and 9 h of HRT, being an aerobic reactor, was responsible for most of the color removal, with efficiencies ranging from 65 to 92%, while UASB showed 30–52% of color removal. The low color removal performance of UASB was attributed to the high salinity and sulfate concentration of the wastewater, which both inhibited the development of organisms in the anaerobic reactor and caused competition between sulfate and azo dye for the electron donors. The treating system was able to almost completely remove toxicity, as evaluated by ecotoxicity assays using *Daphnia magna* as bioindicator. This study highlighted the robustness of the SAB reactor, using simple support material, for treating real, highly varying textile wastewater.

Sequencing Batch Reactors (SBRs) have also been widely employed in the studies on azo dye degradation; however, they are mostly used as one-stage systems. Therefore, there are few studies that have used SBR in two-stage systems [12, 65, 66]. These reactors are versatile, since they can be manufactured in different volumes, from microcosms to full scale, they can be employed with suspended or granular biomass or with carriers for biofilm growth.

SBR can be used as an anaerobic or aerobic reactor in two-stage systems. Bonakdarpour et al. [12] used two SBR, one anaerobic and one aerobic, for the degradation of a synthetic textile wastewater containing glucose at the concentration of 2 g/L of organic matter and the azo dye Reactive Black 5 at a concentration ranging from 100 to 500 mg/L. SBR was operated at 30 °C for 45 h for each anaerobic and aerobic reaction. In this study, more than 80% of color and organic matter removal were observed in the anaerobic SBR, which is in agreement with most studies. It was reported that increasing azo dye concentration from 100 to 500 mg/L resulted in a slight decrease in anaerobic organic matter removal, indicating a possible toxicity of the azo dye and its metabolites to the anaerobic microbial community. The authors also reported 40–44% of aromatic amine removal in the aerobic SBR and reappearance of color, indicating autoxidation of aromatic amines. In this case, it is possible that the long HRT used in this system was favorable for the first stage, although the long exposure to oxygen might have resulted in the autoxidation of aromatic amines.

Koupaie et al. [66] used Moving Bed Biofilm Reactors (MBBR) as the second stage in a system used to degrade synthetic textile wastewater containing the azo dye Acid Red 18 (AR18) and a mix of glucose and lactose as electron donors. MBBR uses carriers for the development of microbial growth by biofilm formation, similarly to the SAB reactor. The major difference between MBBR and SAB is that in MBBR, the media support moves freely inside the reactor, while it in the SAB, it is fixed. Biofilm reactors have the advantage of maintaining slow-growing organisms and an oxygen gradient inside the biofilm that will allow the development of diversified organisms. In this study, an anaerobic SBR was used as the first stage, operating with 66 h of HRT and with the temperature controlled at 35 °C. The reactors were fed with wastewater containing glucose and lactose (1.5 g/L of each) and 100–1000 mg/L of AR18. More than 98% of decolorization was reported, and organic matter removal was higher than 80% in the anaerobic SBR. Increasing dye concentration did not cause a significant change in color and organic matter removal efficiency. MBBR operated with 66 h of HRT and was responsible for the total color removal of the system and almost total residual organic matter removal. The authors identified the degradation of 1-naphthylamine-4-sulfonate ranging from 88.7 to 100%, indicating the system efficiency for the degradation of aromatic amines. The concentration of biomass attached to the carriers decreased with the increase in dye concentration, indicating that the metabolites might have inhibited the development of biomass in the MBBR. The combination of SBR-MBBR resulted in excellent azo dye decolorization, organic matter removal, and metabolite removal, while treating a synthetic textile wastewater.

Anaerobic Baffled Reactor (ABR) has also been used in textile wastewater treating systems. ABR is a continuous flux reactor in which multiple compartments are grouped together, sequentially, where the effluent passes through biomasses in each compartment, being gradually treated. ABR compartments turn these reactors into systems of several stages, which results in the development of multiple specialized microbial communities, in addition to being able to avoid long-term contact between toxic compounds and more sensitive organisms, such as methanogenic archaea and the metabolites from azo dye degradation [88, 126, 127].

Zhu et al. [126] used two baffled reactors, each with six identical compartments in which the three initial compartments were anaerobic, followed by two aerobic and one for sedimentation, operating with a total HRT of 24 h. Each compartment had a baffle that made the wastewater follow a downflow and an up-flow while in contact with the biomass. The authors evaluated the degradation of 30–60 mg/L of the azo dyes Acid Orange 7 (AO7) and Methyl Orange (MO) while using 2 g/L of glucose as electron donor. Organic matter was gradually removed along the compartments, without distinct removal efficiency among them, until reaching a total removal of 83.3–83.8%, in the reactors fed with AO7 and MO, respectively. While the aerobic compartments were efficient for aromatic amine removal, the anaerobic compartments also showed the ability to partially degrade aromatic amines. The authors reported dissolved oxygen concentrations higher than 1 mg/L, indicating that they were not effectively hermetic.

A variety of aerobic reactors were successfully used as second stage for mineralization of azo dyes and Fig. 6 shows a didactic scheme of the most used. Table 3 presents the studies that have used different two-stage system configurations for the degradation of azo dyes in synthetic wastewater or real textile wastewater.

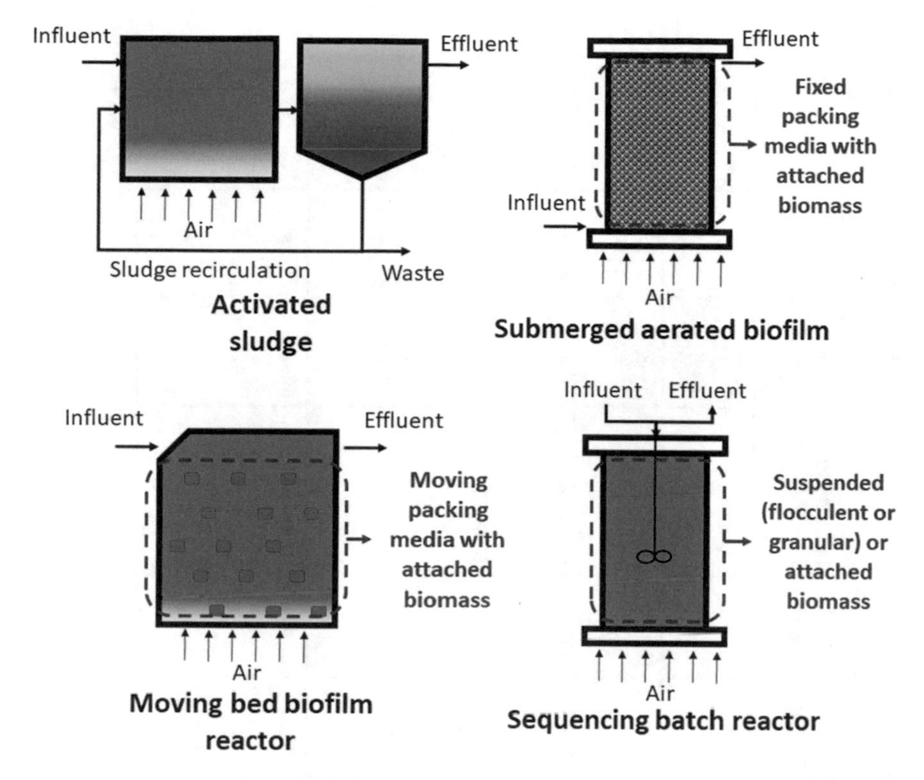

Fig. 6 Schematics of common aerobic reactors used as second stage for degradation of azo dyes by-products

4.2 *One-Stage Systems*

SBRs were pioneering reactors in the study of azo dye degradation using one-stage systems [75, 76, 120]. In these reactors, the possibility of applying multiple environments along the same reaction cycle (anaerobic, aerobic, microaerophilic or intermittently aerobic/anaerobic) makes them ideal for use in the degradation of azo dyes.

Lourenço et al. [76] studied the degradation of the dye Remazol Brilliant Violet 5R (RBV5R) at concentrations from 60 to 100 mg/L and Remazol Black B (RBB) at the concentration of 30 mg/L, using the starch-derived compound Emsize E1 at a concentration that resulted in 750 mg COD/L. The reactors operated in a 24-h cycle, divided into 21 h of reaction and 2 h for volumetric exchange. The reaction was divided into 9–13 h of anaerobic reaction followed by 8–9 of aerobic reaction. The authors reported the efficiencies of organic matter and color removal of 90 and 80%, respectively, in the reactor fed with RBV5R. The authors also reported that the aromatic amines formed from the anaerobic degradation of RBV5R were partially mineralized during the aerobic reaction, but that increasing aerobic reaction

Table 3 Main characteristics and efficiencies of different two-stage sequential anaerobic and aerobic reactor configurations for degradation of azo dyes

Reactor type		Fisrt Stage			Second Stage		Overall		References
First stage	Second stage	Azo dye and concentration (mg/L)	Electron donors (mg COD/L)	HRT (hours)	HRT (hours)	Degradation of aromatic amines (%)	Color removal (%)	COD removal (%)	
Upflow anaerobic sluge blanket	Activated sludge	Procion Red H-E7B (150–750)	Starch and acetate	24	16	–	45–77	66–88	[90]
Upflow anaerobic sluge blanket	Completely stirred tank	Reactive Black 5 (100)	Glucose (3000)	3–30	10–107	–	73–84	45–96	[114]
Upflow anaerobic sluge blanket	Sumberged aerated biofilter	Real textile wastewater[a]	Real textile wastewater (880)	12–24	9–18	b	61–86	69–77	[36]
Upflow anaerobic sluge blanket	Sumberged aerated biofilter	Real textile wastewater[c]	Real textile wastewater (1045–1143)	8–12	6–9	b	65–96	56–71	[2]
Upflow anaerobic sluge blanket	Sumberged aerated biofilter	Real textile wastewater[d]	Real textile wastewater (629)	4–16	12–3	b	55–65	64–75	[1]
Upflow anaerobic sluge blanket	Sequencing batch reactor	Acidic Orange II (50), Methyl Orange (50), Congo Red (50), Amino Black (50)	Glucose and yeast extract (1000–5000 total)	12–24	24–120	Aniline, naphthylamine (73–100)	54–100	78–90	[125]

(continued)

Table 3 (continued)

Reactor type		Fisrt Stage			Second Stage		Overall		References
First stage	Second stage	Azo dye and concentration (mg/L)	Electron donors (mg COD/L)	HRT (hours)	HRT (hours)	Degradation of aromatic amines (%)	Color removal (%)	COD removal (%)	
Expanded granular sludge bed	Activated sludge	Mordant Yellow 10 (100–200)	Ethanol (1000–2000)	33–17	7–50	SA (1–98); 5-ASA (88–100)[e]	87–100	100	[120]
Sequencing batch	Sequencing batch	Reactive Black 5 (100–500)	Glucose (2000)	45	45	36–44	82–87[f]	>80[f]	Bonakdarpour et al. (2011)
Sequencing batch	Moving bed biofilm	Acid Red 18 (100–1000)	Glucose (1600) and Lactose (1695)[g]	66	66	64–71	100	>80	[66]
Sequencing batch	Moving bed biofilm	Acid Red 18 (100–1000)	Glucose (1600) and Lactose (1695)[g]	66	66	1 N-4S (>80)[g]	>97	>81	[65]
Anaerobic baffled	Aerobic baffled	Acid Orange 7 and Methyl Orange (30–60)	Glucose (2134)[g]	12	6		~100	~83	[126]
Anaerobic baffled	Aerobic baffled	Acid Orange 7, Congo Red, Methyl Orange and Amino Black (30–60)	Glucose (2134)[g]	12	6		~100	83–85	[127]

(continued)

Table 3 (continued)

Reactor type		Fisrt Stage			Second Stage		Overall		References
First stage	Second stage	Azo dye and concentration (mg/L)	Electron donors (mg COD/L)	HRT (hours)	HRT (hours)	Degradation of aromatic amines (%)	Color removal (%)	COD removal (%)	
Anaerobic baffled	Downflow-hanging sponge (DHS)	Hellozol Reactive Black HSR (20)	Starch (300)	18	4	–	48	78	[88]

[a] Real textile wastewater was used, and 7 different azo dyes were used by the industry where the study took place
[b] Qualitative aromatic amine production and removal was observerd through UV–Vis spectrum analysis
[c] Authors identified that Direct Black 22 was the predominant azo dye in the wastewater
[d] Authors identified that Direct Black 22 was the predominant azo dye in the wastewater at mean concentration of 30,4 mg/L
[e] Authors identified Sulfanilic Acid - SA and 5–5-Aminosalicylic acid as the aromatic amines produced by the anaerobic degradation of the azo dye analyzed
[f] Values estimated from original work's graphs
[g] Values calculated from substrate concentration and its theoretical COD demand
[h] Authors reported the formation and degradation of the aromatic amine 1-naphthylamine-4-sulfonate, 1 N-4S
[i] Qualitative aromatic amine production and removal was observerd through UV–Vis spectrum, FTIR and GC–MS analysis

time from 8 to 10 and 12 h did not result in improvements. The low removal of aromatic amines was attributed to the absence of organisms capable of degrading these compounds. The reactor fed with RBB showed lower color removal efficiency, which was due to the RBB molecular structure.

Çinar et al. [25] evaluated different reaction cycles in SBR for the degradation of the azo dye RBV5R. The study used 100 mg/L of the azo dye and 1000 mg/L of glucose in a reactor inoculated with biomass previously adapted to the degradation of real textile wastewater. The reactor was operated with three different reaction times: 48 h (24 anaerobic, followed by 23.9 aerobic), 24 h (12 anaerobic followed by 11.9 aerobic), and 12 h (6 anaerobic, followed by 5.6 aerobic). Over 75% of organic matter removal was reported in all reaction times, and approximately 70% of this removal occurred during the anaerobic reaction. Reducing reaction time resulted in a decrease in organic matter removal efficiency, from 92% in cycles of 48 h to 87% in 12 h. The change in reaction time was not related to the change in color removal, which showed 72, 89, and 86% efficiency for 48, 24, and 12 h of total reaction, respectively. The removal of aromatic amines derived from benzene was identified in all conditions, but only the reaction time of 24 h was efficient for the removal of amines derived from naphthalene, with 64% of removal. This study indicated that 24-h cycles showed the best results in terms of organic matter, color and aromatic amine removal.

More recently, [79] studied the use of intermittent microaeration as the aerobic reaction in a sequential anaerobic-aerobic SBR, used to degrade the azo dye DB22. The reactors operated with cycles of 24 h, divided into 12 h of anaerobic reaction, 11 h of aeration, and 1 h of volumetric exchange. Similar organic matter, color, and toxicity removal were verified comparing a reactor that used 30 min of microaeration every 2 h and a reactor with continuous microaeration. The reactor that operated with intermittent microaeration showed 76.8% of DB22 removal, while the reactor with continuous microaeration showed 74.5%. Organic matter removal was 81.4 and 79.5% for the intermittently microaerated and continuously microaerated reactors, respectively. The authors also found similar results for aromatic amine and toxicity removal between both reactors, showing that intermittent microaeration is a great strategy for the aerobic reaction in an anaerobic-aerobic SBR used to treat synthetic textile wastewater containing the azo dye DB22.

Manavi et al. [77] used SBR for the development of aerobic granules acclimated for the degradation of real textile wastewater. The use of aerobic granules is interesting because it can provide an oxygen gradient inside the granules, allowing the development of anaerobic, facultative, and aerobic organisms in the same compartment. The reactors operated with reaction cycles of 6, 8, 12, and 24 h, divided into anaerobic and aerobic reactions, following a ratio of 2 (for cycles with 6, 8, and 12 h of reaction) and 3 (for cycles with 24 h of reaction). The authors developed granules able to remove 73% of color and 68% of organic matter at the end of the operational period with 24-h cycles, which showed the best results. Organic matter was consumed during the 6 initial hours of anaerobic reaction, but it increased during the last 12 h of anaerobic reaction. This behavior was explained by the hydrolysis of extracellular polymeric substances produced by organisms in the granules. Therefore, the aerobic reaction was responsible for most of the organic matter removal. Although most of the color was removed during the anaerobic reaction, color removal during the aerobic reaction was also reported, possibly due to the formation of an anaerobic region inside the granules. At the end of the operation, the authors verified color reappearance during the aerobic reaction, possibly due to autoxidation of aromatic amines.

Few studies have evaluated the possibility of using aerobic biofilm reactors for the degradation of azo dyes. Schoeberl et al. [108] used a membrane reactor to degrade real textile wastewater, adopting HRT between 2.9 and 5 h. Membrane reactors are highly efficient treating systems because they use a mix of physical and biological mechanisms to treat wastewater, resulting in effluents of high quality,nevertheless, their operation and maintenance are costly. This reactor showed organic matter removal efficiency from 89 to 94% and color removal efficiency ranging from 65 to 91%. It was reported that, in general, increasing backwash and aeration intensity helped to reduce membrane fouling. Sahinkaya et al. [104] used membrane reactors for the degradation of 50 mg/L of the azo dye Remazol Brilliant Violet 5R. The authors found better color removal efficiencies using intermittent aeration and the efficiency increased with the increase in the anaerobic periods, reaching 97% of removal for the ratio of 1/15 min of aerobic/anaerobic reaction.

Carvalho et al. [16] used a UASB reactor with a microaeration system placed immediately above the sludge blanket layer for the degradation of synthetic textile wastewater containing 32.5 mg/L of the azo dye Direct Black 22 and 1200 mg COD/L of starch as an electron donor. Color removal between 69 and 79% was reported, as well as organic matter between 59 and 78%, similarly to the completely anaerobic UASB reactor, which was used as control. Nonetheless, the microaerated reactor showed high efficiency for the removal of aromatic amines, reducing almost totally the toxicity of the effluent evaluated by toxicity assays in *Vibrio fischeri*, while the anaerobic reactor did not remove aromatic amines and toxicity. The microaerated UASB system showed good color, organic matter, and toxicity removal efficiencies without causing sufficient autoxidation to impair color removal, indicating this system as a great alternative for continuous flow systems.

Sequencing batch reactors, microaerated UASB, and membrane bioreactors are popular among one-stage systems, and Fig. 1.x shows a schematic of these reactors. Table 4 summarizes the studies that have used different one-stage system reactors (Fig. 7).

5 Conclusions

The chapter elucidated the main processes of azo dye removal and biodegradation. Specifically, the biological treatment of azo dyes revealed the need of using sequential anaerobic-aerobic environments for the mineralization of azo dyes employing systems that use one or two stages or by one-stage systems, which allow anaerobic and aerobic environments in the same compartment.

Table 4 Main characteristics and efficiencies of different one-stage sequential anaerobic and aerobic reactor configurations for degradation of azo dyes

Reactor type	Anaerobic reaction (hours)	Aerobic reaction (hours)	Azo dye and concentration (mg/L)	Electron donors (mg COD/L)	Color removal (%)	COD Removal (%)	Degradation of aromatic amines	Reference
Sequencing batch	13	8	Remazol Brilliant Violet 5R (60–100)	Emsize E1 (862)	30–90	60–80	Only benzene-based aromatic amines	[75]
Sequencing batch	9–13	8–12	Remazol Brilliant Violet 5R (60–100), Remazol Black B (30)	Emsize E1 (750)	75–90	70–85	No degradation	[76]
Sequencing batch	6–24	6–24	Remazol Brilliant Violet 5R (100)	Glucose (1067)[g]	72–86	> 75	Benzene-based (64–92%) and naphthalene-based (25–64%)	[25]
Sequencing batch (aerobic granules)	4–18	2–6	Real textile wastewater	Real textile wastewater (290–1190)	50–73	70–95	–	[77]
Sequencing batch (microaerated)	12	11	Direct Black 22 (32,5)	Ethanol (1200)	74–76	79–81	a	
Aerobic biofilm reactor	–	2	Acid Orange 8, Acid Orange 10 and Acid Red 14 (<5)	–	20–90	40–90	–	
Membrane bioreactor	–	3–5	Real textile wastewater	Real textile wastewater (1606–2997)	65–91	89–94	–	
Membrane bioreactor	–	17–96	Real textile wastewater	Real textile wastewater (1380–6033)	30–99	74–90	–	

(continued)

Table 4 (continued)

Reactor type	Anaerobic reaction (hours)	Aerobic reaction (hours)	Azo dye and concentration (mg/L)	Electron donors (mg COD/L)	Color removal (%)	COD Removal (%)	Degradation of aromatic amines	Reference
Membrane bioreactor	–	12	Remazol Brilliant Violet 5R (50)	Glucose (500)	> 10–97	~95[b]	–	
Microaerated upflow anaerobic sludge blanket	24–82	Siriusgelb GG	Ethanol (820)	56–71	~ 100	Low removal of 5-ASA and 1,4-PDA[c]		
Microaerated upflow anaerobic sludge blanket	–	–	Direct Black 22 (32,5)	Starch (1200)	69–80	59–78	a	

[a]Qualitative aromatic amine production and removal was observerd through UV–Vis spectrum analysis
[b]Mean value calculated based on mean COD concentrations from untreated and treated wastewater
[c]Authors reported that most of 5-aminosalicylic acid (5-ASA) and 1,4-phenylenediamine (1,4-PDA) were rapidly autoxidized to polymeric products

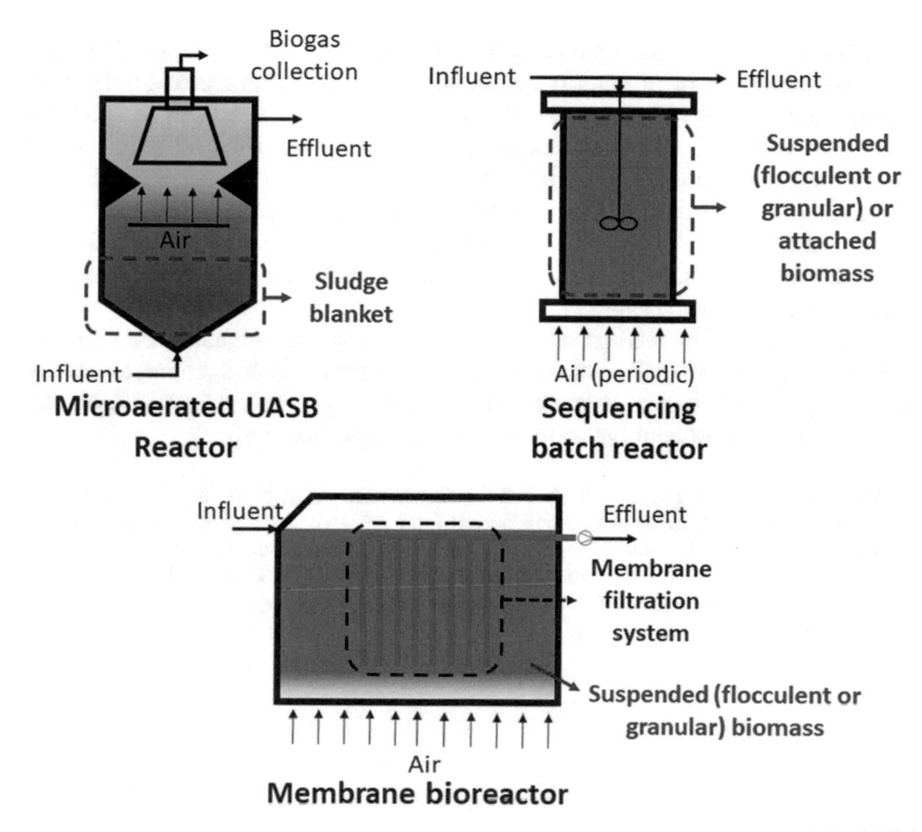

Fig. 7 Schematics of common one-stage reactors used for degradation of azo dyes

A variety of reactor configurations and their combinations have been used with good color and organic matter removal efficiency, from simpler configurations, such as UASB and SBR, to more sophisticated, such as MBR. The reactors RBBS and UASB and their variations, such as EGSB, are commonly used and have proved to be good alternatives as anaerobic reactors, because they are relatively simple technologies, widespread in the treatment of municipal and industrial wastewater, and have shown good results for azo dye degradation in studies that have worked with synthetic and real textile wastewater. The results obtained with aerobic reactors indicated that these systems are not optimal for the degradation of azo dyes, mostly because of the competition for electron donors between oxygen and azo dyes. Nonetheless, they are necessary for the mineralization of the aromatic amines formed during the anaerobic reductive cleavage of azo bonds.

The aerobic reactors BAS and MBBR, which use support material for promoting the growth of biofilms, have shown good results of residual organic matter, color, and aromatic amine removal. The organisms necessary for the mineralization of aromatic amines might be slow growers and the biofilm growth in the support material promotes the development and maintenance of these organisms.

SBRs are widely used as a one-stage system for the treatment of azo dyes, and they have also shown good color, organic matter, and aromatic amine removal performance; nonetheless, they have mostly been used for the treatment of synthetic wastewater. SBRs have the advantage of being versatile, with ability to operate in different environments (anaerobic, aerobic, microaerophilic, or intermittently aerated) in the same reaction cycle, using different types of biomass, such as granular, flocculent of biofilm. Microaerated UASB reactors have shown promising results, yet there are not many studies with this system and with similar microaerated reactors working with continuous flow.

Anaerobic reactors used for the degradation of azo dyes and organic matter are well consolidated; however, the aerobic stage still needs further investigations, especially regarding its optimization in continuous or intermittent aeration or microaeration systems used to treat both synthetic and real textile wastewater.

Acknowledgements The authors thank the following funding agencies for the financial support: FACEPE (Science and Technology Foundation of the State of Pernambuco, Brazil—process numbers IBPG-0553301/21 and APQ-0456-3.07/20), CAPES (Coordination for the Improvement of Higher Education Personnel—process number 88887.609069/2021-00 and Proap and PrInt programs), and CNPq (National Council for Scientific and Technological Development—grant number 304862/2018-5).

References

1. Amaral FM, Florêncio L, Kato MT, Santa-Cruz PA, Gavazza S (2017) Hydraulic retention time influence on azo dye and sulfate removal during the sequential anaerobic-aerobic treatment of real textile wastewater. Water Sci Technol 76(12):3319–3327. https://doi.org/10.2166/wst.2017.378
2. Amaral FM, Kato MT, Florêncio L, Gavazza S (2014) Color, organic matter and sulfate removal from textile effluents by anaerobic and aerobic processes. Bioresour Technol 163:364–369. https://doi.org/10.1016/j.biortech.2014.04.026
3. Amin MSA, Stüber F, Giralt J, Fortuny A, Fabregat A, Font J (2021) Comparative anaerobic decolorization of azo dyes by carbon-based membrane bioreactor. Water 13(8):1060. https://doi.org/10.3390/w13081060
4. Amorim SM, Kato MT, Florencio L, Gavazza S (2013) Influence of redox mediators and electron donors on the anaerobic removal of color and chemical oxygen demand from textile effluent. Clean - Soil, Air, Water 41:928–933. https://doi.org/10.1002/clen.201200070
5. Arora PK (2015) Bacterial degradation of monocyclic aromatic amine. Front Microbiol 6:1–14. https://doi.org/10.3389/fmicb.2015.00820
6. Arora S, Saini HS, Singh K (2011) Biological decolorization of industrial dyes by Candida tropicalis and Bacillus firmus. Water Sci Technol 63:761–768. https://doi.org/10.2166/wst.2011.305
7. Balapure K, Bhatt N, Madamwar D (2015) Mineralization of reactive azo dyes present in simulated textile waste water using down flow microaerophilic fixed film bioreactor. Biores Technol 175:1–7. https://doi.org/10.1016/j.biortech.2014.10.040
8. Balapure KH, Jain K, Chattaraj S, Bhatt NS, Madamwar D (2014) Co-metabolic degradation of diazo dye—reactive blue 160 by enriched mixed cultures BDN. J Hazard Mater 279:85–95. https://doi.org/10.1016/j.jhazmat.2014.06.057
9. Banat IM, Nigam P, Singh D, Marchant R (1996) Microbial decolorization of textile-dye-containing effluents: a review. Bioresour Technol 58:217–227. https://doi.org/10.1016/S0960-8524(96)00113-7
10. Barros AR, Amorim ELC, Reis CM, Shida GM, Silva ED (2010) Biohydrogen production in anaerobic fluidized bed reactors: effect of support material and hydraulic retention time. Int J Hydrogen Energy 35:3379–3388. https://doi.org/10.1016/j.ijhydene.2010.01.108
11. Beydilli MI, Pavlostathis SG, Tincher WC (1998) Decolorization and toxicity screening of selected reactive azo dyes under methanogenic conditions. Water Sci Technol 38:225–232. https://doi.org/10.1016/S0273-1223(98)00531-9
12. Bonakdarpour B, Vyrides I, Stuckey DC (2011) Comparison of the performance of one stage and two stage sequential anaerobic-aerobic biological processes for the treatment of reactive-azo-dye-containing synthetic wastewaters. Int Biodeterior Biodegrad 65:591–599. https://doi.org/10.1016/j.ibiod.2011.03.002
13. Brás R, Gomes A, Ferra MIA, Pinheiro HM, Gonçalves IC (2005) Monoazo and diazo dye decolorisation studies in a methanogenic UASB reactor. J Biotechnol 115:57–66. https://doi.org/10.1016/j.jbiotec.2004.08.001
14. Brik M, Schoeberl P, Chamam B et al (2006) Advanced treatment of textile wastewater towards reuse using a membrane bioreactor. Process Biochem 41:1751–1757. https://doi.org/10.1016/j.procbio.2006.03.019
15. Buitrón G, Quezada M, Moreno G (2004) Aerobic degradation of the azo dye acid red 151 in a sequencing batch biofilter. Biores Technol 92(2):143–149. https://doi.org/10.1016/j.biortech.2003.09.001
16. Carvalho JRS, Amaral FM, Florencio L, Kato MT, Delforno TP, Gavazza S (2020) Microaerated UASB reactor treating textile wastewater: the core microbiome and removal of azo dye Direct Black 22. Chemosphere 242:125157. https://doi.org/10.1016/j.chemosphere.2019.125157

17. Cervantes FJ, Dos Santos AB (2011) Reduction of azo dyes by anaerobic bacteria: microbiological and biochemical aspects. Environ Sci Biotechnol 10:125–137.https://doi.org/10.1007/s11157-011-9228-9
18. Cervantes FJ, Gómez R, Alvarez LH, Martinez CM, Hernandez-Montoya V (2015) Efficient anaerobic treatment of synthetic textile wastewater in a UASB reactor with granular sludge enriched with humic acids supported on alumina nanoparticles. Biodegradation 26:289–298. https://doi.org/10.1007/s10532-015-9734-5
19. Cervantes FJ, Van Der Velde S, Lettinga G, Field JA (2000) Competition between methanogenesis and quinone respiration for ecologically important substrates in anaerobic consortia. FEMS Microbiol Ecol 34:161–171. https://doi.org/10.1016/S0168-6496(00)00091-X
20. Cervantes FJ, Van Der Zee FP, Lettinga G, Field JA (2001) Enhanced decolourisation of acid orange 7 in a continuous UASB reactor with quinones as redox mediators. Water Sci Technol 44:123–128. https://doi.org/10.2166/wst.2001.0198
21. Chengalroyen MD, Dabbs ER (2013) The microbial degradation of azo dyes: minireview. World J Microbiol Biotechnol 29:389–399. https://doi.org/10.1007/s11274-012-1198-8
22. Chinwetkitvanich S, Tuntoolvest M, Panswad T (2000) Anaerobic decolorization of reactive dyebath effluents by a two-stage UASB system with tapioca as a co-substrate. Water Res 34:2223–2232. https://doi.org/10.1016/S0043-1354(99)00403-0
23. Chong S, Sen TK, Kayaalp A, Ang HM (2012) The performance enhancements of upflow anaerobic sludge blanket (UASB) reactors for domestic sludge treatment-A State-of-the-art review. Water Res 46:3434–3470. https://doi.org/10.1016/j.watres.2012.03.066
24. Cirik K, Dursun N, Sahinkaya E, Çinar Ö (2013) Effect of electron donor source on the treatment of Cr(VI)-containing textile wastewater using sulfate-reducing fluidized bed reactors (FBRs). Bioresour Technol 133:414–420. https://doi.org/10.1016/j.biortech.2013.01.064
25. Çinar Ö, Yaşar S, Kertmen M et al (2008) Effect of cycle time on biodegradation of azo dye in sequencing batch reactor. Process Saf Environ Prot 86:455–460. https://doi.org/10.1016/j.psep.2008.03.001
26. Coughlin MF, Kinkler BK, Tepper A, Bishop PL (1997) Characterization of aerobic azo dye-degrading bacteria and their activity in biofilms. Water Sci Technol 36:215–220. https://doi.org/10.1016/S0273-1223(97)00327-2
27. Daeshvar N, Ayazloo M, Khataee AR, Pourhassan M (2007) Biological decolorization of dye solution containing Malachite Green by microalgae Cosmarium sp. Bioresour Technol 98:1176–1182. https://doi.org/10.1016/j.biortech.2006.05.025
28. Donlon B, Razo-Flores E, Luijten M, Swarts H, Lettinga G, Field J (1997) Detoxification and partial mineralization of the azo dye mordant orange 1 in a continuous upflow anaerobic sludge-blanket reactor. Appl Microbiol Biotechnol 47:83–90. https://doi.org/10.1007/s002530050893
29. dos Santos AB, Cervantes FJ, van Lier JB (2007) Review paper on current technologies for decolourisation of textile wastewaters: perspectives for anaerobic biotechnology. Bioresour Technol 98:2369–2385
30. Dos Santos AB, Bisschops IAE, Cervantes FJ, Van Lier JB (2004) Effect of different redox mediators during thermophilic azo dye reduction by anaerobic granular sludge and comparative study between mesophilic (30°C) and thermophilic (55°C) treatments for decolourisation of textile wastewaters. Chemosphere 55:1149–1157. https://doi.org/10.1016/j.chemosphere.2004.01.031
31. Dos Santos AB, Cervantes FJ, Van Lier JB (2004) Azo dye reduction by thermophilic anaerobic granular sludge, and the impact of the redox mediator anthraquinone-2,6-disulfonate (AQDS) on the reductive biochemical transformation. Appl Microbiol Biotechnol 64:62–69. https://doi.org/10.1007/s00253-003-1428-y
32. Dos Santos AB, Cervantes FJ, Yaya-Beas RE, Van Lier JB (2003) Effect of redox mediator, AQDS, on the decolourisation of a reactive azo dye containing triazine group in a thermophilic anaerobic EGSB reactor. Enzyme Microb Technol 33:942–951. https://doi.org/10.1016/j.enzmictec.2003.07.007

33. Ekici P, Leupold G, Parlar H (2001) Degradability of selected azo dye metabolites in activated sludge systems. Chemosphere 44:721–728. https://doi.org/10.1016/S0045-6535(00)00345-3
34. Farabegoli G, Chiavola A, Rolle E (2009) The Biological Aerated Filter (BAF) as alternative treatment for domestic sewage. Optimization of plant performance. J Hazard Mater 171:1126–1132. https://doi.org/10.1016/j.jhazmat.2009.06.128
35. Fernando E, Keshavarz T, Kyazze G (2014) Complete degradation of the azo dye Acid Orange-7 and bioelectricity generation in an integrated microbial fuel cell, aerobic two-stage bioreactor system in continuous flow mode at ambient temperature. Bioresour Technol 156:155–162. https://doi.org/10.1016/j.biortech.2014.01.036
36. Ferraz ADN, Kato MT, Florencio L, Gavazza S (2011) Textile effluent treatment in a UASB reactor followed by submerged aerated biofiltration. Water Sci Technol 64:1581–1589. https://doi.org/10.2166/wst.2011.674
37. Ferraz ERA, Grando MD, Oliveira DP (2011) The azo dye Disperse Orange 1 induces DNA damage and cytotoxic effects but does not cause ecotoxic effects in Daphnia similis and Vibrio fischeri. J Hazard Mater 192:628–633. https://doi.org/10.1016/j.jhazmat.2011.05.063
38. Field JA, Stams AJM, Kato M, Schraa G (1995) Enhanced biodegradation of aromatic pollutants in cocultures of anaerobic and aerobic bacterial consortia. Antonie Van Leeuwenhoek 67:47–77. https://doi.org/10.1007/BF00872195
39. Florêncio TM, Godoi LAG, Rocha VC, Oliveira JMS, Motteran F, Gavazza S, Vicentine KFD, Damianovic MHRZ (2021) Anaerobic structured-bed reactor for azo dye decolorization in the presence of sulfate ions. J Chem Technol Biotechnol 96:1700–1708. https://doi.org/10.1002/jctb.6695
40. Foresti L (2002) Anaerobic treatment of domestic sewage: established technologies and perspectives. Water Sci Technol 45:181–186. https://doi.org/10.2166/wst.2002.0324
41. Forgacs E, Cserhati T, Oros G (2004) Removal of synthetic dyes from wastewaters: a Review. Environ Int 30:953–571. https://doi.org/10.1016/j.envint.2004.02.001
42. Fuchs G (2008) Anaerobic metabolism of aromatic compounds. Ann N Y Acad Sci 1125:82–99. https://doi.org/10.1196/annals.1419.010
43. Fuchs G, Boll M, Heider J (2011) Microbial degradation of aromatic compounds- from one strategy to four. Nat Rev Microbiol 9:803–816. https://doi.org/10.1038/nrmicro2652
44. Gavazza S, Guzman JJL, Angenent LT (2015) Electrolysis within anaerobic bioreactors stimulates breakdown of toxic products from azo dye treatment. Biodegradation 26:151–160. https://doi.org/10.1007/s10532-015-9723-8
45. Ghoropade A, Spencer HT (1993) Azo dyes metabolism by Pseudomonas putida. In: Wukasch RF (Ed.), Proceedings of 48th purdue industrial waste conference. Lewis Publishers Inc., Chelsea, Michigan, pp 699–714
46. Gomi N, Yoshida S, Matsumoto K, Okudomi M, Konno H, Hisabori T, Sugano Y (2011) Degradation of the synthetic dye amaranth by the fungus Bjerkandera adusta Dec 1: inference of the degradation pathway from an analysis of decolorized products. Biodegrad 22:1239–1245. https://doi.org/10.1007/s10532-011-9478-9
47. Gottlieb A, Shaw C, Smith A, Wheatley A, Forsythe S (2003) The toxicity of textile reactive azo dyes after hydrolysis and decolourisation. J Biotechnol 101:49–56. https://doi.org/10.1016/S0168-1656(02)00302-4
48. Grand View Research (2021) Textile Market Size Worth $1412.5 Billion By 2028: CAGR: 4.4%. https://www.grandviewresearch.com/press-release/global-textile-market. Accessed 25 Jun 25 2021
49. Haroun M, Idris A (2009) Treatment of textile wastewater with an anaerobic fluidized bed reactor. Desalination 237:357–366. https://doi.org/10.1016/j.desal.2008.01.027
50. Haug W, Schmidt A, Nortemann B, Hempel DC, Stolz A, Knackmuss HJ (1991) Mineralization of the sulfonated azo dye Mordant Yellow 3 by a 6-aminonaphthalene-2-sulfonate-degrading bacterial consortium. Appl Environ Microbiol 57:3144–3149. https://doi.org/10.1128/aem.57.11.3144-3149.1991
51. Hayaishi O (2008) From oxygenase to sleep. J Biol Chem 283(28):19165–19175. https://doi.org/10.1074/jbc.X800002200

52. Ibrahim Z, Amim MFM, Yahya A, Aria A, Muda K (2010) Characteristics of developed granules containing selected decolourising bacteria for the degradation of textile wastewater. Water Sci Technol 61(5):1279–1288. https://doi.org/10.2166/wst.2010.021
53. Işik M, Sponza DT (2008) Anaerobic/aerobic treatment of a simulated textile wastewater. Sep Purif Technol 60:64–72. https://doi.org/10.1016/j.seppur.2007.07.043
54. Jiang H, Bishop PL (1994) Aerobic biodegradation of azo dyes in biofilms. Water Sci Technol 29:525–530
55. Katuri KP, Mohan SV, Sridhar S, Pati BR, Sarma PN (2009) Laccase membrane reactors for decolorization of an acid azo dye in aqueous phase: process optimization. Water Res 43:3647–3658. https://doi.org/10.1016/j.watres.2009.05.028
56. Kalyuzhnyi S, Sklyar V (2000) Biomineralisation of azo dyes and their breakdown products in anaerobic-aerobic hybrid and UASB reactors. Water Sci Technol 41:23–30. https://doi.org/10.2166/wst.2000.0233
57. Khan MZ, Mondal PK, Sabir S (2011) Bioremediation of 2-chlorophenol containing wastewater by aerobic granules-kinetics and toxicity. J Hazard Mater 190:222–228. https://doi.org/10.1016/j.jhazmat.2011.03.029
58. Khan MZ, Singh S, Sultana S, Sreekrishnan TR, Ahammad SZ (2015) Studies on the biodegradation of two different azo dyes in bioelectrochemical systems. New J Chem 39(7):5597–5604. https://doi.org/10.1039/c5nj00541h
59. Khan R, Bhawana P, Fulekar MH (2013) Microbial decolorization and degradation of synthetic dyes: a review. Rev Environ Sci Biotechnol 12:75–97. https://doi.org/10.1007/s11157-012-9287-6
60. Khandegar V, Saroha AK (2013) Electrocoagulation for the treatment of textile industry effluent–a review. J Environ Manage 128:949–963. https://doi.org/10.1016/j.jenvman.2013.06.043
61. Khelifi E, Gannoun H, Touhami Y, Bouallagui H, Hamdi M (2008) Aerobic decolorization of the indigo dye containing textile wastewater using continuous reactors. J Hazard Mater 152:683–689. https://doi.org/10.1016/j.jhazmat.2007.07.059
62. Knowles R (1982) Denitrification. Microbiol Rev 46:43–70. 0146-0749/82/010043-28$02.00/0
63. Kolekar YM, Nemade HN, Markad VL, Adav SS, Patole MS, Kodam KM (2012) Decolorization and biodegradation of azo dye, reactive blue 59 by aerobic granules. Bioresour Technol 104:818–822. https://doi.org/10.1016/j.biortech.2011.11.046
64. Koupaie EH, Moghaddam MRA, Hashemi SH (2013) Evaluation of integrated anaerobic/aerobic fixed-bed sequencing batch biofilm reactor for decolorization and biodegradation of azo dye Acid Red 18: comparison of using two types of packing media. Bioresour Technol 127:415–421. https://doi.org/10.1016/j.biortech.2012.10.003
65. Koupaie EH, Moghaddam MRA, Hashemi SH (2012) Investigation of decolorization kinetics and biodegradation of azo dye Acid Red 18 using sequential process of anaerobic sequencing batch reactor/moving bed sequencing batch biofilm reactor. Int Biodeterior Biodegrad 71:43–49. https://doi.org/10.1016/j.ibiod.2012.04.002
66. Koupaie EH, Moghaddam MRA, Hashemi SH (2011) Post-treatment of anaerobically degraded azo dye Acid Red 18 using aerobic moving bed biofilm process: enhanced removal of aromatic amines. J Hazard Mater 195:147–154. https://doi.org/10.1016/j.jhazmat.2011.08.017
67. Kudlich M, Hetheridge MJ, Knackmuss HJ, Stolz A (1999) Autoxidation reactions of different aromatic o-aminohydroxynaphthalenes that are formed during the anaerobic reduction of sulfonated azo dyes. Environ Sci Technol 33:896–901. https://doi.org/10.1021/es9808346
68. Kulla HG (1981) Aerobic bacterial degradation of azo dyes. In: Leisinger T (ed) Microbial degradation of xenobiotic and recalcitrant compounds. Academic Press, London, pp 387–399
69. Kulla HG, Klausener F, Meyer U, Lüdeke B, Leisinger T (1983) Interference of aromatic sulfo groups in the microbial degradation of the azo dyes Orange I and Orange II. Arch Microbiol 135:1–7. https://doi.org/10.1007/BF00419473

70. Lettinga G (1995) Anaerobic digestion and wastewater treatment systems. Antonie Van Leeuwenhoek 67:3–28. https://doi.org/10.1007/BF00872193
71. Lettinga G, Velsen ASM, Hobma SW, Zeeuw W, Klapwijk A (1980) Use of the upflow sludge blanket (USB) reactor concept for biological wastewater treatment. Biotechnol Bioeng 22:699–734. https://doi.org/10.1002/bit.260220402
72. Li N, Wei D, Wang S, Hu L, Xu W, Du B, Wei Q (2017) Comparative study of the role of extracellular polymeric substances in biosorption of Ni (II) onto aerobic/anaerobic granular sludge. J Colloid Interface Sci 490:754–761. https://doi.org/10.1016/j.jcis.2016.12.006
73. Libra JA, Borchert M, Vigelahn L, Storm T (2004) Two stage biological treatment of a diazo reactive textile dye and the fate of the dye metabolites. Chemosphere 56:167–180. https://doi.org/10.1016/j.chemosphere.2004.02.012
74. Liu Y-Q, Maulidiany N, Zeng P, Heo S (2021) Decolourization of azo, anthraquinone and triphenylmethane dyes using aerobic granules: acclimatization and long-term stability. Chemosphere 263: 128312. https://doi.org/10.1016/j.chemosphere.2020.128312
75. Lourenço ND, Novais JM, Pinheiro HM (2000) Reactive textile dye colour removal in a sequencing batch reactor. Water Sci Technol 42:321–328. https://doi.org/10.2166/wst.2000.0531
76. Lourenço ND, Novais JM, Pinheiro HM (2001) Effect of some operational parameters on textile dye biodegradation in a sequential batch reactor. J Biotechnol 89:163–174. https://doi.org/10.1016/S0168-1656(01)00313-3
77. Manavi N, Kazemi AS, Bonakdarpour B (2017) The development of aerobic granules from conventional activated sludge under anaerobic-aerobic cycles and their adaptation for treatment of dyeing wastewater. Chem Eng J 312:375–384. https://doi.org/10.1016/j.cej.2016.11.155
78. Mata AMT, Pinheiro HM, Lourenço ND (2015) Effect of sequencing batch cycle strategy on the treatment of a simulated textile wastewater with aerobic granular sludge. Bioch Eng Jour 104:106–114. https://doi.org/10.1016/j.bej.2015.04.005
79. Menezes O, Brito R, Hallwass F, Florêncio L, Kato MT, Gavazza S (2019) Coupling intermittent micro-aeration to anaerobic digestion improves tetra-azo dye Direct Black 22 treatment in sequencing batch reactors. Chem Eng Res Des 146:369–378. https://doi.org/10.1016/j.cherd.2019.04.020
80. Miran W, Jang J, Nawaz M, Shahzad A, Lee DS (2018) Sulfate-reducing mixed communities with the ability to generate bioelectricity and degrade textile diazo dye in microbial fuel cells. J Hazard Mater 352:70–79. https://doi.org/10.1016/j.jhazmat.2018.03.027
81. Mishra S, Maiti A (2018) The efficacy of bacterial species to decolourise reactive azo, anthroquinone and triphenylmethane dyes from wastewater: a review. Environ Sci Pollut Res 25:8286–8314. https://doi.org/10.1007/s11356-018-1273-2
82. Modi HA, Rajput G, Ambasana C (2010) Decolorization of water soluble azo dyes by bacterial cultures, isolated from dye house effluent. Bioresour Technol 101:6580–6583. https://doi.org/10.1016/j.biortech.2010.03.067
83. Moddor Intelligence (2020) Global Home Textile Market - Growth, Trends, Covid-19 Impact, And Forecasts (2021 - 2026). https://www.mordorintelligence.com/industry-reports/global-home-textile-market. Accessed 22 Jun 2021
84. Moreira FC, Boaventura RAR, Brillas E, Vilar VJP (2017) Electrochemical advanced oxidation processes: a review on their application to synthetic and real wastewaters. Appl Catal B Environ 202:217–261. https://doi.org/10.1016/j.apcatb.2016.08.037
85. Motteran F, Braga JK, Sakamoto IK, Silva EL, Varesche MBA (2014) Degradation of high concentrations of nonionic surfactant (linear alcohol ethoxylate) in an anaerobic fluidized bed reactor. Sci Total Environ 481:121–128. https://doi.org/10.1016/j.scitotenv.2014.02.024
86. Motteran F, Nadai BM, Braga JK, Silva EL, Varesche MBA (2018) Metabolic routes involved in the removal of linear alkylbenzene sulfonate (LAS) employing linear alcohol ethoxylated and ethanol as co-substrates in enlarged scale fluidized bed reactor. Sci Total Environ 640–641:1411–1423. https://doi.org/10.1016/j.scitotenv.2018.05.375

87. Murali V, Ong S, Ho L, Wong Y (2013) Evaluation of integrated anaerobic–aerobic biofilm reactor for degradation of azo dye methyl orange. Biores Technol 143:104–111. https://doi.org/10.1016/j.biortech.2013.05.122
88. Nguyen TH, Watari T, Hatamoto M, Sutani D, Setiadi T, Yamaguchi T (2020) Evaluation of a combined anaerobic baffled reactor–downflow hanging sponge biosystem for treatment of synthetic dyeing wastewater. Environ Technol Innov 19:100913.https://doi.org/10.1016/j.eti.2020.100913
89. Ooi T, Shibata T, Sato R, Ohno H, Kinoshita S, Thuoc TL, Taguchi S (2007) An azoreductase, aerobic NADH-dependent flavoprotein discovered from Bacillus sp.: functional expression and enzymatic characterization. Appl Microbiol Biotechnol 75:377–386. https://doi.org/10.1007/s00253-006-0836-1
90. O'Neill C, Hawkes FR, Hawkes DL, Esteves S, Wilcox SJ (2000) Anaerobic-aerobic biotreatment of simulated textile effluent containing varied ratios of starch and azo dye. Water Res 34:2355–2361. https://doi.org/10.1016/S0043-1354(99)00395-4
91. Oliveira JMS, Lima MR De, Issa CG, Corbi JJ, Damianovic MHRZ, Foresti E (2019) Intermittent aeration strategy for azo dye biodegradation: A suitable alternative to conventional biological treatments? J Hazard Mater 121558.https://doi.org/10.1016/j.jhazmat.2019.121558
92. Oliveira LL, Duarte ICS, Sakamoto IK, Varesche MBA (2009) Influence of support material on the immobilization of biomass for the degradation of linear alkylbenzene sulfonate in anaerobic reactors. J Environ Manage 90:1261–1268. https://doi.org/10.1016/j.jenvman.2008.07.013
93. Pagga U, Brown D (1986) The degradation of dyestuffs: Part II Behaviour of dyestuffs in aerobic biodegradation tests. Chemosphere 15:479–491. https://doi.org/10.1016/0045-6535(86)90542-4
94. Pandey A, Singh P, Iyengar L (2007) Bacterial decolorization and degradation of azo dyes. Int Biodeterior Biodegrad 59(2):73–84. https://doi.org/10.1016/j.ibiod.2006.08.006
95. Pearce CI, Lloyde JR, Guthrie JT (2003) The removal of colour from textile wastewater using whole bacterial cells: a review. Dyes Pigm 58:179–196. https://doi.org/10.1016/S0143-7208(03)00064-0
96. Pereira R, Pereira L, van der Zee FP, Madalena Alves M (2011) Fate of aniline and sulfanilic acid in UASB bioreactors under denitrifying conditions. Water Res 45:191–200. https://doi.org/10.1016/j.watres.2010.08.027
97. Pinheiro HM, Touraud E, Thomas O (2004) Aromatic amines from azo dye reduction: status review with emphasis on direct UV spectrophotometric detection in textile industry wastewaters. Dyes Pigm 61:121–139. https://doi.org/10.1016/j.dyepig.2003.10.009
98. Punzi M, Anbalagan A, Aragão Börner R, Svensson B-M, Jonstrup M, Mattiasson B (2015) Degradation of a textile azo dye using biological treatment followed by photo-Fenton oxidation: Evaluation of toxicity and microbial community structure. Chem Eng J 270:290–299. https://doi.org/10.1016/j.cej.2015.02.042
99. Punzi M, Nilsson F, Anbalagan A, Svensson B-M, Jönsson K, Mattiasson B, Jonstrup M (2015) Combined anaerobic-ozonation process for treatment of textile wastewater: removal of acute toxicity and mutagenicity. J Hazard Mater 292:52–60. https://doi.org/10.1016/j.jhazmat.2015.03.018
100. Puvaneswari N, Muthukrishnan J, Gunasekaran P (2006) Toxicity assessment and microbial degradation of azo dyes. Indian J Exp Biol 44:618–626
101. Qu Y, Shi S, Ma F, B, (2010) Decolorization of reactive dark blue K-R by the synergism of fungus and bacterium using response surface methodology. Biores Technol 101:8016–8023. https://doi.org/10.1016/j.biortech.2010.05.025
102. Rau J, Knackmuss HJ, Stolz A (2002) Effects of different quinoid redox mediators on the anaerobic reduction of azo dyes by bacteria. Environ Sci Technol 36:1497–1504. https://doi.org/10.1021/es010227+
103. Razo-Flores E, Luijten M, Donlon BA, Lettinga G, Field JA (1997) Complete biodegradation of the azo dye azodisalicylate under anaerobic conditions. Environ Sci Technol 31:2098–2103. https://doi.org/10.1021/es960933o

104. Sahinkaya E, Yurtsever A, Çınar Ö (2016) Treatment of textile industry wastewater using dynamic membrane bioreactor: impact of intermittent aeration on process performance. Sep Purif Technol 174:445–454. https://doi.org/10.1016/j.seppur.2016.10.049
105. Saratale RG, Saratale GD, Chang JS, Govindwar SP (2009) Decolorization and biodegradation of textile dye Navy Blue HER by Trichosporon beigelii NCIM-3326. J Hazard Mater 166:14211428. https://doi.org/10.1016/j.jhazmat.2008.12.068
106. Saratale RG, Saratale GD, Chang JS, Govindwar SP (2011) Bacterial decolorization and degradation of azo dyes: a review. J Taiwan Inst Chem 42:138–157. https://doi.org/10.1016/j.jtice.2010.06.006
107. Sarayu K, Sandhya S (2010) Aerobic biodegradation pathway for Remazol Orange by Pseudomonas aeruginosa. Appl Biochem Biotechnol 160:1241–1253. https://doi.org/10.1007/s12010-009-8592-1
108. Schoeberl P, Brik M, Bertoni M, Braun R, Fuchs W (2005) Optimization of operational parameters for a submerged membrane bioreactor treating dyehouse wastewater. Sep Purif Technol 44:61–68. https://doi.org/10.1016/j.seppur.2004.12.004
109. Seghezzo L, Zeeman G, Van Lier JB, Hamelers HVM, Lettinga G (1998) A review: The anaerobic treatment of sewage in UASB and EGSB reactors. Bioresour Technol 65:175–190. https://doi.org/10.1016/S0960-8524(98)00046-7
110. Selvaraj V, Swarna T, Mansiya C, Alagar M (2021) An over review on recently developed techniques, mechanisms and intermediate involved in the advanced azo dye degradation for industrial applications. J Mol Struct 1224:129195.https://doi.org/10.1016/j.molstruc.2020.129195
111. Shaul GM, Holdsworth TJ, Dempsey CR, Dostal KA (1991) Fate of water soluble azo dyes in the activated sludge process. Chemosphere 22:107–119. https://doi.org/10.1016/0045-6535(91)90269-J
112. Shoukat R, Khan SJ, Jamal Y (2019) Hybrid anaerobic-aerobic biological treatment for real textile wastewater. J Water Process Eng 29:100804.https://doi.org/10.1016/j.jwpe.2019.100804
113. Solis M, Solis A, Perez HI, Manjarrez N, Flores M (2012) Microbial decolouration of azo dyes: a review. Process Biochem 47:1723–1748. https://doi.org/10.1016/j.procbio.2012.08.014
114. Sponza DT, Işik M (2002) Decolorization and azo dye degradation by anaerobic/aerobic sequential process. Enzyme Microb Technol 31:102–110. https://doi.org/10.1016/S0141-0229(02)00081-9
115. Starling MCVM, dos Santos PHR, de Souza FAR, Oliveira SC, Leão MMD, Amorim CC (2017) Application of solar photo-Fenton toward toxicity removal and textile wastewater reuse. Environ Sci Pollut Res 24:12515–12528. https://doi.org/10.1007/s11356-016-7395-5
116. Stolz A (2001) Basic and applied aspects in the microbial degradation of azo dyes. Appl Microbiol Biotechnol 56:69–80. https://doi.org/10.1007/s002530100686
117. Sun W, Li Y, McGuinness LR, Luo S, Huang W, Kerkhof LJ, Mack EE, Häggblom MM, Fennell DE (2015) Identification of anaerobic aniline-degrading bacteria at a contaminated industrial site. Environ Sci Technol 49:11079–11088. https://doi.org/10.1021/acs.est.5b02166
118. Tan L, Li H, Ning S, Xu B (2014) Aerobic decolorization and degradation of azo dyes by suspended growing cells and immobilized cells of a newly isolated yeast Magnusiomyces ingens LH-F. Bioresour Technol 158:321–328. https://doi.org/10.1016/j.biortech.2014.02.063
119. Tan L, Ning S, Zhang X, Shi S (2013) Aerobic decolorization and degradation of azo dyes by growing cells of a newly isolated yeast Candida tropicalis TL-F1. Biores Technol 138:307–313. https://doi.org/10.1016/j.biortech.2013.03.183
120. Tan NCG, Borger A, Slenders P, Svitelskaya A, Lettinga G, Field JA (2000) Degradation of azo dye Mordant Yellow 10 in a sequential anaerobic and bioaugmented aerobic bioreactor. Water Sci Technol 42:337–344. https://doi.org/10.2166/wst.2000.0533
121. Van Der Zee FP, Villaverde S (2005) Combined anaerobic-aerobic treatment of azo dyes-A short review of bioreactor studies. Water Res

122. Van der Zee FP, Lettinga G, Field JA (2001) Azo dye decolourisation by anaerobic granular sludge. Chemosphere 44:1169–1176. https://doi.org/10.1016/S0045-6535(00)00270-8
123. Waghmode TR, Kurade MB, Khandare RV, Govindwar SP (2011) A sequential aerobic/microaerophilic decolorization of sulfonated mono azo dye Golden Yellow HER by microbial consortium GG-BL. Int Biodeterior Biodegrad 65:1024–1034. https://doi.org/10.1016/j.ibiod.2011.08.002
124. Wang X, Cheng X, Sun D, Ren Y, Xu G (2014) Fate and transformation of naphthylamine-sulfonic azo dye Reactive Black 5 during wastewater treatment process. Environ Sci Pollut Res 21:5713–5723. https://doi.org/10.1007/s11356-014-2502-y
125. Zhu Y, Wang W, Ni J, Hu B (2020) Cultivation of granules containing anaerobic decolorization and aerobic degradation cultures for the complete mineralization of azo dyes in wastewater. Chemosphere 246:125753.https://doi.org/10.1016/j.chemosphere.2019.125753
126. Zhu Y, Xu J, Cao X, Cheng Y, Zhu T (2018) Performances and structures of functional microbial communities in the mono azo dye decolorization and mineralization stages. Chemosphere 210:1051–1060. https://doi.org/10.1016/j.ibiod.2018.03.006
127. Zhu Y, Xu J, Cao X, Cheng Y (2018) Characterization of functional microbial communities involved in different transformation stages in a full-scale printing and dyeing wastewater treatment plant. Biochem Eng J 137:162–171

Fundamental of Aerobic and Anaerobic Processes in Dye Wastewater

Pallavi Jain, A. Geetha Bhavani, Prashant Singh, and Madhur Babu Singh

Abstract One of the critical challenges of the textile industry is the removal and degradation of dyes. While designing a wastewater treatment plant, the design, environment, and toxicity concerns associated with releasing textile effluents into a water body must be considered. The various traditional methods like coagulation, adsorption, oxidation, and flocculation are used to remove the contamination. Compared to all oxidation and physicochemical methods, the biological process, either aerobic or anaerobic is easy and demonstrates substantial benefits. This chapter covers the different aerobic and anaerobic processes employed to remove contaminants from textile industries.

Keywords Decolorization · Textile effluent · Biological treatment · Aerobic · Anaerobic treatments

1 Introduction

The wastewaters left from textile industries are heavily polluted and possess higher temperatures with pH, strong colour, and a higher percentage of chemical oxygen demand (COD) concentration. It also demonstrates a lower ability of biodegradation [11, 16, 38]. These wastewaters need to be treated and disposed of properly, otherwise lead to grim health and environmental concerns. The capacity of reoxygenation and penetration of sunlight will be stopped by the wastewater containing coloured contaminates and pollute the water bodies which degrade the aquatic life

P. Jain (✉) · M. B. Singh
Department of Chemistry, SRM Institute of Science and Technology, Delhi-NCR Campus, Modinagar, Ghaziabad 201204, U.P., India

A. G. Bhavani
Department of Chemistry, School of Sciences, Noida International University, Greater Noida 2013086, U.P., India

P. Singh
Department of Chemistry, Atma Ram Sanatan Dharma College, University of Delhi, New Delhi, India

A. Khadir et al. (eds.), *Biological Approaches in Dye-Containing Wastewater*, Sustainable Textiles: Production, Processing, Manufacturing & Chemistry,
https://doi.org/10.1007/978-981-19-0545-2_2

[35]. The various traditional methods like coagulation, adsorption, oxidation, and flocculation [18, 53] are used to remove the contamination. Compared to all oxidation and physicochemical methods, the biological process is easy and demonstrates substantial benefits. Among all the treatments, the frequently employed biological methods [23] in the form of aerobic and anaerobic ways, are the potential methods to convert the complex pollutants [42]. The activated sludge usage in aerobic biological treatment is quite regular to treat effluents from the textile industry. Even though activated sludge is not up to the mark to degrade the complex dye molecules, however, researchers are finding alternatives with the combination of bacteria and fungi for effective dye degradation [34]. The anaerobic process significantly cleaves the azo dye bonds, which leads to aromatic amine formation after the degradation [37, 55]. For any type of anaerobic process, the highly reactive bioreactors are deployed for increasing the efficiency to run for a longer time [33]. The textile wastewater can undergo an aerobic process before the anaerobic process as a post-treatment in the case of coloured wastewater [9]. The chapter aims to intend to provide insights into dye removal from industrial effluents left in wastewaters through anaerobic and aerobic treatments.

2 Impact on Environment

The existence of the untreated dye in water bodies is the biggest concern to the environment, which hinders the photosynthesis of underwater plants by preventing sunlight [57]. Few reports showed that untreated dyes led to toxic aromatic amine affecting the bladder, liver, spleen, and causing deformities in the chromosomes of mammalian cells [2]. Many issues are related to wastewater treatment with the degradation process of dye molecules. The industrial effluents are highly corrosive, possess an unpleasant smell and the capacity to increase the insolubility of the materials in water. These insoluble materials in water with chemical contaminants merge with fresh waters and increases the turbid formation [25], which are unsuitable for irrigation and human consumption, and other needs. The inorganic contaminants like sulphur trioxide and chromium are dangerous to aquatic life and to the beneficial microorganisms contributing to clean water bodies naturally. The range and level of the contamination may vary with the textile effluent and the chemicals used in the process and may lead to the variation in the pH levels that affect the aquatic life.

Some researchers [3, 14, 39] reported the anaerobic conditions for the decolorization mechanism of azo dye using bacteria. This bacterium allowed the electron transfer by the enzymatic process to link ATP generation. Although other studies [26] showed that Fe^{2+} or H_2S generated from inorganic compounds were far more effective to degrade the azo dyes compared to the synthesis of ATP generation through the anaerobic process. The inorganic contaminants (nitrites, nitrates, sulphides, starch) and organic contaminants (cotton, starch) are easily degraded by bacteria, which improves the BOD of the water. The consequences of the benzidine and its analogues were studied over hamsters, canines, and mice as a model. The outcome of the research showed the presence of N-acetylated and aromatic amines and their derivatives in their urine and was quite tumorigenic. The findings strongly indicated that certain textile dye chemicals may lead to cancer and certain other illnesses in individuals and other aquatic animals.

3 Biological Processes

Every industrial and commercial wastewater is highly polluted and contains biodegradable constituents. Thus, textile effluents can be effectively treated by the aerobic and anaerobic processes. Apart from the degradation of organic matter, the adsorption process and the conversion of the complex molecules to simple substances using microbes played a vital role in reducing pollutants like heavy metal complexes [50]. Figure 1 shows the two ways of the biological treatments, i.e., aerobic and anaerobic processes depending on the microorganism's growth.

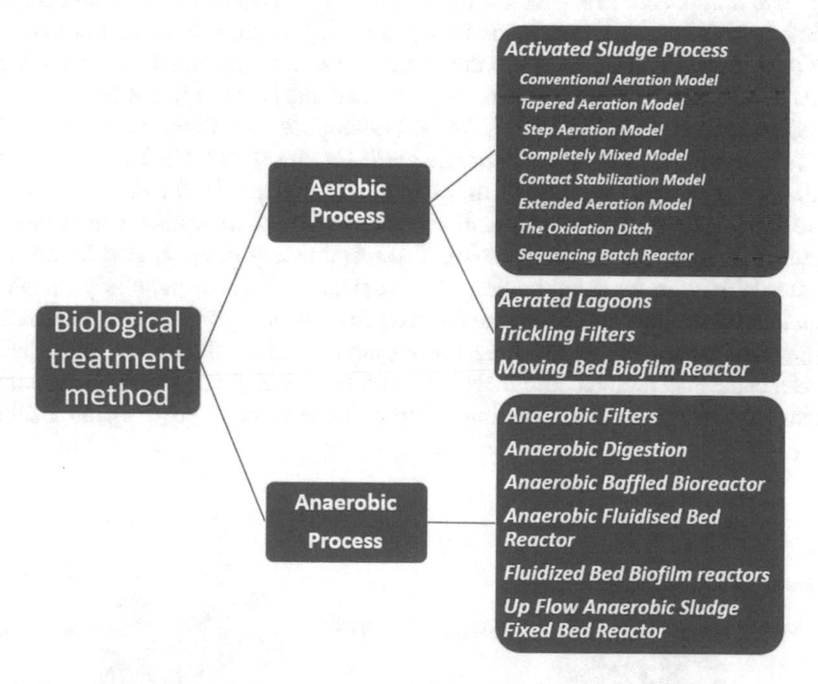

Fig. 1 Flow chart of the biological processing of the textile effluent in the treatment process [5]

4 Aerobic and Anaerobic Process

These processes are classified based on the presence of air in the treatment process. Figure 2a, b demonstrates oxygen as an electron acceptor in the aerobic system, whereas Fig. 2b shows that the organic material acts as an electron acceptor in the anaerobic system (oxygen is absent). In the aerobic system, the oxygen converts the organic compounds into CO_2, H_2O, and new cells. However, in the anaerobic system, CH_4, CO_2, and new cells are produced at the end of the treatment in the absence of oxygen.

Both the processes are quite effective for textile wastewater treatments with substantial benefits and are proved to be eco-friendly as chemical usage is not required. The chemical content and volumes of the sludge produced because of cell manufacturing are modest and nontoxic in both the processes. Additionally, the water consumption is on the lower side in comparison to physicochemical treatments. The mixed processes lead to better results due to the synergetic metabolic activity compared to a single process, i.e., aerobic or anaerobic system [13]. Further, every dye compound will have an interaction with various kinds of independent strains through bonds, which leads to a proper degradation process. Bacteria like *B.subtilis, A.hydrophilia, B.cereus* are quite effective in decomposing the azo dye bonds in wastewaters [4]. Many reports confirmed that bacteria like *A.faecalis 6132, B.megaterium, B. licheniformis LS04, R.erythropolis 24, P.desmolyticum NCIM 2112, R.aquatilis, M. aerodenitrificans, A.guillouiae* are efficient in decolourizing the textile wastewaters [19, 24, 40]. On the other hand, fungi like *D.flavida, H.larincinus, D.squalens, I.hispidus, P.tremellosa, P.sanguineus, C.versicolor* also seem to be quite helpful in treating the wastewaters [6]. The membrane process is also found to be commendable with the combination of biological treatments, popularly known as membrane bioreactor. When compared to a traditional biological process followed by sedimentation, the membrane bioreactor has numerous benefits like the high quality of water in the final product, minimal sludge production, requires less area, and highly effective in biological decomposition. Microorganisms decompose organic material

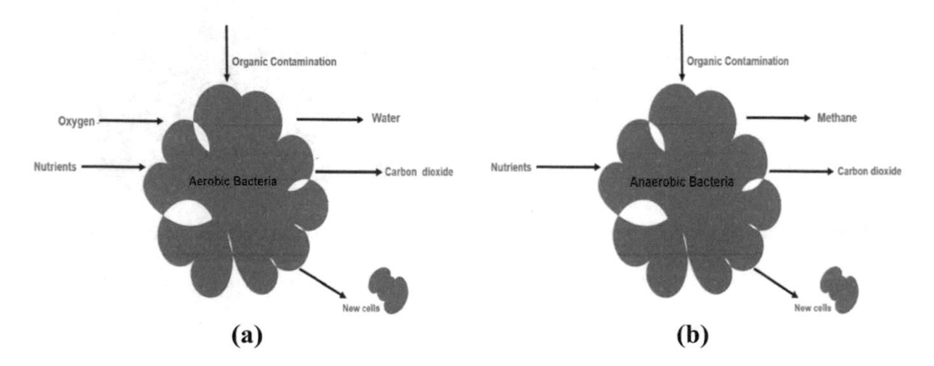

Fig. 2 **a** Aerobic and **b** Anaerobic treatment process [43]

in biological treatment processes, resulting in the formation of new cells. For the system's stability, the produced sludge containing old microbes should be eliminated. The separation of the sludge is carried out by a conventional system, i.e., sedimentation. In the sedimentation system, sediment sludge is removed from the bottom of the tank. In few cases, flocculants are added to initiate the sedimentation process. While the membrane approach separates sludge from the liquid mixture utilizing a membrane with a large pore width. The above-discussed benefits of membrane bioreactor are encouraging the textile industries for effluent treatment.

5 Aerobic Processes

5.1 Activated Sludge Process (ASP)

This process is aerobic with continuous flow system having large quantities of microorganisms (protozoa and bacteria) for stable organic matter decomposition to produce CO_2, H_2O, and new cells with few other products [49]. The aerobic basin is mechanically aerated to mix the reactor liquor completely for a specific time. The 15 Gram-negative bacteria are mainly used in activated sludge systems, and also include nitrogen-carbon oxidizers, facultative anaerobes and aerobes, non-floc formers [7]. As shown in Fig. 3, the mixed liquor from the basin is transferred to the secondary sedimentation tank (SST)/clarifier to settle the sludge and remove it from the SST. For effective removal of organic matters, many operational parameters like (a) frequent checking of dissolved oxygen in tanks, (b) to control the activated sludge returning, and (c) to regulate the activated sludge waste, must be taken care of. Recycling is essential for any activated sludge system. The different types of ASP are discussed below.

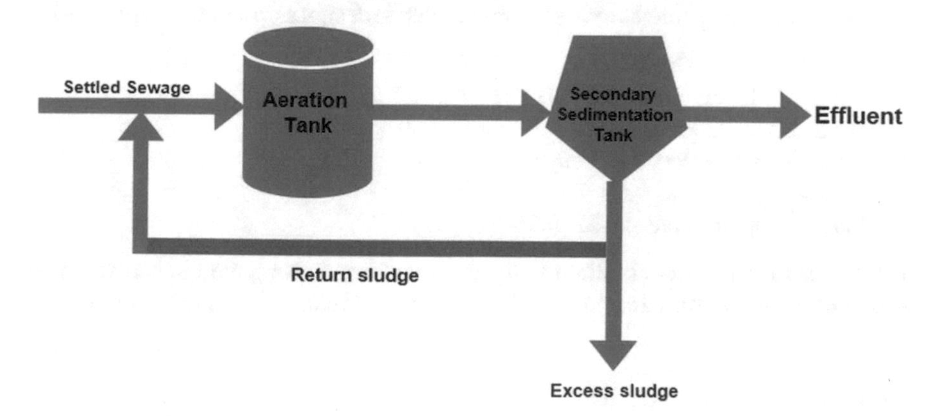

Fig. 3 Conventional activated sludge process [28]

5.1.1 Conventional Aeration Model

The conventional process has a plug flow type aeration tank. The recycled sludge and industrial wastewaters enter at the top of the tank and get mixed with mechanical aerators/diffusers. The mixture is aerated for 5–6 h at a constant rate. During the period of constant aeration, the organic matter undergoes the flocculation, adsorption, and oxidation processes. The following parameters are recommended to design conventional aeration type of ASP:

- Food to microorganism ratio (F/M): 0.2–0.4 kg BOD kg^{-1} VSS.d
- Volumetric loading rate: 0.3–0.6 kg BOD m^{-3} .d
- Mixed liquor suspended solids (MLSS): 1500–3000 mg L^{-1}
- Cell residence period: 5–15 days
- Hydraulic retention time (HRT): 4–8 h
- Sludge circulation ratio: 0.25–0.5

5.1.2 Tapered Aeration Model

Tapered aeration is a similar process of plug flow model as the effluent moves from the top of the tank to the end. In the tapered aeration model, maximum air (aeration) is used at the top of the tank and gradually reduces to the bottom of the tank. The rest of the operational conditions are similar to conventional ASP. The tapered aeration model is highly economic as it increases the efficiency of the aeration unit.

5.1.3 Step Aeration Model

In this process, the wastewater is loaded at more than one point along with aeration channels. The aeration is uniform throughout the tank and lowers the returned sludge load. The following parameters are recommended to design step aeration type of ASP:

- F/M ratio: 0.2–0.4 kg BOD kg^{-1} VSS.d
- Volumetric loading rate: 0.6–1.0 kg BOD m^{-3} .d
- MLSS: 2000–3500 mg L^{-1}
- Cell residence period: 5–15 days
- HRT: 3–5 h
- Sludge circulation ratio: 0.25–0.75

The step aeration process is slightly higher productive as compared to conventional ASP with better performance due to the reduction of returned sludge load [48].

5.1.4 Completely Mixed Model

In this model, the mixed flow system is adopted in the aeration tank. The wastewater is loaded uniformly from one side through channels with return sludge and the final products are collected from the other side of the tank. The following parameters are recommended to design a completely mixed type of ASP:

- F/M ratio: 0.2–0.6 kg BOD kg^{-1} VSS.d
- Volumetric loading rate: 0.8–2.0 kg BOD m^{-3} .d
- MLSS: 3000–6000 mg L^{-1}
- Cell residence period: 5–15 days
- HRT: 3–5 h
- Sludge circulation ratio: 0.25–1.0

Completely mixed ASP can manage flocculation in organic matter concentrations, and its functionality will not be adversely impacted if any toxic substance (minor concentration) exists in the wastewater for a short period of time. Owing to this property, this model is favoured in treating the wastewater with flocculation issue.

5.1.5 Contact Stabilization Model

This ASP process takes the benefits of activated sludge's absorptive capabilities and removes BOD in two phases. The first step is the absorption phase, during which the activated sludge absorbs many of the colloidal, fine suspended particles, and dissolved organic materials. The second step is the oxidation of organic matter. The model runs in two separate tanks, and finally, the settled products are mixed and aerated in a contact tank for 30–90 min with re-aerated activated sludge. The re-aerated process allows absorbing of the organic matter over sludge flocs and finally separated in SST. The aeration volume required in this model is roughly half that of a conventional/traditional ASP. The existing plant can also be converted to a contact stabilization process by upgrading the plant for potential effect. If the organic material is in liquid form only, then this model is not efficient enough to treat such wastewater [36].

5.1.6 Extended Aeration Model

In this model, the aeration time is quite long with low F/M rate to run for endogenous respiration phase of growth curve. The low excess sludge is produced due to the endogenous respiration of cells. The produced sludge is dried over a sand bed for aerobic decomposition. This model involves a completely mixed type aeration tank and is appropriate for small-scale plants. The following parameters are recommended to design an extended aeration type of ASP:

- F/M ratio: 0.05–0.15 kg BOD kg^{-1} VSS.d
- Volumetric loading rate: 0.1–0.4 kg BOD m^{-3} .d
- MLSS: 3000–6000 mg L^{-1}
- Cell residence period: 20–30 days
- HRT: 18–36 h
- Sludge circulation ratio: 0.75–1.5

5.1.7 Oxidation Ditch

It is a modified ASP that uses lengthy solids retention periods (SRTs) to eliminate biodegradable organics. In this model, the tank is in ditch (oval) shape and consists of 1.0–1.5 m deep ring-shaped channels with appropriate width that are ideal for trapezoidal/rectangular cross section. A Kessener brush aeration rotor is installed crosswise the ditch that offers aeration and wastewater circulation (0.3–0.6 m s^{-1}). The ditch operating conditions are:

(a) shutting the inflow valve and permitting the wastewater aeration for the same duration as of the design detention,
(b) turning off the aeration and circulation mechanism and permitting the deposition of the sludge in the tank itself,
(c) enabling the entering wastewater to replace the cleared effluent by unlocking the inlet and outlet valves.

A continuous wastewater entering was observed in case of continuous operation and is operated as a flow-through system. The vertical aerators are utilized for abundant air supply while also maintaining adequate horizontal velocity to prevent the cells from settling at the ditch's bottom. The generated sludge in the oxidation ditch is lower than the conventional ASP and is dried over the sand bed [31].

5.1.8 Sequencing Batch Reactor (SBR)

The SBR is employed in small-scale plants and centralized sewage treatment. It involves a single complete mixed type reactor that performs all ASP steps (Fig. 4). The reactor basin is filled and aerated for a specific duration. After 0.5 h, cell will settle down and decanted from top that requires around 30 min. The emptied tanks are refilled with wastewater and this process will be repeated until the idle step is achieved. An idle phase exists between the fill and decant step. In this phase, a little portion of activated sludge from the SBR basin's bottom is squandered. The treatment parameter is attuned with anaerobic, aerobic, and anoxic, for effective denitrification, nitrification, and removal of phosphorus. The SBR requires only one tank with less space and left small footprint [44].

5.2 Aerated Lagoons

Aerated lagoons are one of the main biological treatments of textile industries. The large tanks coated with polythene or rubber are utilized for the aeration of primary treated effluent for 2–6 days. After the aeration, the produced sludge is separated. The process of ammonia nitrification is occurred in lagoons and effectively removes

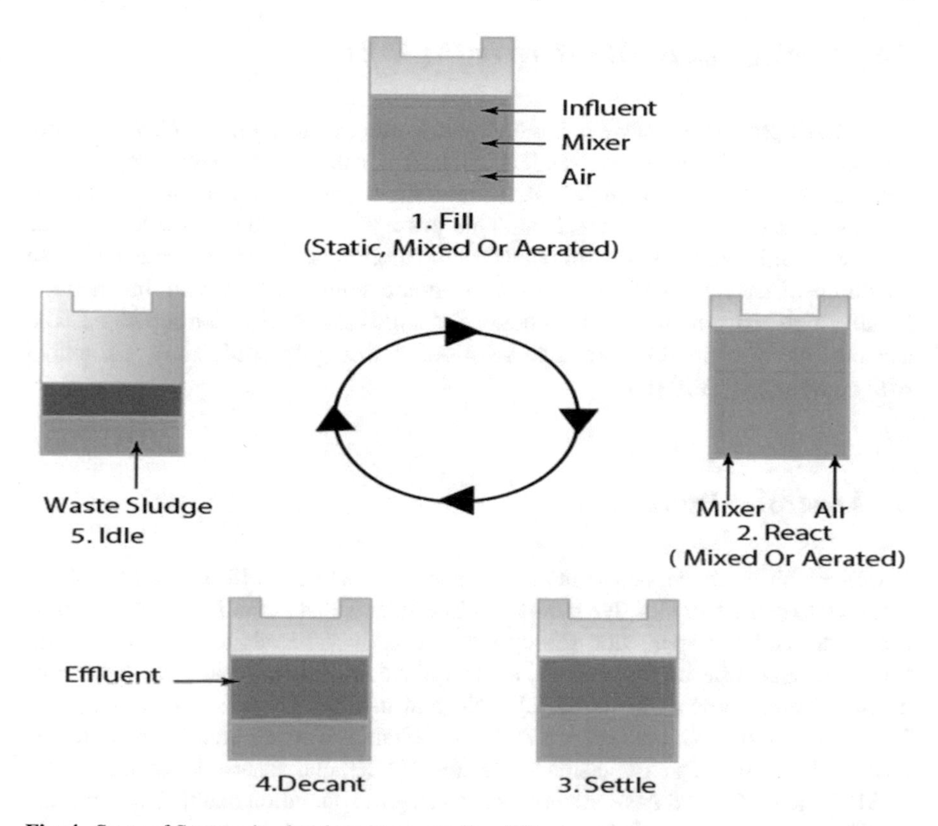

Fig. 4 Steps of Sequencing batch reactor operation [22]

99% of BOD and 15 to 25% of phosphorous. The algae usage in lagoons successfully removes the total suspended solids. The limitation of this process is bacterial contamination and the requirement of large space [12].

5.3 *Trickling Filters*

Trickling filters are used as a secondary treatment in the presence of air. After the primary process, the wastewater is trickled/sprayed over the filters. The filters are prepared with a combination of gravel, coal, broken stones, polyvinyl chloride, and synthetic resins. The filter medium is gel-based formed with microorganisms to oxidize the organic matter to CO_2 and water. The benefit of the trickling filters is requirement is quite less space, whereas limitations are emitting odour and huge capital investment [17].

5.4 Moving Bed Biofilm Reactor (MBBR)

The MBBR process is a new technology with cutting-edge solutions on wastewater treatment with microbiology [61]. The MBBR is suitable for every simple treatment process involving the high BOD/COD removal, nitrogen reduction from challenging industrial effluents treatment. This process is operated over a large space to grow the microorganisms with polyethylene biofilm carriers to operate in precise conditions. Carriers float in the aerobic or anoxic zone to attach with the bacteria. Finally, high active biomass settled down and solids are timely been separated. The usage of biofilms provides effective solutions with highly productivity compared with conventional ASP [60].

6 Anaerobic Process

The anaerobic treatment occurs in the absence of air, which is effective for decolourization and degradation of dye molecules in effluents that come from textile industries. The oxidation–reduction reaction of the azo bond leads to the decolourization under anaerobic conditions, which is a quite economic process [15]. Anaerobic decolourization process in industrial scale and medium scale is very efficacious. The mono azo dye (CI acid orange 7) decolourization was reported in an anaerobic reactor like up-flow anaerobic sludge blanket (UASB) and anaerobic baffled reactor (ABR). The UASB and ABR are used by varying the operation conditions [56]. The UASB reactor was found to be appropriate for the thermophilic system compared to the mesophilic system. Moreover, both the systems are efficient to decolourize the wastewaters [59]. The anaerobic conditions were observed for the azo dye, Black 2 HN, Orange II of 400 mg/l decolourization using biofilter leading to 99% efficiency. The decolourization of the azo dye converted into aromatic amines and found to be carcinogenic [30].

6.1 Anaerobic Filters

Anaerobic filters are practising for the last 100 years. The septic tank is filled with contaminated wastewaters with controlled temperature with activated biomass in an airtight tank. The anaerobic filter, fluidized bed, and growth technique are successfully used for wastewater treatment. This process has much more benefits like easy to load the organic matter, high contact time with microbes, high performance, low cost, reduction of contaminates in wastewater in the small reactor.

6.2 *Anaerobic Digestion*

This sludge is digested in presence of an anaerobic digester for organic matters. The reaction rate and reaction duration in presence of microbes is crucial for the bioaugmentation phase. This anaerobic digestion takes stable operating conditions with long time duration, industrial scale reactor size to complete operation. After the anaerobic process, the released biogas is cached at a high rate. The ideal wastewater digestion required adequate mixing with heat, which is equally proportional to highly effective digestion. The hydraulic retention time of digestion needs 15 days, whereas 30 to 60 days is the commonly used duration while doing the anaerobic process [51].

6.3 *Anaerobic Baffled Bioreactor (ABR)*

The ABR technique plays a vital role in the treatment of textile effluent owing to better stabilization towards toxicity, excessive time for contact, less requirement of energy, non-requirement of fixed media and sludge/gar separators, less velocity of flowing liquid, lesser washout of bacteria, possess an anaerobic filter [58]. ABR is also seen as a sequence of UASB reactors.

ABRs can be used as an activated sludge reactor, trickling fixed reactor, UASB reactors, fluidized bed reactor, and septic tanks, among other things. It is made up of several compartments and baffles through which the influent passes. The remarkable finding of ABR is the induction of capable bioreactors with high solid retention capability. As a result, for treating a slurry of high solids, vertical baffled to a plug flow was used.

The ABR's architecture provides enough time for influents to degrade and discharge appropriate outflow, as well as the ability to handle wastewater in a single unit with a circulating pattern and reduced capital expenditures. For increasing the efficiency of the bioreactor, it does not require a granulation technique [21].

Since 1987, the hybrid anaerobic baffled bioreactors (HABRs) are in process. The suspension and growth are instigated by flowing liquid with the 0.46 mh^{-1} velocities of the flocculent and granular biomass, 0.97kg COD kg^{-1} VSS l^{-1} of low. initial loading. rate, and 4.01g VSS l^{-1} [8]. The formation of stable 0.5 mm granules presented in all. HABRs takes 30 days duration. Despite the flocs formed during this process being found less than 1.5 mm and weak, however, after 90 days, they reached 3.5 mm and this was determined by the kind of substrate taken in the process. The composition of granules is the acetoclastic methanogens (Methanosarcina cluster) that upraise the surface due to the cavities being full of gas and having low density.

6.4 Anaerobic Fluidized Bed Reactor (AFBR)

For the treatment of wastewater at relatively less hydraulic retention times (HRT), several high-rate anaerobic reactor configurations have been developed. These AFBR successfully works in a broad spectrum of treating wastewater. This one has been known as one of the breakthroughs in technology. The brief information of the performance of different anaerobic systems for the treatment of biodegradable and synthetic wastewater is reported in the literature, respectively [20]. The functioning of AFBR gives negligible issues in plugging, gas holding, and channelling matters. Besides this, it also provides substantial surface area for adsorption and requires low HRT for treating highly polluted wastewater and to handle high OLR. The efficiency of this reactor can be upgraded using better bed material that will enhance the whole process [45].

6.5 Fluidized Bed Biofilm reactors (FBBR)

These reactors are in use from an age ago to treat textile wastewater [52]. This technology is ten times more efficient than ASP and only takes up 10% of the space needed by stirred tank reactors of comparable capacity. Unlike the suspended growth process, which involves sedimentation and biomass recirculation, this process does not require any of the special measures to preserve the biomass in culture. These FBBR are preferred because the carrier particles are small, allowing a wide surface for biomass attachment and development [32]. Furthermore, the fluidization of these reactors reduces the risk of blockage, electrical conduction, and, as a result, large pressure losses in packed towers [41]. The fluidized bed's superiority over many of the suspended and attached growth biological systems is interpreted based on the pilot scale and experimental evidence [47].

6.6 Up Flow Anaerobic Sludge Blanket Reactor (UASB)

The extremely apt method for the treatment of textile dye house outflow and sago discharge is the UASB reactor. These reactors come in the class of high-rate anaerobic reactors composed of a sludge bed. Sandy biomass has high methanogenic properties and outstanding settling down properties cultures in these. reactors. But the washout of biomass limited the function of UASB reactors [10, 46].

Wastewater enters the USAB reactors from the bottom of the reactor and drifts ascending through "sludge blanket" that consist of a molecular sludge bed. This configuration provides effective anaerobic decomposition due to the efficacious mixing of biomass and wastewater [54].

The functioning of this reactor systematically rotates on its sludge bed that gets expanded and made the wastewater drift vertically rising up through it. The granular sludge bed consists of microflora film to the sludge particles removes the pollutant in wastewater, the main factors directing the UASB efficiency are the quality and intimacy of the biofilm of the sludge-wastewater contact [1]. The upper part of the reactors consists of the phase (gas -liquid–solid) separator that allows the release of biogas. This released biogas allows the mixing and the contact between residue and wastewater, separating it from liquid outflow and residual sediment particles [54]. The upward flow with a velocity of 0.5–1.0 m/h and diameter ratio of 0.2–0.5 geometry attributes the operation of UASB reactors [27]. This reactor is well known for the treatment of wastewater with high content of BOD and COD efficiency and the ratio of BOD and COD [29].

6.7 Up flow Anaerobic Sludge Fixed Bed Reactor (UASFB)

The UASFB is made to answer the setback with high performance with much more table benefits. The compensations of UASFB are the retention of biomass in the reactor, elimination of the clog, great retention time, and ability to reduce OLR with stability. The UASFB is suitable for brewery, sugar wastewaters treatment within the acceptable ranges of CH4 release. The UASFB process is not suitable for wood fibre wastewater treatment, whereas other effluents are efficiently removed by COD level, still, the process will vary depending upon operating conditions of treatment.

7 Conclusions

The wastewaters left from textile industries are heavily polluted. The wastewaters show higher temperature with pH, lower ability of biodegradation, strong colour, and higher percentage of chemical oxygen demand concentration. The various traditional methods used are biologically treated, coagulation, adsorption, oxidation, and flocculation. The reactor is used to remove the contamination. Among all the treatments, the biological methods frequently used in the form of anaerobic or/add-on aerobic ways are potential methods to convert the complex pollutants. The textile wastewater undergoes an aerobic process prior to an anaerobic process as a post-treatment in case of coloured wastewater to degrade the organic matters. The adsorption process plays a vital role in converting the complex molecule into simple substances using microorganisms like heavy metal complexes. The oxygen acts as an electron acceptor in the aerobic system, whereas organic material acts as an electron acceptor in the anaerobic system. Both processes end degradable products lefts new cells with low quantities of the chemicals with less water consumption and no toxicity. In this chapter, fundamentals and different types of aerobic and anaerobic treatments were discussed. For the membrane process on an industrial scale, membrane bioreactors are used to treat wastewater followed by sedimentation using various types of anaerobic and aerobic reactions, models using biofilms, and membranes to achieve greater treatment.

References

1. Abbasi T, Abbasi SA (2012) Formation and impact of granules in fostering clean energy production and wastewater treatment in Upflow Anaerobic Sludge Blanket (UASB) reactors. Renew Sustain Energy Rev 16:1696–1708. https://doi.org/10.1016/j.rser.2011.11.017
2. Ali H (2010) Biodegradation of synthetic dyes-a review. Water Soil Pollut 213:251–273. https://doi.org/10.1007/s11270-010-0382-4
3. Amoozegar MA, Hajighasemi M, Hamedi J, Asad S, Ventosa A (2011) Azo dye decolorization by halophilic and halotolerant microorganisms. Ann Microbiol 61:217–230. https://doi.org/10.1007/s13213-010-0144-y
4. Anjaneyulu Y, Chary NS, Raj SSD (2005) Decolourization of industrial effluents–available methods and emerging technologies-a review. Envirn Sci Biotechnol 4:245–273. https://doi.org/10.1007/s11157-005-1246-z
5. Arslan S, Eyvaz M, Gurbulak E, Yuksel E (2015) A review of state of the art technologies in dye-containingwastewater treatment–the textile industry case. In: Kumbasar EPA, Körlü AE, IntechOpen. Textile Wastewater Treatment. https://doi.org/10.5772/64140
6. Banat ME, Nigam P, Singh D, Marchant R (1996) Microbial decolorization of textile dye containing effluents, a review. Biores Technol 58:217–227. https://doi.org/10.1016/S0960-8524(96)00113-7
7. Bhargava A (2016) Activated sludge treatment process–concept and system design. Int J Eng Dev Res 4:890–896
8. Boopathy R, Tilche A (1991) Anaerobic digestion of high strength molasses wastewater using hybrid anaerobic baffled reactor. Water Res 25:785–790. https://doi.org/10.1016/0043-1354(91)90157-L
9. Brown D (1987) Effects of colorants in the aquatic environment. Ecotoxicol Environ Saf 13:139–147. https://doi.org/10.1016/0147-6513(87)90001-7
10. Buzzini AP, Sakamoto IK, Varesch MB, Pires EC (2006) Evaluation of the microbal diversity in an UASB reactor treating from an unbleached pulp plant. Process Biochem 41:168–176. https://doi.org/10.1016/j.procbio.2005.06.009
11. Cattaneo C, Nicolella C, Rovatti M (2003) Denitrification performance of Pseudomonas denitricanc in a fluidized-bed biofilm reactor and in a stirred tank reactor. Eng Life Sci 3(4):187–192. https://doi.org/10.1002/elsc.200390026
12. Chandran D (2016) A review of the textile industries waste water treatment methodologies. Int J Sci Eng Res 7:392–403
13. Chen BY, Chang JS (2007) Assessment upon species evolution of mixed consortia for azo dye decolorization. J Chin Inst Chem Eng 38:259–266. https://doi.org/10.1016/j.jcice.2007.04.002
14. Doble M, Kumar A (2005) Biotreatment of industrial effluents. Butterworth-Heinemann, pp111–132
15. Erkurt HA (2010) Biodegradation of azo dyes. Springer, Heidelberg Dordrecht, New York. https://doi.org/10.1007/978-3-642-11847-0
16. Gao BY, Wang Y, Yue QY, Wei JC, Li Q (2007) Color removal from simulated dye water and actual textile wastewater using a composite coagulant prepared by polyferric chloride and polydimethyldiallylammonium chloride. Sep Purif Technol 54:157–163. https://doi.org/10.1016/j.seppur.2006.08.026
17. Ghaly AE, Ananthashankar R, Alhattab M, Ramakrishnan VV (2014) Production, characterization and treatment of textile effluents: a critical review. J Chem Eng Process Technol 5:1000182. https://doi.org/10.4172/2157-7048.1000182

18. Ghorbani M, Eisazadeh H (2013) Removal of COD, color, anions and heavy metals from cotton textile wastewater by using polyaniline and polypyrrole nanocomposites coated on rice husk ash. Compos B Eng 45:1–7. https://doi.org/10.1016/j.compositesb.2012.09.035
19. Han JL, Ng IS, Wang Y, Zheng X, Chen WM, Hsueh CC (2012) Exploring new strains of dye-decolorizing bacteria. J Biosci Bioeng 113:508–514. https://doi.org/10.1016/j.jbiosc.2011.11.014
20. Haroun M, Idris A (2009) Treatment of textile wastewater with an anaerobic fluidized bed reactor. Desalination 237:357–366. https://doi.org/10.1016/j.desal.2008.01.027
21. Hassan SR, Dahlan I (2013) Anaerobic wastewater treatment using anaerobic baffled bioreactor: a review. Cent Eur J Eng 3:389–399. https://doi.org/10.2478/s13531-013-0107-8
22. Irvine RL, Ketchum LH Jr, Asano T (2009) Sequencing batch reactors for biological wastewater treatment. Crit Rev Environ Control 18:225–294. https://doi.org/10.1080/10643388909388350
23. Joshi M, Purwar R (2004) Developments in new processes for colour removal from effluent. Color Technol 34:58–71. https://doi.org/10.1111/j.1478-4408.2004.tb00152.x
24. Kalme SD, Parshetti GK, Jadhav SU, Govindwar SP (2007) Biodegradation of benzidine based dye Direct Blue-6 by Pseudomonas desmolyticum NCIM 2112. Bioresour Technol 98:405–1410. https://doi.org/10.1016/j.biortech.2006.05.023
25. Kemker C (2014) Turbidity, total suspended solids and water clarity. Fondriest Environmental Inc., Fundamentals of Environmental Measurements, p 13
26. Kuhad RC, Sood N, Tripathi KK, Singh A, Ward OP (2004) Developments in microbial methods for the treatment of dye effluents. Adv Appl Microbiol 56:185–213. https://doi.org/10.1016/s0065-2164(04)56006-9
27. Lim SJ, Kim TH (2014) Applicability and trends of anaerobic granular sludge treatment processes. Biomass Bioenergy 60:189–202. https://doi.org/10.1016/j.biombioe.2013.11.011
28. Liu X, Yuan W, Di M, Li Z, Wang J (2019) Transfer and fate of microplastics during the conventional activated sludge process in one wastewater treatment plant of China. Chem Eng J 362:176–182. https://doi.org/10.1016/j.cej.2019.01.033
29. Mainardis M, Buttazzoni M, Goi D (2020) Up-Flow anaerobic sludge blanket (UASB) technology for energy recovery: a review on state-of-the-art and recent technological advances. Bioengineering 7:43. https://doi.org/10.3390/bioengineering7020043
30. Manu B, Chaudhari S (2002) Anaerobic deodorization of simulated textile wastewater containg azo dyes. Biores Technol 82:225–231. https://doi.org/10.1016/s0960-8524(01)00190-0
31. Mostafa M (2015) Waste water treatment in textile industries- the concept and current technologies. J Biodivers Environ Sci 7:501–525
32. Mulcahy LT, Shieh WK, LaMotta EJ (1980) Kinetic model of biological denitrification in a fluidized bed biofilm reactor (FBBR). Prog Water Technol 12:143–157. https://doi.org/10.1016/B978-1-4832-8438-5.50015-7
33. Naimabadi A, Movahedian AH, Shahsavani A (2009) Decolorization and biological degradation of azo dye reactive red2 by anaerobic/aerobic sequential process. Iran J Environ Health Sci Eng 6:67–72
34. Naresh B, Jaydip J, Prabhat B, Rajkumar P (2013) Recent biological technologies for textile effluent treatment. Int Res J Biological Sci 2:77–82
35. Nassar MM, Magdy YH (1997) Removal of different basic dyes from aqueous solutions by adsorption on palm-fruit bunch particles. Chem Eng J 66:223–226. https://doi.org/10.1016/S1385-8947(96)03193-2
36. Nncy VS, Jenny YRV, Patrica TL, Carlos MP (2011) Performance of a contact stabilization process for domestic wastewater treatment of Cali, Colombia. Dyna 78:98–107
37. O'Neill C, Hawkes FR, Hawkes DW, Esteves S, Wilcox SJ (2000) Anaerobic-aerobic biotreatment of simulated textile effluent containing varied ratios of starch and azo dye. Wat Res 34:2355–2361. https://doi.org/10.1016/S0043-1354(99)00395-4
38. O'Neill FR, Hawkes DL, Lourenco ND, Pinheiro HM, Delee W (1999) Colour in textile effluents-sources, measurement, discharge consents and simulation: a review. Chem Technol Biotechnol 74:1009–1018. https://doi.org/10.1002/(SICI)1097-4660(199911)

39. Pearce CI, Lloyd JR, Guthrie JT (2003) The removal of colour from textile wastewater using whole bacterial cells: a review. Dyes Pigm 58:179–196. https://doi.org/10.1016/S0143-7208(03)00064-0
40. Praveen Kumar GN, Bhat SK (2012) Decolorization of Azo dye Red 3BN by Bacteria. Inter Res J Biol Sci 1:46–52
41. Rabah FKJ, Dahab MF (2004) Nitrate removal characteristics of high performance fluidized-bed biofilm reactors. Water Res 38:3719–3728. https://doi.org/10.1016/j.watres.2004.07.002
42. Rai HS, Bhattacharyya MS, Singh J, Bansal TK, Vats P, Banerjee UC (2005) Removal of dyes from the effluent of textile and dyestuff manufacturing industry: a review of emerging techniques with reference to biological treatment. Environ Sci Technol 35:219–238. https://doi.org/10.1080/10643380590917932
43. Ranade V, Bhandari V (2014) Industrial Wastewater Treatment, recycling and reuse, 1st edn. Butterworth-Heinemann, Oxford-UK, p 576
44. Sandhy S, Padmavathy S, Swaminathan K, Subrahmanyam YV, Kaul SN (2005) Microaerophilic–aerobic sequential batch reactor for treatment of azo dyes containing simulated wastewater. Process Biochem 40:885–890. https://doi.org/10.1016/j.procbio.2004.02.015
45. Şen S, Demirer G (2003) Anaerobic treatment of real textile wastewater with a fluidized bed reactor. Water Res 37:1868–1878. https://doi.org/10.1016/s0043-1354(02)00577-8
46. Senthil kumar M, Gnanapragasam G, Arutchelvan V, Nagarajan S, (2011) Treatment of textile dyeing wastewater using two-phase pilot plant UASB reactor with sago wastewater as co-substrate. Chem Eng J 166:10–14. https://doi.org/10.1016/j.cej.2010.07.057
47. Shieh WK, Keenan JD (1986) Fluidized bed biofilm reactor for wastewater treatment. In: Biproducts. Adv Biochem Eng/Biotechnol. Springer, Berlin, Heidelberg. https://doi.org/10.1007/BFb0002455
48. Shivaranjani SK, Thomas LM (2007) Performance study for treatment of institutional wastewater by activated sludge process. Int J Civ Eng Technol 8:376–382
49. Simpson JR (1998) Activated sludge process. In: Encyclopedia of hydrology and lakes. Encyclopedia of Earth Science, Springer, Dordrecht. https://doi.org/10.1007/1-4020-4497-6_8
50. Singh R, Gautam N, Mishra A, Gupta R (2011) Heavy metals and living systems: an overview. Ind J Pharmacol 43:246–253. https://doi.org/10.4103/0253-7613.81505
51. Stanbury PF, Whitaker A, Hall SJ (2017) Principles of fermentation technology, 3rd edn. Butterworth-Heinemann, pp 687–723
52. Sutton PM, Mishra PN (1994) Activated carbon based biological fluidized beds for contaminated water and wastewater treatment: a state-of-the-art review. Water Sci Technol 29(10–11):309–317. https://doi.org/10.2166/wst.1994.0774
53. Szpyrkowicz LC, Juzzolino C, Kaul SN (2001) A comparative study on oxidation of disperse dyes by electrochemical process, ozone, hypochlorite and fenton reagent. Water Res 35:2129–2136. https://doi.org/10.1016/s0043-1354(00)00487-5
54. Tauseef SM, Abbasi T, Abbasi SA (2013) Energy recovery from wastewaters with high-rate anaerobic digesters. Renew Sustain Energy Rev 19:704–741. https://doi.org/10.1016/j.rser.2012.11.056
55. Van der Zee FP, Bisschops IAE, Blanchard VG, Bouwman RHM, Letting G, Field JA (2003) The contribution of biotic and abiotic processes during azo dye reduction in anaerobic sludge. Water Res 37:3098–3109. https://doi.org/10.1016/s0043-1354(03)00166-0
56. Van der Zee FP, Lettinga G, Field JA (2000) The role of (auto)catalysis in the mechanism of anaerobic azo reduction. Water Sci Technol 42:301–308. https://doi.org/10.2166/wst.2000.0528
57. Wang H, Su JQ, Zheng XW, Tian Y, Xiong XJ, Zheng TL (2009) Bacterial decolorization and degradation of the reactive dye Reactive Red 180 by Citrobacter sp. CK3. Int Biodeterior Biodegradation 63:395–399. https://doi.org/10.1016/j.ibiod.2008.11.006
58. Wang J, Huang Y, Zhao X (2004) Performance and characteristics of an anaerobic baffled reactor. Bioresour Technol 93:205–208. https://doi.org/10.1016/j.biortech.2003.06.004

59. Willetts JR, Ashbott NJ, Mossburgger RE, Asaim MR (2000) The use of a thermophilic anaerobic system for pretreatment of textile dye wastewater. Water Sci Technol 42:309–316. https://doi.org/10.2166/wst.2000.0529
60. Yang X, Crespi M, López-Grimau V (2018) A review on the present situation of wastewater treatment in textile industry with membrane bioreactor and moving bed biofilm reactor. Desalin Water Treat 103:315–322. https://doi.org/10.5004/dwt.2018.21962
61. Yang X, López-Grimau V, Vilaseca M, Crespi M (2020) Treatment of textile wastewater by CAS, MBR and MBBR: a comparative study from technical, economic, and environmental perspectives. Water 12:1306. https://doi.org/10.3390/w12051306

Aerobic Biological Units in Dye Removal

Aiza Azam, Gulzar Muhammad, Muhammad Arshad Raza, Muhammad Mudassir Iqbal, Muhammad Shahbaz Aslam, Adnan Ashraf, and Tania Saif

Abstract Over the last few decades, the production of dye effluent is gaining a lot of concern owing to the rapid increase in industrialization. The textile industry is consuming a large number of dyes (azo dyes) and is responsible for the production of dye wastewater full of hazardous chemicals, which pose serious threats to both aquatic life and humans. Therefore, there is a dire need for dye wastewater treatment before discharging into the environment using either physiochemical or biological methods. A biological process is the best to treat wastewater due to being inexpensive and producing less amount of sludge. Among biological processes, aerobic biological process is preferred, which can degrade toxic aromatic amines. The chapter focuses on the treatment of wastewater dye effluents through the aerobic biological process using various aerobic microbes such as bacteria, fungi, algae, and yeast. Bacteria are proved to be more efficient for the treatment of dyed wastewater due to the fast growth and high degradation rate than other microbes.

Keywords Dye effluent · Wastewater treatment · Azo dyes · Aerobic biological process · Biocatalysts · Bacteria · Fungi · Algae · Yeast

1 Introduction

Different industries such as leather, textile, plastic, food, paints, and pharmaceuticals use dyes due to color-giving properties [23, 35]. The textile industry mainly uses synthetic azo dyes in producing 60–70% dyestuff [9, 12, 14, 37]. Once the dyeing

A. Azam · G. Muhammad (✉) · M. A. Raza · T. Saif
Department of Chemistry, GC University Lahore, Lahore 54000, Pakistan
e-mail: mgulzar@gcu.edu.pk

M. M. Iqbal · M. S. Aslam
School of Biochemistry and Biotechnology, University of the Punjab, Lahore, Lahore 54000, Pakistan

A. Ashraf
Department of Chemistry, University of Lahore, Lahore 54000, Pakistan

A. Khadir et al. (eds.), *Biological Approaches in Dye-Containing Wastewater*, Sustainable Textiles: Production, Processing, Manufacturing & Chemistry,
https://doi.org/10.1007/978-981-19-0545-2_3

process is finished, dye wastewater (effluent), a mixture of a lot of hazardous chemicals (Fig. 1) is discharged into the water without proper management and is a major concern because dye effluent is very harmful to both aquatic and human life [23, 35].

Dye wastewater (dye effluent) is the left-over of dye mixture taken at the start of the dyeing process [23]. Refractory dyes and some other organic chemicals are normally present in dye wastewater [5]. Among five major industries (Fig. 2), the textile manufacturing units are mainly liable for dye wastewater, a major source of aromatic amines and color into the environment [23, 26, 51]. Water is very important for every life but due to the increase in industrialization is being spoiled [13].

The textile industry is the part of the largest manufacturing industries and a great contributor to Pakistan's economy [26]. The textile industry uses more than 10,000 dyes and discharged 280,000 tons of dyes annually during operations such as desizing, sizing, bleaching, scouring, dyeing, and mercerizing [47]. Dye wastewater discharged by the textile industry has 72 toxic chemicals out of which major contaminants are coloring agents like tannin and lignin, acids-bases, and organic–inorganic solid materials [9]. Each process in the textile industry produces a different amount of wastewater in which the dyeing process is responsible for 85% dye effluent (Fig. 3) [23]. Major pollutants, chemical types, and processes of origin are enumerated in Table 1 [29].

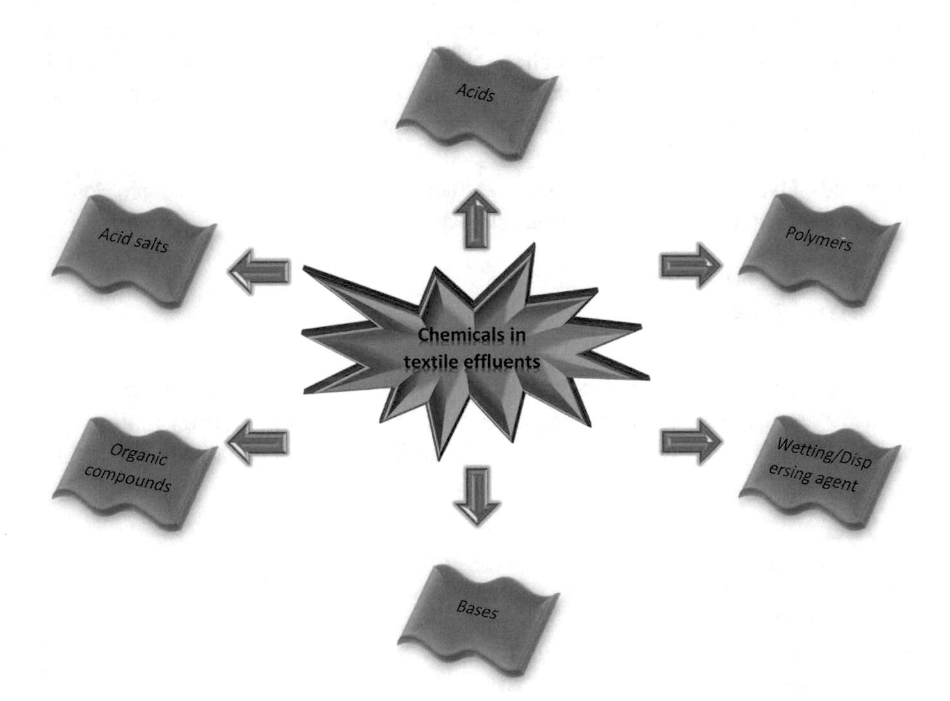

Fig. 1 In dye-utilizing industries, hazardous chemicals present in dye effluents along with dyes

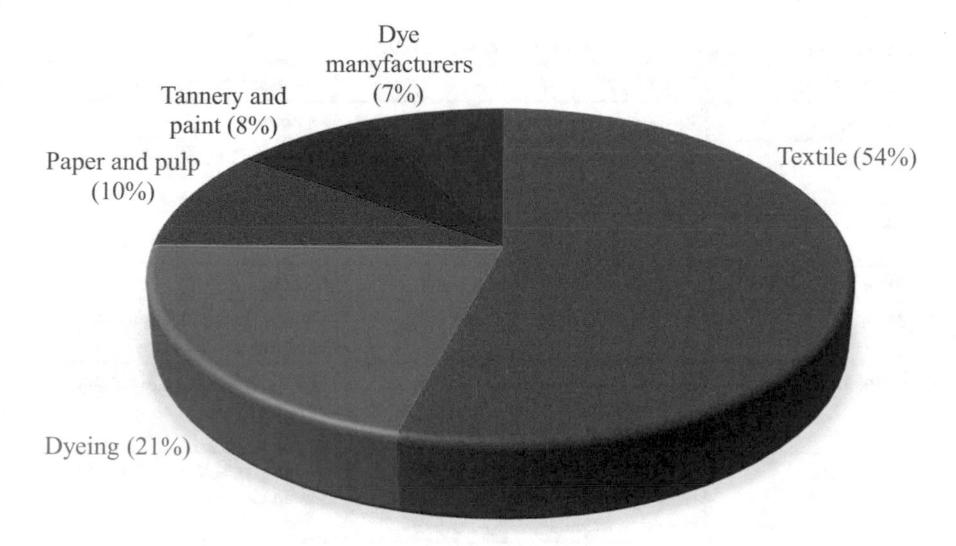

Fig. 2 Amount of dye effluents produced by five major industries [23]

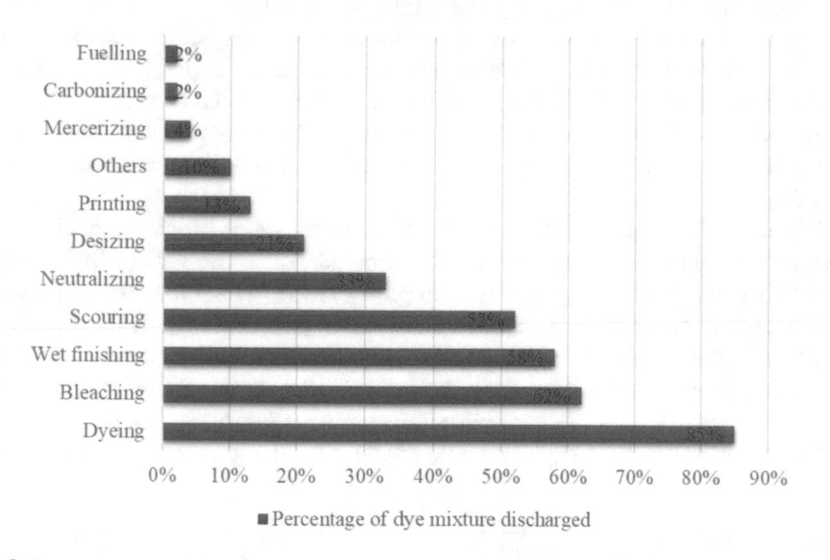

Fig. 3 Percentage of dye mixture discharged from every textile process [23]

Dyed wastewater treatment is crucial because wastewater has negative impacts on almost everything on earth such as animals, aquatic life, and soil. Dyed wastewater causes chronic diseases in humans such as skin irritation, eye burns, breathing problems, sweating, mouth burns, confusion, vomiting, or nausea. Due to the dark

Table 1 Major pollutants present in textile wastewater [29]

Pollutant	Types of chemicals	Processes of origin
Nutrients	Buffers, ammonium salts, sequestrants	Dyeing
Organic load	Surfactants, waxes, fats, starches, enzymes, acetic acid, grease	Dyeing, bleaching, desizing
Sulfur	Sulfide, sulfuric acid, sulfate, hydrosulfide salts	Dyeing
Color	Scoured wool impurities, dyes	Dyeing
pH and salt effects	Sulfate, NaOH, NaCl, carbonate, silicate, mineral/organic acids	Dyeing, bleaching, scouring, desizing, mercerizing
Refractory organics	Resins, carrier organic solvents, surfactant dyes, chlorinated organic compounds	Dyeing, bleaching, desizing, finishing
Toxicants	Reducing agents, biocides, heavy metals, oxidizing agents, quaternary ammonium salts	Dyeing, bleaching, desizing, finishing

color of the polluted water, sunlight penetration is disturbed, which results in photosynthesis prevention and causes hyper-trophication of water bodies. Textile wastewater is rich in very harmful pollutants such as dissolved salts (total dissolved solids and total suspended solids), chemical oxygen demand (COD), colors, biological oxygen demand (BOD), and hazardous chemicals. The only permissible amount of dye wastewater can be discharged into the environment as summarized in Table 2 [23, 29].

The degradation process's driving force is the presence of microorganisms and favorable conditions for the growth of microorganisms. The factors (pH, nutrient availability, temperature, food to microorganism ratio, aeration or oxygen transfer rate, hydraulic retention time, and hydraulic and organic loading rates) influence the textile effluent biodegradation, microorganism's survival, and biochemical oxidation rate. For cell synthesis and survival, various microbes require optimal temperature (25–35 °C) and pH (6.0–7.5). Bacteria mainly grow in slightly alkaline pH while algae and fungi grow mainly in slightly acidic conditions. The growth of microorganisms

Table 2 International standard/permissible amount of dye effluent into the environment [23]

Factor	Standard allowed
COD	<50 mg/L
BOD	<30 mg/L
pH	6–9
Color	<1 ppm
Temperature	<42 °C
Suspended solids	<20 mg/L
Toxic pollutants	Not allowed

can be enhanced using the nutrients like nitrogen and phosphorus. Hydraulic retention time is the average time spent within the bioreactor by the contaminants. Aeration or oxygen transfer rate is the mixing of air with textile wastewater. Aerobic bioreactor specification depends upon both organic and hydraulic loading rates. The food to microorganism ratio helps in determining the number of microorganisms in relevance to the aeration chamber [29].

To treat the dyed wastewater, three different conventional methods are used such as physical, chemical, and biological (Fig. 4) [11]. Physio-chemical methods are expensive with limited versatility and low efficiency and are associated with hazardous metabolites. Physio-chemical methods produce a large amount of sludge, require a large amount of energy and proper handling of waste [14, 33].

In contrast, a biological method is advantageous because it is ecofriendly and in-expensive with minimum sludge and hazardous metabolites [30]. A biological process is the secondary treatment process as shown in the system of traditional wastewater treatment (Fig. 5) and is further divided into aerobic and anaerobic treatment or a combination [23, 24]. Aerobic treatment is preferred because anaerobic treatment produces carcinogenic and mutagenic aromatic amines [43].

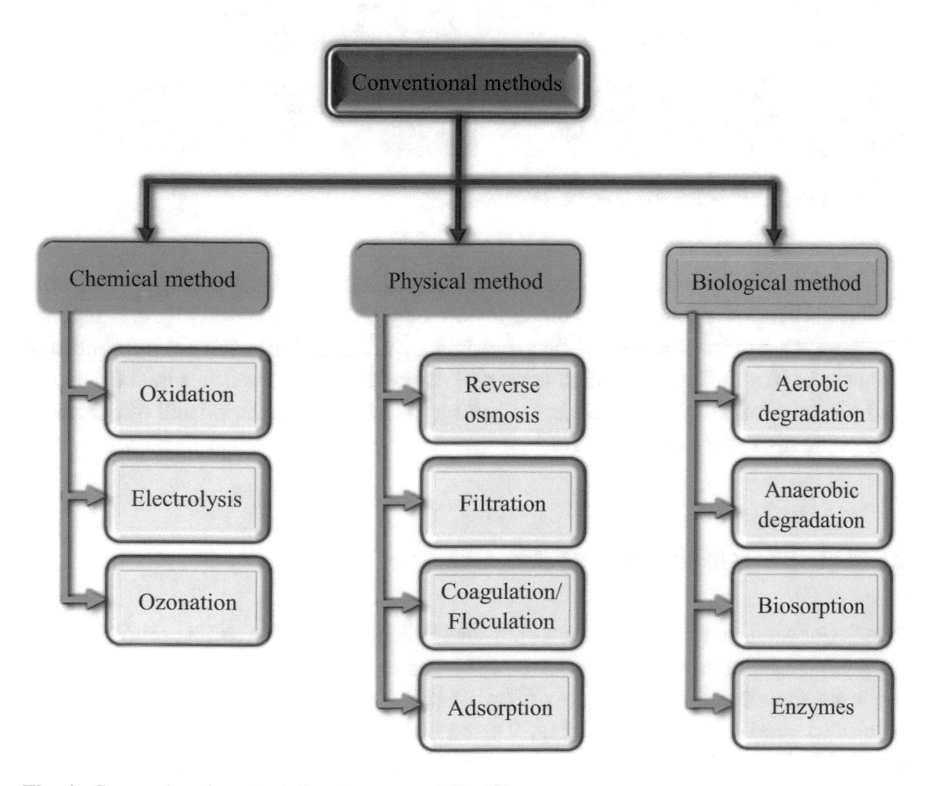

Fig. 4 Conventional methods for dye removal [2, 40]

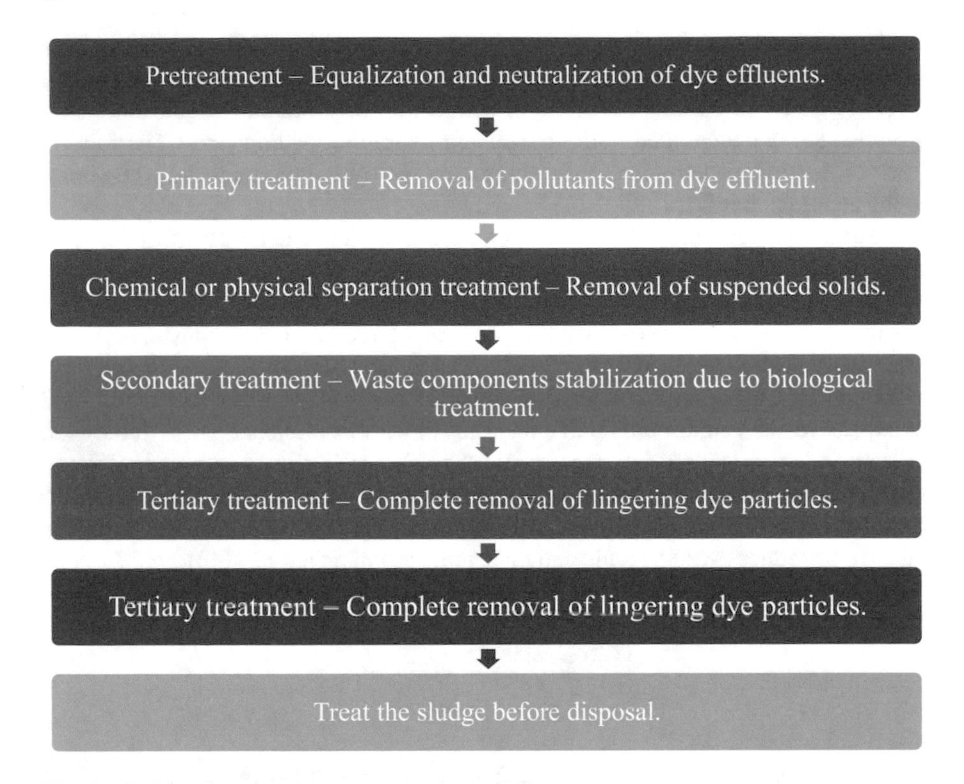

Fig. 5 Traditional wastewater treatment systems [23]

The chapter emphasizes the dyed wastewater treatment with the help of an aerobic biological process using various biocatalysts such as bacteria, fungi, algae, and yeast. The process and mechanism of dye removal by aforesaid microorganisms are also discussed. Effect of different factors on dye degradation such as dye's initial concentrations, time taken, and different conditions applied for dye removal and dye removal results are also taken into account.

2 Aerobic Biological Treatment

Treatment processes using oxygen are metabolically active in which cellular energy is generated using aerobic respiration and more residual solids are generated as cell mass are called aerobic processes (Fig. 6) [1]. Pagga and Brown [32] studied the aerobic degradability of dyes (Fig. 7) and checked 87 dyestuffs to aerobic treatment [35, 37]. Aerobic treatment includes different processes or systems such as activated sludge treatment methods, oxidation, aerobic digestion, trickling filtration, and lagoons [11]. Owing to its greater productivity and wide range of applications, aerobic biological treatment is a conventional biological treatment [22].

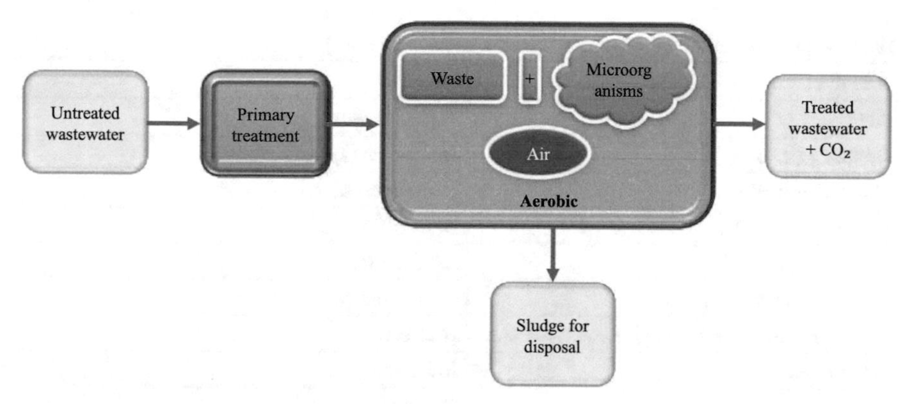

Fig. 6 Schematic representation of aerobic biological wastewater treatment process [38]

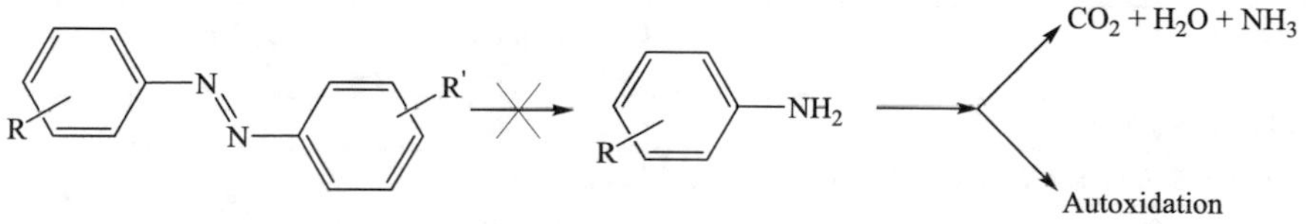

$CO_2 + H_2O + NH_3$

Autoxidation

Fig. 7 Aerobic biological treatment of azo dyes [3]

3 Aerobic Biological Units

For aerobic degradability of dyes, different aerobic microorganisms (aerobic biological units) such as bacteria, fungi, yeast, and algae are used [35]. The two most common microorganisms used for dye removal are bacteria and fungi [37].

3.1 Bacteria

Many microorganisms act as biocatalysts to treat organic pollutants present in dye wastewater. Among biocatalysts, bacteria are more advantageous owing to easy culturing, rapid growth, and dye degradation rate than fungi and other microbes [7, 9]. Aerobic treatment using bacteria is also economical with minimum sludge creation and greater decolorization and mineralization rate [29]. Besides, bacterial activity can be increased easily by molecular genetic manipulation, and several bacterial strains are being isolated from ecological niches like water, contaminated food materials, soil, and excreta of humans and animals [9, 33, 48]. For dyes or some toxic effluent degradation, bacteria of genera *Bacillus, Pseudomonas, Proteus,* and *Aeromonas* are mostly used [29]. Various aerobic bacteria used to decolorize different azo dyes are listed in Table 3 [41, 45].

Table 3 Docolorization of dyes using aerobic bacteria [41, 45]

Azo dye	Bacteria
Acid Yellow	*P. fluorescens*
Remazol Black B	*Halomonas sp.*
Navitan Fast Blue	*P. aeruginosa*
Disperse Blue 79, Acid Orange 10	*B. fusiformis kmk 5*
Disperse dye	*Bacillus sp.*
Reactive Black 5, Basic Blue 41	*Klebsiella sp.*
Reactive Blue, Reactive Red, Reactive Black	*P. luteola*
Congo Red	*S. aureus*
Acid blue 113	*B. subtilis*
Orange II	*X, azovorans*
Acid Orange 7	*Pseudomonas, Enterobacter, Morganella*
Methyl Red	*Bacillus sp., Enterobacter agglomerans, P. nitroreductase, V. logei*
Direct Black 38	*E. gallinarum*
Methyl Red and sulfonated azo dyes	*E. faecalis*
Remazol Orange	*B. lastrosporus*
Reactive Black 5, Acid Red 88, Disperse Orange 3 and Direct Red 81	*S. putrefaciens*
Azo, Triphenylmethane and Anthraquinone	*A. hydrophila*

The aerobic bacteria responsible for the organic compounds biodegradation are *Acetobacter liquefaciens, Aeromonas hydrophilia, Bacillus cereus, Klebsiella pneumoniae, Bacillus subtilis, Sphingomonas,* and *Pseudomonas species* [39]. The principle involved is the intake of oxygen by aerobes to degrade organic contaminants resulting in several byproducts like CO_2, biomass, water, and inorganic nitrogen (Fig. 8) [50]. Bacterial strains need organic carbon sources for growth, obtained by utilizing amines produced by the azo bond cleavage [18, 33]. Azo dyes provide carbon, nitrogen, and energy for many bacteria. For example, *Xenophilus azovorans KF46* and *Pigmentiphaga Kullae K24* can grow on carboxy-orange I and II aerobically and strain 1CX of *Sphingomonas sp.* uses an azo dye, acid orange 7 for its growth [33].

Aerobic bacteria have an oxidoreductase enzyme responsible for degrading dye molecules either symmetrically or asymmetrically and can also carry out desulfonation, deamination, and hydroxylation, etc. [18]. Degradation of dyes with aerobic bacteria started in 1970s when *B. subtilis* was reported followed by some other aerobic bacteria like *A. hydrophilia* and *Bacillus cereus* [6, 27]. Similarly, *A. hydrophilia, Pseudomonas strains, Pseudomonas cepacia,* and *B. subtilis IFO 13,719* were used for the simple azo compounds aerobic reduction, sulfonated azo dyes degradation, reductive cleavage of azo dyes, and five tri-phenyl methane dyes degradation, respectively [37].

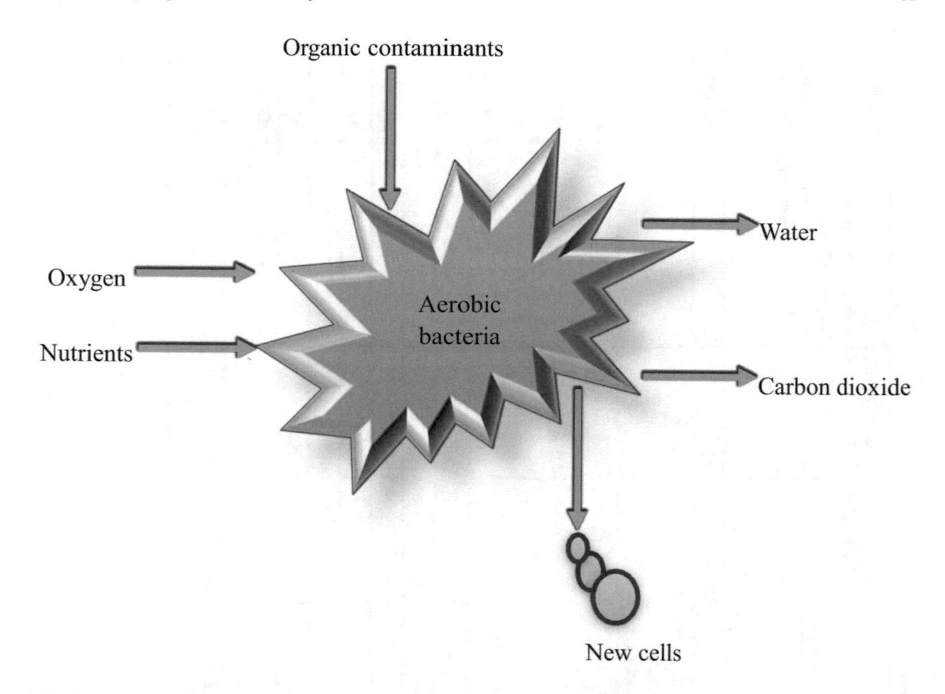

Fig. 8 Aerobic biological principle for organic contaminants degradation [50]

Many bacterial strains such as *Bacillus megaterium, Rhodococcus erythropolis 24, Alcaligenes faecalis 6132, Rahnella aquatilis, Bacillus licheniformis LS04, Acinetobacter guillouiae, Pseudomonas desmolyticum,* and *Microvirgule aerodenitrificans* decolorize or degrade azo dyes [15, 21, 36]. Another bacterium, *Pseudomonas luteola* isolated from the sludge of dyed wastewater depicted 33, 93, 88, and 93% decoloring efficiency for red G, RP_2B, V_2RP, and RBP, respectively after 48 h shaking incubation followed by 48 h static incubation [17]. To degrade and oxidize sulfur-based textile dyes into sulfuric acid, sulfate-reducing bacteria are used [34, 37, 52].

Some triphenylmethane dyes such as malachite green, crystal violet, magenta, brilliant green, ethyl violet, and pararosaniline were decolorized using a strain of *Bacillus sp. MTCC B0006.* and *Kurtbia sp.* cells. Using *Kurtbia sp.*, cells, COD reduction was more than 88% in triphenylmethane dyes like in cell-free dyes but it was almost 70% in ethyl violet after biotransformation. Crystal violet decolorization with *Escherichia coli DH5α, Schizosaccharomyces pombe,* and *Saccharomyces cerevisiae* was found to be toxic but was not toxic when *Bacillus sp. MTCC B0006* was used [37].

Various dyes are mineralized under aerobic conditions using bacterial strains under conditions as listed in Table 4. *Hydrogenophaga Palleronii S1* derived *bacterial strain S5*, aerobic film reactor isolated *Bacterial strain MI2,* and *Sphingomonas sp. strain 1CX* were used to decolorize and mineralize sulfonated azo dye, which can be used as an energy and a sole carbon source like acid orange 7, acid orange 8, acid

Table 4 Dye removal from wastewater using various bacteria under different conditions

Sr No	Bacteria	Dye	Initial dye concentration (mg/L)	Conditions				Time (h)/incubation period	Decolorization (%)	Reaction mechanism/types of enzymes involved	References
				pH	Temperature (°C)	Agitation (rpm)	Mode of operation				
1	*Acinetobacter calcoaceticus NCIM-2890*	Direct Brown MR	50	7.0	30	–	Static	48	91.3	Oxidative and reductive	[29, 40]
2	*Aeromonas hydrophilia*	Red RBN	3000	5.5–10.0	20–35	–	–	8	90	–	[29, 40]
		Reactive Red 141	3800	7.0	30	200	Chemostat pulse technique	48	100	–	[40]
		Reactive Yellow 84, Reactive Red 141, Reactive Red 198, Reactive Blue 171, Reactive black 5	300	7.5	30	125	–	–	33.7, 66.5, 60.2, 36.0, 80.9,	Reductive	[40]
		Crystal Violet	50	–	–	–	–	10	>90	–	[4]
		Reactive Red 198	3000	–	–	–	–	192	>90	–	[4]
		Acid Amaranth	50	–	–	–	–	36	>85	–	[4]
		Brilliant Green	50	–	–	–	–	10	>90	–	[4]
		Basic Fuchsine	50	–	–	–	–	10	>90	–	[4]
		Malachite green	50	–	–	–	–	10	>90	–	[4]
		Reactive Red K3-3B	50	–	–	–	–	36	>85	–	[4]
		Reactive brilliant blue K-GR	50	–	–	–	–	36	>85	–	[4]

(continued)

Table 4 (continued)

Sr No	Bacteria	Dye	Initial dye concentration (mg/L)	Conditions				Time (h)/incubation period	Decolorization (%)	Reaction mechanism/types of enzymes involved	References
				pH	Temperature (°C)	Agitation (rpm)	Mode of operation				
		Great Red GR	50	–	–	–	–	36	>85	–	[4]
3	*Aeromonas hydrophilia var 24 B*	Various azo dyes	10–100	–	–	–	–	24	50–90	Reductive	[40]
4	*Arthrobacter, Pseudomonas, Rhizobium*	AO7	–	–	–	–	–	–	–	Multistage packed bed BAC reactor	[33]
5	*Bacillus aryabhattai DC100*	Coomassie Brilliant Blue G-250	150	–	–	–	–	72	100	–	[44]
6	*Bacillus cereus*	Cibacron Red P4B, Cibacron Black PSG	–	7.0	35	–	–	120	81, 67	Aerobic biodegradation	[48]
7	*Bacillus fusiformis KMK 5*	Acid Orange 10 and Disperse Blue 79	1500	9.0	37	–	Anoxic	48	100	–	[29, 40]
8	*Bacillus Halodurans MTCC 865*	Acid Black 24	–	9.0	37	–	Static	6	90	Aerobic degradation	[48]
9	*Bacillus sp.*	Congo Red	–	7.0	40	–	–	24	85	Aerobic	[48]
10	*Bacillus sp. VUS*	Navy Blue 2GL	50	7.0	40	–	Static	18	94	Oxidative and reductive	[29, 40]
11	*Bacillus velezensis*	Direct Red 28	–	–	–	–	–	–	–	Azo-reductase isolation	[33]

(continued)

Table 4 (continued)

Sr No	Bacteria	Dye	Initial dye concentration (mg/L)	Conditions				Time (h)/incubation period	Decolorization (%)	Reaction mechanism/types of enzymes involved	References
				pH	Temperature (°C)	Agitation (rpm)	Mode of operation				
12	*Bacteria consortium SKB-II*	Congo Red	–	–	–	–	–	–	90	–	[54]
13	*Bacteria consortium SKB-I*	Blue BCC	–	–	–	–	–	–	74	–	[54]
14	*Bacteroides fragilis*	Tartrazine, Amaranth, and Orange II	0.1 mM	8.0	35	–	Static	–	100	–	[29, 40]
15	*Brevibacillus latersporus MTCC 2298*	Golden Yellow HER	50	7.0	30	–	Static	48	87	Oxidative and reductie	[40]
16	*Citrobacter sp.*	Triphenylmethane and azo dyes such as Gentian Violet, Brilliant Green, Crystal Violet, Malachite Green, Methyl red, Basic Fuchsine, Congo Red	5 μM	7.0–9.0	30	–	Static	1	100	Adsorption	[4, 29, 40]
17	*Comamonas sp. UVS*	Direct Red 5B	1100	6.5	40	–	Static	13	100	Oxidative	[29, 40]

(continued)

Table 4 (continued)

Sr No	Bacteria	Dye	Initial dye concentration (mg/L)	Conditions				Time (h)/incubation period	Decolorization (%)	Reaction mechanism/types of enzymes involved	References
				pH	Temperature (°C)	Agitation (rpm)	Mode of operation				
18	*Enterobacter cloacae and Enterobacter asburiae (consortium)*	Textile effluent	–	1.67	32	–	–	10 min	–	Aerobic	[42]
19	*Enterococcus gallinarum*	Direct Black 38	100	–	–	–	Static	480	100	Reductive	[29, 40]
		Direct Black 38	20–250	–	–	–	–	24	71–85	–	[4]
20	*Exiguobacterium sp. RD3*	Navy Blue HE2R	50	7.0	30	–	Static	48	91	Oxidative and reductive	[29, 40]
21	*Galactomyces geotrichum*	Methyl Red	–	–	–	–	–	–	–	Different enzymes are introduced	[33]
22	*Geobacillus stearothermophilus (UCP 986)*	Orange II	0.050 mM	5.0–6.0	50	–	Aeration	24	96–98	–	[40]
23	*Halomonas sp. strain A55*	Reactive Red 184	150	–	–	–	–	24	96	–	[44]
24	*Klebsiella sp.*	RR198, RY107, DB71	–	–	–	–	–	–	–	Microaerophilic-aerobic	[33]
25	*Klebsiella pneumoniae R5-13*	Methyl Red	100	6.0–8.0	30	200	–	168	100	Reductive	[40]

(continued)

Table 4 (continued)

Sr No	Bacteria	Dye	Initial dye concentration (mg/L)	Conditions				Time (h)/incubation period	Decolorization (%)	Reaction mechanism/types of enzymes involved	References
				pH	Temperature (°C)	Agitation (rpm)	Mode of operation				
26	*Kocuria rosea*	Malachite green	50	–	–	–	–	5	100	–	[4]
27	*Micrococcus glutamicus*	Scarlet R	150	–	–	–	–	36	100	–	[4]
28	*Micrococcus glutamicus NCIM 2168*	Reactive Green 19 A	50	6.8	37	–	Static	42	100	–	[29, 40]
29	*Micrococcus luteus strain SSN2*	Direct Orange 16		8.0, 37, –, static	37	–	Static	6	96	Oxidative and reductive	[48]
30	*Nocardia corallina*	Crystal Violet	2.3 µM	–	–	–	–	24	100	–	[4]
31	*Paenibacillus azoreducens sp. nov*	Remazol Black B	100	–	37	–	Static	24	98	–	[29, 40]
32	*Proteus mirabilis*	Red RBN	1000	6.5–7.5	30–35	–	Static	20	95	Reduction followed by biosorption	[29, 40]
33	*Providencia rettgeri strain HSL 1*	Reactive Blue 172	–	7.0	30 ± 0.2	120	–	20	–	Microaerophilic	[42]
34	*Pseudomonas aeruginosa*	Remazol Orange	–	–	–	–	–	–	–	Microaerophilic	[33]
		Remazol orange	200	7.0	30	–	Static	24	94	Reductive	[40]
35	*Pseudomonas aeruginosa NBAR12*	Reactive Blue 172	500	7.0–8.0	40	–	Static	42	83	Oxidative and reductive	[40]

(continued)

Table 4 (continued)

Sr No	Bacteria	Dye	Initial dye concentration (mg/L)	Conditions				Time (h)/incubation period	Decolorization (%)	Reaction mechanism/types of enzymes involved	References
				pH	Temperature (°C)	Agitation (rpm)	Mode of operation				
36	*Pseudomonas desmolyticum*	Red HE7B	–	–	–	–	–	–	–	Amine degradation, extracellular peroxidase, and some other enzymes	[33]
37	*Pseudomonas extremorientalis*	Congo Red	100	–	–	–	–	24	75	–	[44]
38	*Pseudomonas putida*	Crystal Violet	60 μM	–	–	–	–	168	~80	–	[4]
39	*Pseudomonas putida mt-2*	Acid Yellow 17	–	–	–	–	–	–	–	Amine degradation	[33]
40	*Pseudomonas sp.*	Reactive Blue 13	2000	7.0	35	–	Static	70	83.2	–	[40]
41	*Pseudomonas sp. SU-EBT*	Congo Red	1000	8.0	40	–	Static	12	97	Oxidative	[29, 40]
42	*Pseudomonas sp. SUK1*	Reactive Red 2	5000	6.2–7.5	30	–	Static	6	96	Oxidative and reductive	[40]
		Reactive red 2	1000	–	–	–	–	18	95		
43	*Rhizobium radiobacter MTCC 8161*	Reactive Red 141	50	7.0	30	–	Static	48	90	Oxidative and reductive	[29, 40]
44	*Shewanella decolorationis*	Crystal Violet	50	–	–	–	–	28	~100	–	[4]

(continued)

Table 4 (continued)

Sr No	Bacteria	Dye	Initial dye concentration (mg/L)	Conditions				Time (h)/incubation period	Decolorization (%)	Reaction mechanism/types of enzymes involved	References
				pH	Temperature (°C)	Agitation (rpm)	Mode of operation				
45	*Shewanella decolorationis S12*	Acid red GR	150 μM	–	30	–	–	68	100	Reductive	[40]
46	*Shewanella putrefaciens*	Acid Red 88	100	–	–	–	–	8	100	–	[4]
		Direct Red 81	100	–	–	–	–	8	100	–	[4]
		Reactive Black 5	100	– , –, –, –	–	–	–	6	100	–	[4]
		Disperse Orange 3	100	–	–	–	–	8	100	–	[4]
47	*Shewanella putrefaciens strain AS96*	Four azo dyes	–	–	–	–	–	–	–	Microaerophilic-aerobic and amine degradation	[33]
48	*Shewanella sp.*	Acid Orange 7	–	–	–	–	–	6	98	Aerobic	[48]
49	*Sphingomonas paucimobolis*	Methyl Red	–	9.0	30	–	–	10	98	Aerobic	[48]
50	*Sphingomonas sp. BN6*	Amaranth, acid azo dyes, and direct azo dyes	0.1 μmol	–	–	–	–	–	–	Flavin reductase	[40]
51	*Staphylococcus arlettae*	RR198, DB71, RY107, RB5	–	–	–	–	–	–	–	Microaerophilic-aerobic	[33]
52	*Staphylococcus sp. K2204*	Remazol Brilliant Blue R	100	–	–	–	–	12	100	–	[44]

orange 10, acid red 4, and acid red 88. Acid 151 was successfully degraded using an aerobic sequenced biofilm reactor and the dye provides carbon to microorganisms, and 73% carbon was changed into CO_2 [37].

Pseudomonas aeruginosa can degrade or decolorize many dyes in the presence of oxygen especially commercial tannery and textile dyes in the presence of glucose such as Navitan Fast blue S5R and strain BCH of *P. aeruginosa* was able to degrade 98% Remazol Orange 3R within 15 min [33, 44]. Azo dye degradation using genus *Halomonas* also showed promising results. *Shewanella strain KMK6* reduced COD and color of dyes mixture. Another bacterium, *Bacillus MK-8 strain* in granular form reduced COD and color to 66.7 and 96.9%, respectively [29]. Likewise, *A. hydrophilia* and some other aerobic bacterial strains used oxygen insensitive or aerobic azo-reductases, because oxygen can inhibit the activity of azo-reductase, to reduce azo bond which proved that oxygen-rich environments are needed for dye degradation and decolorization (Fig. 9) [44, 48].

An azo bond is cleaved by azo-dye degrading bacteria with the help of the enzyme, azo-reductase. There are several bacteria species in which azo-reductase activity is identified such as *Xenophilus azovorans KF6F, Enterobacter agglomerans, Pigmentiphaga Kullae K24,* and *Enterococcus faecalis* [46]. The dyes can be decolorized by pure bacterial culture quickly but are also responsible for the formation of aromatic amines (toxic intermediates). Mixed culture is extra effective as compared to pure bacterial culture because synergistic metabolism of microbes converts toxic intermediates to non-toxic by-products. In such a study, a microbial consortium of *Pseudomonas, Sphingobacterium,* and *Bacillus* decolorized contaminated water even more quickly than pure bacterial culture [29].

3.2 Fungi

Filamentous fungi present everywhere have a great influence in degrading and decolorizing the dyes [29, 40]. The isolation of fungi from plants, waste, and soil has also been carried out successfully [40]. To treat the dye wastewater, different fungal species like White rot fungi are used to degrade triphenylmethane and azo dyes (Fig. 10). A fungal species, *Phanerochaete chrysosporium* degrade lignin (a polymer obtained from woody plants), various dyes, xenobiotics, related aromatic compounds, and many harmful pollutants like nitrotoluenes, chlorophenols, and polycyclic aromatic hydrocarbons [4, 9, 27, 29, 37]. A fungal culture metabolism can adapt to changes in the environment with the help of intracellular or extracellular enzymes [30]. On the other hand, *White rot fungi* have limitations of an extensive growth cycle and prolonged hydraulic retention times [29].

In 1990, the very first azo dye aerobic degradation was performed using lignolytic fungi, *P. chrysosporium,* in the extracellular enzymes like laccases, manganese peroxidase, and lignin peroxidase. Using aforesaid enzymes, various dyes were decolorized under aerobic conditions, however, mainly decolorized three azo dyes like tropaeolin O, Congo red, and orange II [9, 27, 37, 49]. Some other dyes such as

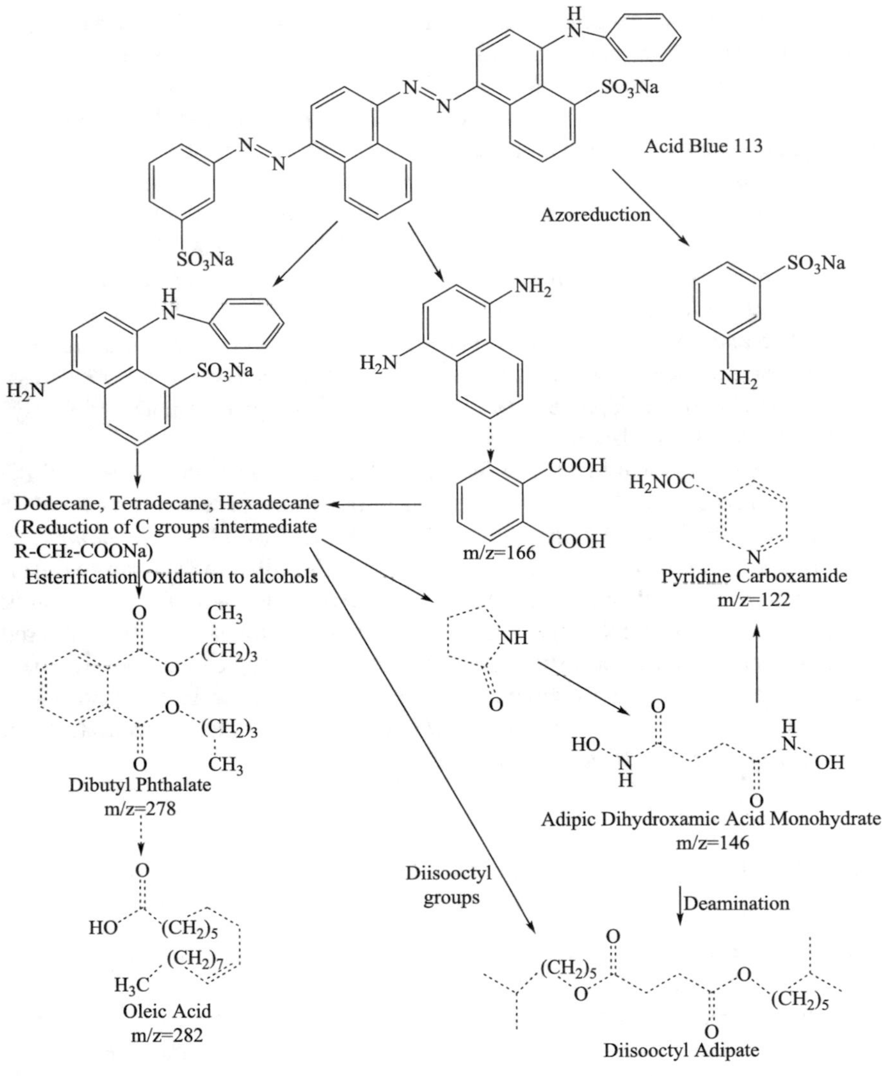

Fig. 9 Scheme showing degradation of acid blue 113 by aerobic bacteria (*S. lentus*) in shade flask [48]

reactive dyes, direct dyes, disperse dyes, acidic dyes, azo dyes (^{14}C labeled azo dyes), polymeric dyes, heterocyclic dyes, and acid green 20 were also decolorized in the presence of lignin peroxidase without using manganese peroxidase at 30 °C [27, 37, 49]. *P. chrysosporium* in the presence of lignin peroxidase also oxidized 3,5-dimethyl-4-hydroxyazobenzene-4′-sulfonate and 1-(4′-acetamidophenylazo)-2-naphthol (Fig. 11–12) [49].

Some other fungal strains such as *Cyathus bulleri, Funalia trogii, Kurthia sp., Coriolus versicolor, P. chrysosporium,* Laetiporous *sulphureus,* and *Streptomyces sp.*

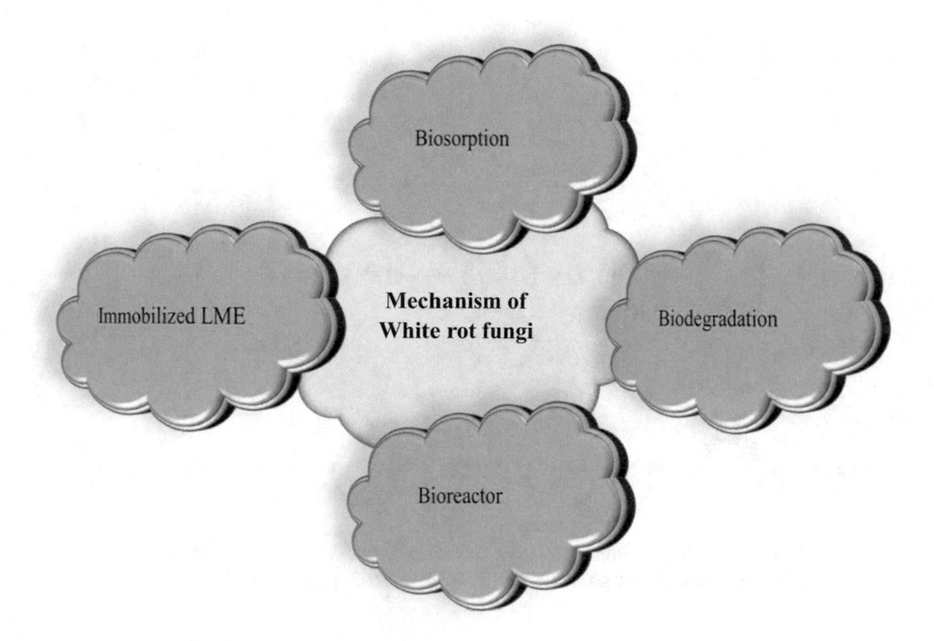

Fig. 10 Process involved in dye removal using *White rot fungi* [19]

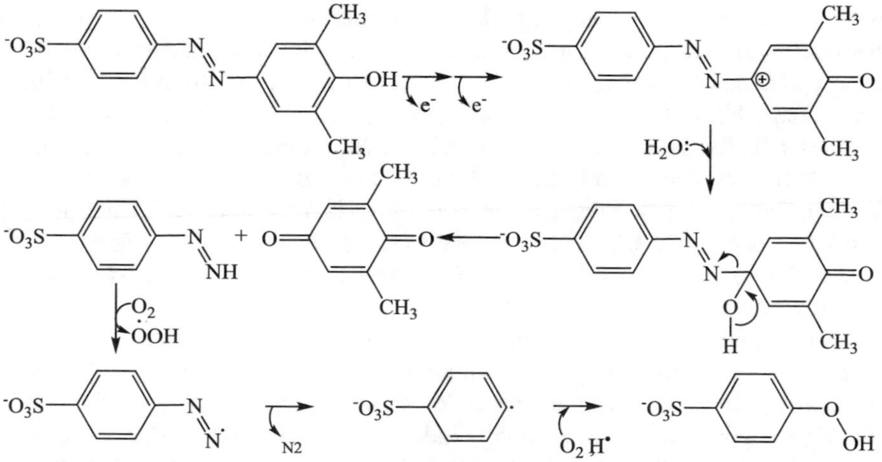

Fig. 11 Mechanism for 1-(4′-acetamidophenylazo)-2-naphthol oxidation by using *P. chrysosporium* bacterium in the presence of an enzyme-lignin peroxidase [49]

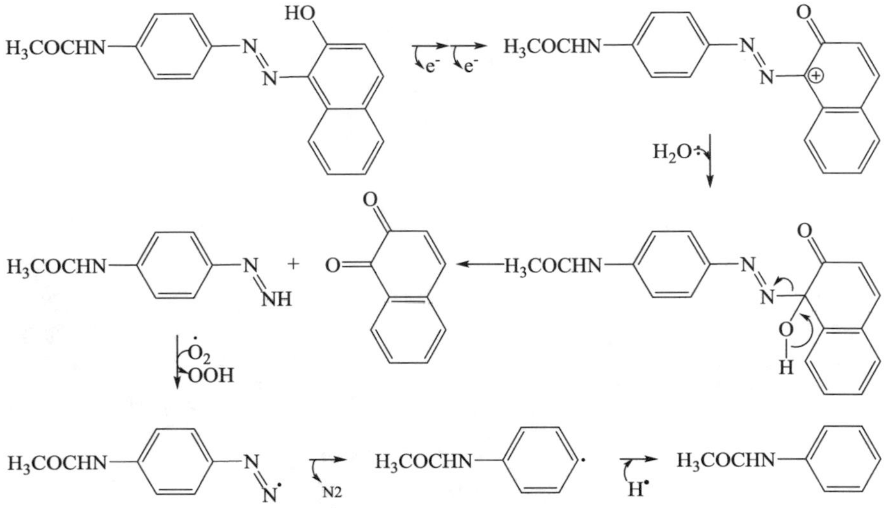

Fig. 12 Mechanism for 3,5-dimethyl-4-hydroxyazobenzene-4′-sulfonate oxidation by using *P. chrysosporium* bacterium in the presence of an enzyme-lignin peroxidase [49]

are also used to decolorize dyes. Among the species mentioned above, *P. chrysosporium* is proved more advantageous due to nonspecific enzymes and can decolorize a diverse range of dyes by catalyzing a lot of reactions with the help of enzyme-lignin peroxidase such as phenol dimerization, benzylic oxidation, carbon–carbon bond cleavage, *O*-demethylation, and hydroxylation. However, the strain has also some limitations while treating the industrial wastewater, if carbon and nitrogen nutrients are present in the industrial wastewater, will inhibit the enzyme release from fungal cells. Moreover, dye degradation with the help of lignin peroxidase needs higher levels of hydrogen peroxide, veratryl alcohol (metabolite released by the fungus) and lignin peroxidase which is not created instantaneously in dyed effluents [37]. Due to the limitations of *P. chrysosporium*, other *White rot fungi* are used such as *Trametes* and *Bjerkandera species* [49].

Trametes versicolor with the help of ligninases catalyzed oxygen was successfully able to decolorize 80% indigo, azo, anthraquinone dyes, and metal complex while manganese peroxidase was not able to decolorize the dyes. It is due to the reason that *T. versicolor* used an extracellular oxidase enzyme, laccase for dye decomposition (indigo, azo, and anthraquinone dyes) [27, 37]. *T. versicolor* has some advantages over *P. chrysosporium* because it can produce an oxidative enzyme, laccase, even in carbon and nitrogen nutrients. Owing to the reasonable redox potential (78 *mV*), *T. versicolor* oxidizes organic contaminants without secondary metabolites and hydrogen peroxide [37].

Unidentified *White rot fungi* were used for the cotton bleaching effluent in the presence of manganese peroxidase while lignin peroxidase was not evidenced effective here [27]. To decolorize Direct Blue 1 and Reactive Black 5, White rot fungi used are *Bjerkandera adusta, Geotrichum candidum, Penicillium sp., Pycnoporus cinnabarinus, Pleurotus ostreatus*, and *Pyricularia oryzae* [49]. Some other White rot fungi like *Daedalea flavida, Dichomitus squalens, Polyporus sanguineus,* and *Irpex Flavus* are used to decolorize and modify many chromophoric groups of dyes present in textile wastewater [10, 16].

Other fungal strains such as *Inonotus hispidus, Hirschioporus larincinus, C. versicolor, Phlebia tremellosa, Umbelopsis isabelline, Aspergillus foetidus, Penicillium geastrivous,* and *Rhizopus oryzae* also biosorb or decolorize many dyes [8, 25, 27]. The reduction of COD and BOD is achieved using mixed fungal cultures of *C. versicolor* and *P. ostreatus*. Various fungal strains used in degrading dyes are summarized in Table 5 [29].

The oxidation of anthraquinone dyes by laccase was achieved directly but azo and indigo dyes (nonsubstrate) decolorization needs mediation by some anthraquinone dye or some synthetic organic compounds. Mediation is important in color removal from textile effluents because of the presence of non-substrate dyes with the substrate dyes. Oxidation of phenolic compounds was done using an enzyme laccase in oxygen and some redox mediators. Fungal laccase isolated from the textile industry was not able to decolorize Remazol brilliant blue R when used without any redox mediator but was able to decolorize when low molecular weight redox mediator violuric acid was used. Some other redox mediators are also used to decolorize various classes of dyes with the help of laccases such as 1-hydroxybenzonitriazole, which was inefficient as compared to violuric acid, acetosyringone, 2-methoxyphenothiaazine, and 4-hydroxybenzenesulfonic acid [37].

3.3 Algae

Like other microbes, algae can treat dyed wastewater containing pigments, dyes, and heavy metals [28]. Algae decolorized dye compounds by using three different mechanisms in which chromophores are utilized to convert colored compounds into uncolored ones, for the algal biomass, water, and carbon dioxide production and dye adsorption on algal biomass. Algae degraded almost 30 azo dyes to aromatic amines,

Table 5 Dye removal from wastewater by using various fungi under different conditions

Sr No	Fungi	Dye	Initial dye concentration (mg/L)	Conditions				Time (h)/Incubation period	Decolorization (%)	Reaction mechanism/ types of enzymes involved	References
				pH	Temperature (°C)	Agitation (rpm)	Mode of operation				
1	*Armillaria sp. F022*	Reactive Black 5	–	4.0	40	–	–	96	<80	Laccase	[48]
2	*Aspergillus niger fungi*	Scarlet Red	–	–	–	–	–	–	80	–	[31]
		Congo Red	–	3.0	–	60	–	36	99	Degradation	[48]
		Direct Red 81	10	–	–	–	–	96	78.3	–	[4]
		Reactive Red 120	40	–	–	–	–	96	74.2	–	[4]
3	*Aspergillus ochraceus*	Reactive Blue-25	–	–	30	–	–	168	100	–	[29]
4	*Bjerkandera adusta OBR105*	Acid and Reactive dyes	200	–	–	–	–	72	99	–	[44]
5	*Coriolus versicolor*	Acid Orange II	–	–	30	–	–	–	85	Degradation	[48]
		Poly S119	100	–	–	–	–	24	15	–	[4]
		Acid Orange II	100	–	–	–	–	24	55	–	[4]
6	*Cunninghamella elegans*	Reactive Black 5, Reactive Orange II, Reactive Red 198	–	5.6	28	–	–	120	98	Adsorption	[48]

(continued)

Table 5 (continued)

Sr No	Fungi	Dye	Initial dye concentration (mg/L)	Conditions				Time (h)/Incubation period	Decolorization (%)	Reaction mechanism/ types of enzymes involved	References
				pH	Temperature (°C)	Agitation (rpm)	Mode of operation				
		Orange II	–	5.8	28	–	–	96	88	–	[29]
7	*Dichomitus squalens*	Coracryl Pink	–	–	–	–	–	–	100	–	[10]
8	*Dietes flavida fungi*	Coracryl Pink	–	–	–	–	–	–	53	–	[10]
9	*Funalia trogii*	Reactive Blue 49	100	–	–	–	–	8	100	–	[4]
		Reactive Blue 19	100	–	–	–	–	8	96.3	–	[4]
		Reactive Black 5	100	–	–	–	–	24	78.3	–	[4]
		Acid Black 52	100	–	–	–	–	24	59	–	[4]
		Reactive Orange 16	100	–	–	–	–	24	22.2	–	[4]
		Acid Violet 43	100	–	–	–	–	8	96.2	–	[4]
10	*Fusarium axysporum*	Yellow GAD	–	–	24	–	–	144	100	Aerobic, degradation	[48]

(continued)

Table 5 (continued)

Sr No	Fungi	Dye	Initial dye concentration (mg/L)	Conditions				Time (h)/Incubation period	Decolorization (%)	Reaction mechanism/ types of enzymes involved	References
				pH	Temperature (°C)	Agitation (rpm)	Mode of operation				
11	*Ganoderma australe, Pleurotus ostreatus sp. 3, Polyporus sp. 2*	Poly R-478	–	6.7	25	–	–	432	93.4, 66.25, 86.53	–	[29]
12	*Ganoderma sp. En 3*	Crystal Violet, Malachite Green, Methyl orange, Bromophenol Blue	–	5.5	28	–	–	72	75, 91, 96.7, 90,	Laccase, degradation	[48]
13	*Ganoderma sp. WR-1*	Amaranth	–	–	28	–	–	8	96	–	[29]
14	*Lentinula edodes*	Poly R-478	–	6.5	30	–	–	264	72 mm decolorized zone	–	[29]
15	*Myrioconium sp. UHH 1–6-18–4*	Disperse Blue 1	–	–	–	–	–	–	43	–	[20]

(continued)

Table 5 (continued)

Sr No	Fungi	Dye	Initial dye concentration (mg/L)	Conditions				Time (h)/Incubation period	Decolorization (%)	Reaction mechanism/types of enzymes involved	References
				pH	Temperature (°C)	Agitation (rpm)	Mode of operation				
16	*Phanerochaete chrysosporium fungi*	Coracryl Violet	–	–	–	–	–	–	100	–	[10]
		Coracryl Pink	–	–	–	–	–	–	100	–	[10]
		Direct Red 180	–	4.5	30	–	–	72	100	Aerobic, lignin peroxidase, Degradation	[48]
		Astrazon Red FBL	–	–	37	–	–	48	87	Degradation	[48]
17	*Phanerochaete chrysosporium fungi ATCC 24,725*	Remazol Red	–	–	–	–	–	–	97	–	[53]
18		Remazol Blue	–	–	–	–	–	–	95	–	[53]
19	*Pleurotus eryngii F032*	Reactive Black 5	–	3.0	40	120	–	72	93.56	Laccase, lignin peroxidase, Manganese peroxidase	[48]
20	*Pleurotus ostreatus*	Disperse Yellow 3	–	–	–	–	–	–	57	–	[29]
		Disperse Orange 3	–	5.0	30	–	–	120	57	–	[29]

(continued)

Table 5 (continued)

Sr No	Fungi	Dye	Initial dye concentration (mg/L)	Conditions				Time (h)/Incubation period	Decolorization (%)	Reaction mechanism/types of enzymes involved	References
				pH	Temperature (°C)	Agitation (rpm)	Mode of operation				
21	*Pleurotus sajor-caju*	Reactive Yellow 15, Reactive Red 198, Reactive Blue 220	–	–	–	–	–	264	100, 100, 100	Adsorption, degradation	[48]
22	*Pleurotus sp. MAK-II*	Remazol Brilliant Blue R	150	–	–	–	–	–	72	–	[44]
23	*Pycnoporus sanguineus fungi*	Coracryl Black	–	–	–	–	–	–	67	–	[10]
		Trypan Blue	20	–	–	–	–	24	70	–	[4]
24	*Pycnoporus sanguineus MUCL 41,582*	Acid Blue 62	–	–	25	–	–	264	99	–	[29]
25	*Schizophyllum commune IBL-06*	Solar Brilliant Red 80	–	4.5	30	–	Shaking	168	100	Manganese peroxidase, laccase, lignin peroxidase	[48]
26	*Stropharia rugosoannulata DSM 11,372*	Reactive Red 4	–	–	–	–	–	–	31	–	[20]

(continued)

Table 5 (continued)

Sr No	Fungi	Dye	Initial dye concentration (mg/L)	Conditions				Time (h)/Incubation period	Decolorization (%)	Reaction mechanism/types of enzymes involved	References
				pH	Temperature (°C)	Agitation (rpm)	Mode of operation				
27	*Trametes sp.*	Brilliant Blue R 250, Orange II	–	–	–	–	–	240	100, 100	Manganese peroxidase, laccase, degradation	[48]
28	*Trametes species CNPR 4783*	Remazol Blue	–	–	–	–	–	–	89	–	[53]
29	*Trametes species CNPR 4801*	Remazol Blue	–	–	–	–	–	–	58	–	[53]
30	*Trametes sp. SQ01*	Fast blue RR	100	–	–	–	–	96	~100	–	[4]
		Bromophenol Blue	100	–	–	–	–	168	100	–	[4]
		Congo Red	100	–	–	–	–	120	~100	–	[4]
		Orange G	100	–	–	–	–	144	~100	–	[4]
		Amido Black 10B	100	–	–	–	–	144	~100	–	[4]
		Crystal Violet	15	–	–	–	–	168	37	–	[4]
		Remazol Brilliant Blue R	200	–	–	–	–	168	~100	–	[4]

(continued)

Table 5 (continued)

Sr No	Fungi	Dye	Initial dye concentration (mg/L)	Conditions				Time (h)/Incubation period	Decolorization (%)	Reaction mechanism/ types of enzymes involved	References
				pH	Temperature (°C)	Agitation (rpm)	Mode of operation				
31	*Trametes trogii*	Indigo Carmine, Remazol Brilliant Blue R, Indigo Carmine	–	4.5–7.0	30	–	–	30 min	84.5, 82, 75	Manganese peroxidase, laccase, degradation	[48]
32	*Trametes versicolor*	Black 5, Blue 49, Orange 12, Reactive Brilliant Blue R, Orange 13	–	3.5–6.5	40	–	–	168	88, 94, 83, 97, 84	Degradation	[48]
		Congo Red	30.5	–	–	–	–	22	100	–	[4]
		Amaranth	33	–	–	–	–	3.5	100	–	[4]
		Remazol Black B	–	4	30	–	–	288	88.4	–	[29]
33	*Trametes versicolor ATCC 20,869*	Remazol Blue	–	–	–	–	–	–	98	–	[53]
		Remazol Red	–	–	–	–	–	–	85	–	[53]
34	*Trametes versicolor CBR43*	Disperse, Acid and Reactive dyes	200	–	–	–	–	216	90	–	[44]

(continued)

Table 5 (continued)

Sr No	Fungi	Dye	Initial dye concentration (mg/L)	Conditions				Time (h)/Incubation period	Decolorization (%)	Reaction mechanism/ types of enzymes involved	References
				pH	Temperature (°C)	Agitation (rpm)	Mode of operation				
35	*Trametes versicolor DSM 11,269*	Disperse Red I	–	–	–	–	–	–	50	–	[20]
36	*Trichoderma tomentosum*	Acid Red 3 R	85.5	–	–	–	–	72	99.20	–	[44]
37	*White-rot fungus Cyathus bulleri*	Kiton Blue A	50	–	–	–	–	6	88	–	[44]

which were further degraded into simple compounds and later into carbon dioxide. Different algal strains such as *Scenedesmus bijugatus* and *Enteromorpha sp.* are used to degrade azo dyes in textile effluents [29]. Likewise, *Chlorella salina* and *Chlorella vulgaris* efficiently reduce BOD, pH, COD, total dissolved solids, nitrate, ammonia, sodium, sulfate, phosphate, magnesium, potassium, calcium, and some other heavy metals (copper, manganese, zinc, cobalt, chromium, nickel, and iron). Both were able to reduce 14–100% heavy metals while *C. vulgaris* was proved more operative than *C. salina* [28]. Algae while growing symbiotically with aerobic microbes will provide oxygen to aerobes in degrading or utilizing aromatic amines produced as a result of dye degradation [29].

3.4 Yeast

Yeast is used in treating dyed wastewater because of yeast biomass's ability to accumulate or absorb chromophores (toxic) or degradation into simpler compounds. For the biosorption of dyes, yeast dead biomass is used as a biosorbent. Degradation or decolorization of dyes with the help of yeast also requires some enzymes and reductases, peroxidases and laccases are the different enzymes used in dye degradation. For the biodegradation of dyes, different types of yeast have been described like *Trichosporon beigelii, Candida krusei, Galactomyces geotrichum,* and S. cerevisiae [29]. Various algae and yeast used for the dye removal under various conditions from wastewater are listed in Table 6.

4 Conclusion

In the last few decades, many different treatments are reported to remove dyes from wastewater, but in a chapter, we focused on dye removal through aerobic biological treatment, which is simple, inexpensive, stable, economical in operation and design, can degrade toxic aromatic amines or aerobic nonbiodegradables, conserve the energy, reduce toxicity levels of the chlorinated organic compound, eliminate off-gas air pollution, and have the proper disposal sites for the disposal of produced excess biomass.

Aerobic treatment is done by using some different aerobic biological units like bacteria, fungi, algae, and yeast. *Bacillus, Pseudomonas,* and *Aeromonas* are the important bacteria used for dye removal. White-rot fungi such as *P. chrysosporium* and *T. versicolor* are used mainly but *T. versicolor* is advantageous as compared with the others. Algae such as *S. bijugatus, Enteromorpha sp.*, and yeast such as *T. Beigelii, C. krusei, G. geotrichum, S. cerevisae* are used to treat the dye wastewater.

Table 6 Dye removal from wastewater by using various algae and yeast under different conditions

Sr No	Strain	Organism	Dye	Initial dye concentration (mg/L)	Conditions					Time (h)/incubation period	Decolorization (%)	Reaction mechanism/types of enzymes involved	References
					pH	Temperature (°C)	Agitation (rpm)	Mode of operation	Medium				
1	Algae	*Brown alga, Stoechospermumma ginatum*	Acid Orange II (AO7) dye	–	–	–	–	–	–	–	–	Adsorption	[30]
2		*Chlorella sorokiniana XJK and Aspergillus sp. XJ-2 (consortia)*	Disperse Red 3B	–	7.0	30	170	Shaking	BG11 and Czapek's medium	96	–	Manganese peroxidase and lignin peroxidase	[42]
3		*Chlorella vulgaris*	Congo Red	–	–	25 ± 3	–	–	Bold's basal mineral medium	72	–	–	[42]
			Direct Blue 71, Reactive Black 5, Disperse Red 1	100 100–500	–	30	–	Static	Mineral salt media	288	–	–	[42]
4		*Chroococcus minutus*	Amido Black 10B	100	–	–	–	–	–	624	55	–	[4]
5		*Cosmarium sp.*	Malachite green, Triphenylmethane dyes	–	–	–	–	–	–	–	–	–	[45]
6		*Elkatothrix viridis, Nostoc linckia, Oscillatoria rubescens, Chlorella vulgaris, Lyngbya lagerlerimi, Volvox aureus*	Orange II, Methyl Red, Basic Cationic, Basic Fuchsin, G- Red (FN-3G)	–	7.0	25 ± 1	–	Continuous light intensity 5000–3000 lx on rotatory shaker	Bold's basal mineral medium	–	–	–	[42]

(continued)

Table 6 (continued)

Sr No	Strain	Organism	Dye	Initial dye concentration (mg/L)	Conditions					Time (h)/incubation period	Decolorization (%)	Reaction mechanism/types of enzymes involved	References
					pH	Temperature (°C)	Agitation (rpm)	Mode of operation	Medium				
7		*Gloeocapsa pleurocapsoides*	Acid Red 97	100	–	–	–	–	–	624	83	–	[4]
8		*Green macroalga, Enteromorpha sp.*	C.I. Basic Red 46 (BR46)	–	–	25	–	–	–	5	83.5	Biodegradation	[29, 30]
9		*Phormidium ceylanicum*	Acid Red 97	100	–	–	–	–	–	624	89	–	[4]
			FF Sky Blue	100	–	–	–	–	–	624	80	–	[4]
10		*Scenedesmus bijugatus*	Azo dyes	–	–	–	–	–	–	144	68	–	[29]
11		*Shewanella algae (SAL)*	Acid Red 27 (AR27)	–	–	–	–	–	–	–	–	Biodegradation	[30]
12		*Spirogyra rhizopus*	Acid red 247	–	–	–	–	–	–	–	–	–	[45]
13		*Xanthophyta alga, Vaucheria species*	Malachite Green (MG)	2.5–17.5	1.5–8.5	5–45	–	–	–	–	–	Adsorption	[30, 42]
			Triphenylmethane dye	–	–	–	–	–	–	–	–	Adsorption	[30]
1	Yeast	*Candida tropicalis TL-FI*	Acid Brilliant Scarlet GR	100	–	–	–	–	–	24	100	–	[44]
2		*Candida krusei*	Basic Violet 3	10	–	–	–	–	–	24	100	–	[44]
			Weak Acid Brilliant Red B	50	–	–	–	–	–	24	94	–	[4]
			Acid Mordant Yellow	50	–	–	–	–	–	24	72	–	[4]

(continued)

Table 6 (continued)

Sr No	Strain	Organism	Dye	Initial dye concentration (mg/L)	Conditions					Time (h)/incubation period	Decolorization (%)	Reaction mechanism/types of enzymes involved	References
					pH	Temperature (°C)	Agitation (rpm)	Mode of operation	Medium				
			Reactive Brilliant Red K-2BP	50	–	–	–	–	–	24	98	–	[4]
			Reactive Black KN-B	50	–	–	–	–	–	24	96	–	[4]
			Acid Mordant Red S-80	50	–	–	–	–	–	24	72	–	[4]
			Reactive Brilliant Blue X-BR	50	–	–	–	–	–	24	78	–	[4]
			Reactive Turquoise Blue KN-G	50	–	–	–	–	–	24	62	–	[4]
			Acid Mordant Light Blue B	50	–	–	–	–	–	24	93	–	[4]
3		*Galactomyces geotrichum*	Malachite Green	50	–	–	–	–	–	9	97	–	[4]
			Orange HE 4B	50	–	–	–	–	–	18	75	–	[4]
			Methyl Red	100	–	–	–	–	–	1	100	–	[4]
			Scarlet RR	50	–	–	–	–	–	18	100	–	[4]
			Amido Black 10B	50	–	–	–	–	–	18	92	–	[4]
4		*Pseudozyma rugulosa*	Weak Acid Brilliant Red B	50	–	–	–	–	–	24	98	–	[4]
			Acid Mordant Yellow	50	–	–	–	–	–	24	94	–	[4]

(continued)

Table 6 (continued)

Sr No	Strain	Organism	Dye	Initial dye concentration (mg/L)	Conditions					Time (h)/incubation period	Decolorization (%)	Reaction mechanism/types of enzymes involved	References
					pH	Temperature (°C)	Agitation (rpm)	Mode of operation	Medium				
			Reactive Brilliant Red K-2BP	50	–	–	–	–	–	24	99	–	[4]
			Reactive Black KN-B	50	–	–	–	–	–	24	96	–	[4]
			Acid Mordant Red S-80	50	–	–	–	–	–	24	22	–	[4]
			Reactive Brilliant Blue X-BR	50	–	–	–	–	–	24	85	–	[4]
			Reactive Turquoise Blue KN-G	50	–	–	–	–	–	24	67	–	[4]
			Acid Mordant Light Blue B	50	–	–	–	–	–	24	89	–	[4]
5		*Kluyveromyces marxianus IMB3*	Ramazol black B	–	–	–	–	–	–			–	[45]
6		*Saccharomyces cerevisiae MTCC 46*	Methyl Red	–	–	–	–	–	–	–	–	–	[45]
7		*Trichosporon akiyoshidainum HP2023*	Reactive Black 5	300	–	–	–	–	–	24	100	–	[44]
8		*Trichosporon beigelii*	Navy Blue HER	50	–	–	–	–	–	24	95	–	[44]

(continued)

Table 6 (continued)

Sr No	Strain	Organism	Dye	Initial dye concentration (mg/L)	Conditions					Time (h)/incubation period	Decolorization (%)	Reaction mechanism/types of enzymes involved	References
					pH	Temperature (°C)	Agitation (rpm)	Mode of operation	Medium				
			Golden Yellow 4BD	50	–	–	–	–	–	24	60	–	[4]
			Navy Blue HER	50	–	–	–	–	–	24	100	–	[4]
			Green HE 4BD	50	–	–	–	–	–	24	70	–	[4]
			Red HE7B	50	–	–	–	–	–	24	85	–	[4]
			Malachite Green	50	–	–	–	–	–	24	90	–	[4]
			Orange HE2R	50	–	–	–	–	–	24	50	–	[4]
			Methyl Violet	50	–	–	–	–	–	24	73	–	[4]
			Crystal Violet	50	–	–	–	–	–	24	57	–	[4]

References

1. Ahammad SZ, Sreekrishnan TR (2016) Energy from wastewater treatment. In: Prasad MNV (eds) Bioremediation and Bioeconomy. Elsevier, pp 523–536. https://doi.org/10.1016/B978-0-12-802830-8.00020-4
2. Al-Sakkaf BM, Nasreen S, Ejaz N (2020) Degradation pattern of textile effluent by using bio and Sono chemical reactor. J Chem https://doi.org/10.1155/2020/8965627
3. Alabdraba WMS, Bayati M (2014) Biodegradation of Azo dyes a review. Int J Environ Eng Nat Resour 1:179–189
4. Ali H (2010) Biodegradation of synthetic dyes—a review. Water Air Soil Pollut 213(1):251–273
5. An H, Qian Y, Gu X, Tang WZ (1996) Biological treatment of dye wastewaters using an anaerobic-oxic system. Chemosphere 33(12):2533–2542. https://doi.org/10.1016/S0045-6535(96)00349-9
6. Anjaneyulu Y, Chary NS, Raj DSS (2005) Decolourization of industrial effluents–available methods and emerging technologies–a review. Rev Environ Sci Bio/Technol 4(4):245–273. https://doi.org/10.1007/s11157-005-1246-z
7. Balamurugan B, Thirumarimurugan M, Kannadasan T (2011) Anaerobic degradation of textile dye bath effluent using Halomonas sp. Bioresour Technol 102(10):6365–6369. https://doi.org/10.1016/j.biortech.2011.03.017
8. Banat IM, Nigam P, Singh D, Marchant R (1996) Microbial decolorization of textile-dyecontaining effluents: a review. Bioresour Technol 58(3):217–227. https://doi.org/10.1016/S0960-8524(96)00113-7
9. Bhatia D, Sharma NR, Singh J, Kanwar R (2017) Biological methods for textile dye removal from wastewater: a review. Crit Rev Env Sci Technol 47(19):1836–1876. https://doi.org/10.1080/10643389.2017.1393263
10. Chander M, Arora D (2007) Evaluation of some white-rot fungi for their potential to decolourise industrial dyes. Dyes Pigm 72(2):192–198. https://doi.org/10.1016/j.dyepig.2005.08.023
11. Chandran D (2016) A review of the textile industries waste water treatment methodologies. Int J Sci Eng Res 7(1):2229–5518
12. García-Montaño J, Domènech X, García-Hortal JA, Torrades F, Peral J (2008) The testing of several biological and chemical coupled treatments for Cibacron Red FN-R azo dye removal. J Hazard Mater 154(1–3):484–490. https://doi.org/10.1016/j.jhazmat.2007.10.050
13. Gupta VK, Mittal A, Gajbe VJJoc, science i (2005) Adsorption and desorption studies of a water soluble dye, Quinoline Yellow, using waste materials. J colloid Interf Sci 284(1):89–98. https://doi.org/10.1016/j.jcis.2004.09.055
14. Hakimelahi M, Moghaddam MRA, Hashemi SH (2012) Biological treatment of wastewater containing an azo dye using mixed culture in alternating anaerobic/aerobic sequencing batch reactors. Biotechnol Bioprocess Bioeng 17(4):875–880. https://doi.org/10.1007/s12257-011-0673-7
15. Han JL, Ng IS, Wang Y, Zheng X, Chen WM, Hsueh CC, Liu SQ, Chen BY (2012) Exploring new strains of dye-decolorizing bacteria. J Biosci Bioeng 113(4):508–514. https://doi.org/10.1016/j.jbiosc.2011.11.014
16. Holkar CR, Jadhav AJ, Pinjari DV, Mahamuni NM, Pandit ABJJoem (2016) A critical review on textile wastewater treatments: possible approaches. J Environ Manage 182:351–366. https://doi.org/10.1016/j.jenvman.2016.07.090
17. Hu T (1994) Decolourization of reactive azo dyes by transformation with Pseudomonas luteola. Bioresour Technol 49(1):47–51. https://doi.org/10.1016/0960-8524(94)90172-4
18. Jamee R, Siddique R (2019) Biodegradation of synthetic dyes of textile effluent by microorganisms: an environmentally and economically sustainable approach. Eur J Microbiol Immunol 9(4):114–118. https://doi.org/10.1556/1886.2019.00018
19. Jebapriya GR, Gnanadoss JJIJoCR, Review, (2013) Bioremediation of textile dye using white rot fungi: A review. Int J Cur Res Rev 5(3):1

20. Junghanns C, Krauss G, Schlosser D (2008) Potential of aquatic fungi derived from diverse freshwater environments to decolourise synthetic azo and anthraquinone dyes. Bioresour Technol 99(5):1225–1235. https://doi.org/10.1016/j.biortech.2007.02.015
21. Kalme SD, Parshetti GK, Jadhav SU, Govindwar SP (2007) Biodegradation of benzidine based dye Direct Blue-6 by Pseudomonas desmolyticum NCIM 2112. Bioresour Technol 98(7):1405–1410. https://doi.org/10.1016/j.biortech.2006.05.023
22. Karthik V, Saravanan K, Bharathi P, Dharanya V, Meiaraj C (2014) An overview of treatments for the removal of textile dyes. J Chem Pharm Sci 7(4):301–307
23. Katheresan V, Kansedo J, Lau SY (2018) Efficiency of various recent wastewater dye removal methods: a review. J Environ Chem Eng 6(4):4676–4697. https://doi.org/10.1016/j.jece.2018.06.060
24. Khouni I, Marrot B, Amar RB (2012) Treatment of reconstituted textile wastewater containing a reactive dye in an aerobic sequencing batch reactor using a novel bacterial consortium. Sep Purif Technol 87:110–119. https://doi.org/10.1016/j.seppur.2011.11.030
25. Kirby N, McMullan G, Marchant R (1997) Bioremediation of textile industry wastewater by white-rot fungi. In: Studies in Environmental Science. Elsevier, pp 711–718. https://doi.org/10.1016/S0166-1116(97)80083-0
26. Lodha B, Chaudhari S (2007) Optimization of Fenton-biological treatment scheme for the treatment of aqueous dye solutions. J Hazard Mater 148(1–2):459–466. https://doi.org/10.1016/j.jhazmat.2007.02.061
27. Miao Y (2005) Biological remediation of dyes in textile effluent: a review on current treatment technologies. Bioresour Technol 58:217–227
28. Muhammad G, Mehmood A, Shahid M, Ashraf RS, Altaf M, Hussain MA, Raza MA (2020) Biochemical Methods for Water Purification. In: Ahamed MI, Lichtfouse E, Asiri AM (eds) Methods for Bioremediation of Water and Wastewater Pollution. Springer, pp 181–212
29. Mullai P, Yogeswari MK, Vishali S, Namboodiri MMT, Gebrewold BD, Rene ER, Pakshirajan K (2017) Aerobic treatment of effluents from textile industry. In: Lee DJ, Hallenbeck PC, Ngo HH, Jegatheesan V, Pandey A (eds) Current Developments in Biotechnology and Bioengineering: Biological Treatment of Industrial Effluents. Elsevier, pp 3–34. https://doi.org/10.1016/B978-0-444-63665-2.00001-1
30. Navin PK, Kumar S, Mathur M (2018) Textile wastewater treatment: a critical review. Int J Eng Res Technol 6(11)
31. Nikhath K, Seshikala D, Charya MAS (2000) Decolourisation of textile dyes by fungi. Indian J Microbiol 40(3):191–197
32. Pagga U, Brown D (1986) The degradation of dyestuffs: Part II behaviour of dyestuffs in aerobic biodegradation tests. Chemosphere. 15(4):479–491
33. Pandey A, Singh P, Iyengar L (2007) Bacterial decolorization and degradation of azo dyes. Int Biodeterior Biodegrad 59(2):73–84. https://doi.org/10.1016/j.ibiod.2006.08.006
34. Parshetti G, Saratale G, Telke A, Govindwar S (2009) Biodegradation of hazardous triphenylmethane dye methyl violet by Rhizobium radiobacter (MTCC 8161). J Basic Microbiol 49(S1):S36–S42. https://doi.org/10.1002/jobm.200800200
35. Popli S, Patel UD (2015) Destruction of azo dyes by anaerobic–aerobic sequential biological treatment: a review. Int J Environ Sci Technol 12(1):405–420. https://doi.org/10.1007/s13762-014-0499-x
36. Praveen Kumar GN, Bhat SK (2012) Decolorization of azo dye Red 3BN by bacteria. Int Res J Biol Sci 1(5):46–52
37. Rai HS, Bhattacharyya MS, Singh J, Bansal T, Vats P, Banerjee U (2005) Removal of dyes from the effluent of textile and dyestuff manufacturing industry: a review of emerging techniques with reference to biological treatment. Crit Rev Env Sci Technol 35(3):219–238. https://doi.org/10.1080/10643380590917932
38. Ranade VV, Bhandari VM (2014) Industrial wastewater treatment, recycling and reuse: an overview. In: Ranade VV, Bhandari VM (eds) Industrial wastewater treatment, recycling and reuse. Butterworth-Heinemann, pp 1–80. https://doi.org/10.1016/B978-0-08-099968-5.00001-5

39. Saini RD (2017) Textile organic dyes: polluting effects and elimination methods from textile waste water. Int J Chem Eng Res 9(1):121–136
40. Saratale RG, Saratale GD, Chang JS, Govindwar SP (2011) Bacterial decolorization and degradation of azo dyes: a review. J Taiwan Inst Chem Eng 42(1):138–157. https://doi.org/10.1016/j.jtice.2010.06.006
41. Sarayu K, Sandhya S (2012) Current technologies for biological treatment of textile wastewater–a review. Appl Biochem Biotechnol 167(3):645–661. https://doi.org/10.1007/s12010-012-9716-6
42. Saxena A, Gupta S (2020) Bioefficacies of microbes for mitigation of Azo dyes in textile industry effluent: a review. Bioresour 15(4):9858. https://doi.org/10.15376/biores.15.4.Saxena
43. Senan RC, Abraham TE (2004) Bioremediation of textile azo dyes by aerobic bacterial consortium aerobic degradation of selected azo dyes by bacterial consortium. Biodegrad 15(4):275–280. https://doi.org/10.1023/B:BIOD.0000043000.18427.0a
44. Sghaier I, Guembri M, Chouchane H, Mosbah A, Ouzari HI, Jaouani A, Cherif A, Neifar M (2019) Recent advances in textile wastewater treatment using microbial consortia. J Text Eng Fash Technol 5(3):134–146
45. Shah K (2014) Biodegradation of azo dye compounds. Int Res J Biochem Biotechnol 1(2):5–13
46. Shiroya AJ, Vaghasiya KK, Ghantala NJ (2012) A review biotechnological removal of color and dye from waste water. PharmaTutor
47. Shoukat R, Khan SJ, Jamal Y (2019) Hybrid anaerobic-aerobic biological treatment for real textile wastewater. J Water Process Eng 29:100804.https://doi.org/10.1016/j.jece.2018.06.060
48. Singh PK, Singh RL (2017) Bio-removal of azo dyes: a review. Int J Appl Sci Biotechnol 5(2):108–126. https://doi.org/10.3126/ijasbt.v5i2.16881
49. Stolz A (2001) Basic and applied aspects in the microbial degradation of azo dyes. Appl Microbiol Biotechnol 56(1):69–80. https://doi.org/10.1007/s002530100686
50. Tantak N, Chandan N, Raina P (2014) An introduction to biological treatment and successful application of the aqua EMBR system in treating effluent generated from a chemical manufacturing unit. In: Ranade VV, M BV (eds) industrial wastewater treatment, recycling and reuse. Butterworth-Heinemann, pp 369–397. https://doi.org/10.1016/B978-0-08-099968-5.00009-X
51. Tantak NP, Chaudhari S (2006) Degradation of azo dyes by sequential Fenton's oxidation and aerobic biological treatment. J Hazard Mater 136(3):698–705. https://doi.org/10.1016/j.jhazmat.2005.12.049
52. Togo CA, Mutambanengwe CCZ, Whiteley CG (2008) Decolourisation and degradation of textile dyes using a sulphate reducing bacteria (SRB)–biodigester microflora co-culture. Afr J Biotechnol 7(2)
53. Toh Y-C, Yen JJL, Obbard JP, Ting YP (2003) Decolourisation of azo dyes by white-rot fungi (WRF) isolated in Singapore. Enzyme Microb Technol 33(5):569–575. https://doi.org/10.1016/S0141-0229(03)00177-7
54. Tony BD, Goyal D, Khanna S (2009) Decolorization of textile azo dyes by aerobic bacterial consortium. Int Biodeter Biodegr 63(4):462–469. https://doi.org/10.1016/j.ibiod.2009.01.003

Anaerobic Processes in Dye Removal

Ashutosh Vashisht, Rohit Rai, Sapna Thakur, Satish Kondal, Kumud Ashish Singh, Manju, Diksha Sharma, and Vishakha Gilhotra

Abstract Dyes are synthetic organic chemicals that are widely employed in a variety of sectors, including textiles, leather, plastics, food, pharmaceuticals, and paints. Untreated industrial effluents have accumulated resistant azo dyes, which have had a negative influence on the ecology. Discharging textile effluents in a water body has aesthetic, environmental, and toxicological issues that should be considered. Biological treatment, particularly anaerobic treatment, is usually considered to be one of the most effective ways for eliminating the majority of contaminants from complex or high-strength organic wastewater. The purpose of this book chapter is to describe the effect of biological treatment on dye removal, taking into account both anaerobic and aerobic treatments separately as well as a hybrid approach.

Keywords Anaerobic process · Dye removal · Hybrid dye removal method · Aerobic process · Azo dyes · Synthetic dyes · Wastewater · Biodegradation · Colour removal · Industrial waste treatment

---

A. Vashisht
Department of Molecular Biology and Biochemistry, Guru Nank Dev University, Amritsar, Punjab 143005, India
e-mail: ashutosh.mob@gndu.ac.in

R. Rai · K. A. Singh · Manju
Faculty of Applied Medical Sciences, Lovely Professional University, Phagwara, Punjab 144411, India

S. Thakur · S. Kondal
Department of Microbiology, Guru Nanak Dev University Amritsar, Amritsar, Punjab 143005, India
e-mail: sapnamicro.rsh@gndu.ac.in

S. Kondal
e-mail: satishmicro.rsh@gndu.ac.in

D. Sharma
Department of Biotechnology, CT Institute of Pharmaceutical Sciences, Jalandhar, Punjab 144020, India

V. Gilhotra (✉)
Department of Botanical and Environmental Sciences, Guru Nanak Dev University, Amritsar, Punjab 143005, India

A. Khadir et al. (eds.), *Biological Approaches in Dye-Containing Wastewater*, Sustainable Textiles: Production, Processing, Manufacturing & Chemistry,
https://doi.org/10.1007/978-981-19-0545-2_4

1 Introduction

The textile business is both one of the largest and one of the most polluting industries in the world. Textile manufacture entails a number of procedures, the majority of which produce highly polluted wastes [21]. One of the most important steps in the textile industries is dyeing. About 15% of colours used in the dyeing process are expected to be released in wastewater [40]. Among the many dye classes, azo dyes are the most popular, accounting for 70% of colours used in textile wastewater [110]. Under anaerobic conditions, spontaneous reduction of the azo bond or photocatalytic degradation may occur, resulting in extremely carcinogenic and mutagenic colourless aromatic amines [3]. The remaining dyes (10–15%) have a considerable impact on sunlight penetration, changing the natural water body [85]. The presence of various amounts of dyes with high nitrogen and organic content causes eutrophication, which causes the aquatic ecology to be disrupted [62].

Different treatment techniques for removing dyes and their related metabolites from textile effluents have been proposed such as adsorption [108], coagulation [29], electrocoagulation [6], filtering [50], chemical processes such as oxidation (Fenton's oxidation) [28], ozonation [96], and biological methods [12]. The main disadvantages of these technologies are their high prices and the vast amount of sludge produced. The elimination of the principal contaminants from complicated or high-strength organic wastewater is typically thought to be most effective via biological treatment, either aerobic or anaerobic [79].

Biological approaches are the most basic, natural, efficient, and cost-effective. Using biological techniques large amounts of effluents can be treated without causing significant environmental harm. Activated sludge and microaerophilic modes are accessible in a variety of bioreactors that can function in anaerobic, anoxic, and aerobic settings [8, 63]. Although some attempts to decolorize dyes under aerobic circumstances have proven successful, most azo dyes are still thought to be non-biodegradable [89]. Various studies found that under aerobic circumstances, dyestuff did not have substantial biodegradation rates. Because most aerobic bacteria are incapable of destroying dye molecules, the traditional activated sludge technique is ineffective in the treatment of textile wastewater in terms of colour. Under anaerobic conditions, however, in contrast to aerobic treatment, reductive cleavage of azo bonds is common. The development of high-rate bioreactors, which achieve a high reaction rate per unit reactor volume by retaining the biomass in the reactor for long periods of time, is critical for the successful application of anaerobic technology for the treatment of industrial wastewaters [7].

In this chapter, we will be discussing the importance of dye removal strategies, different aspects of using anaerobic method as an efficient dye removal process and how it is better than the present technologies and how it can be made more efficient.

2 Classification of Dyes

Dyes are categorised in a number of ways. The classification of a single parameter is exceedingly complicated and useless from a practical view. Each class of dye is distinguished by its source of material, chemical structure, chemical characteristics, and mode of interaction with substrate molecules [39]. Figure 1 classifies the dyes on the basis of source of material, ionic nature and chemical structure.

2.1 Classification Based on the Source of Material

2.1.1 Natural Dyes

Natural dyes are primarily derived from plants, animals, and minerals. The majority of natural dyes are extracted from plant materials i.e., leaves, stem, roots, flower and seeds. Natural dyes are biodegradable, non-toxic, and have a greater environmental compatibility than synthetic dyes [54]. The most of natural dyes have a negative charge, and positive charge dyes are very rare. Natural dyes are categorised based on their chemical constitution and colour (Fig. 2).

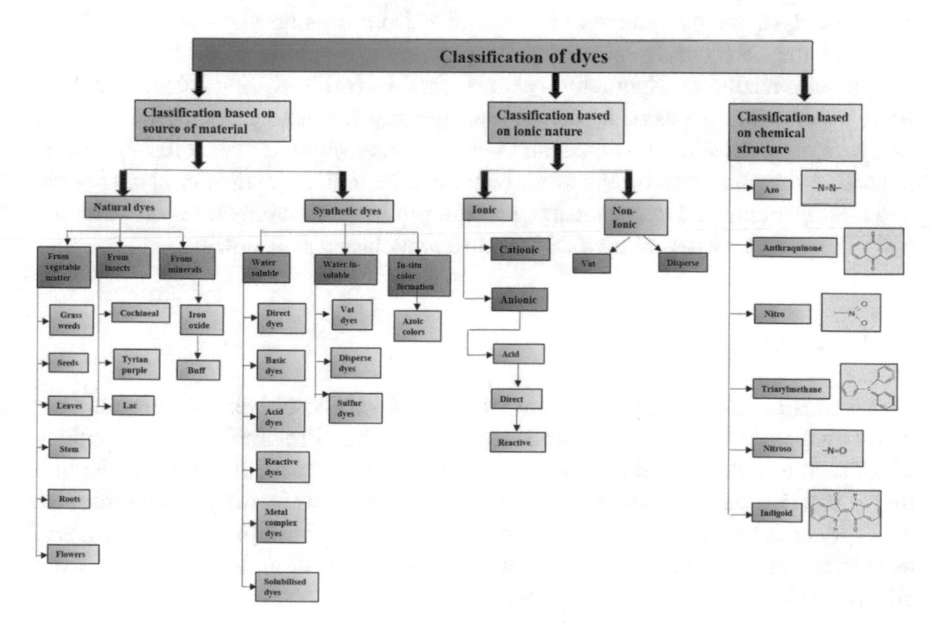

Fig. 1 Broad classification of dyes [66]

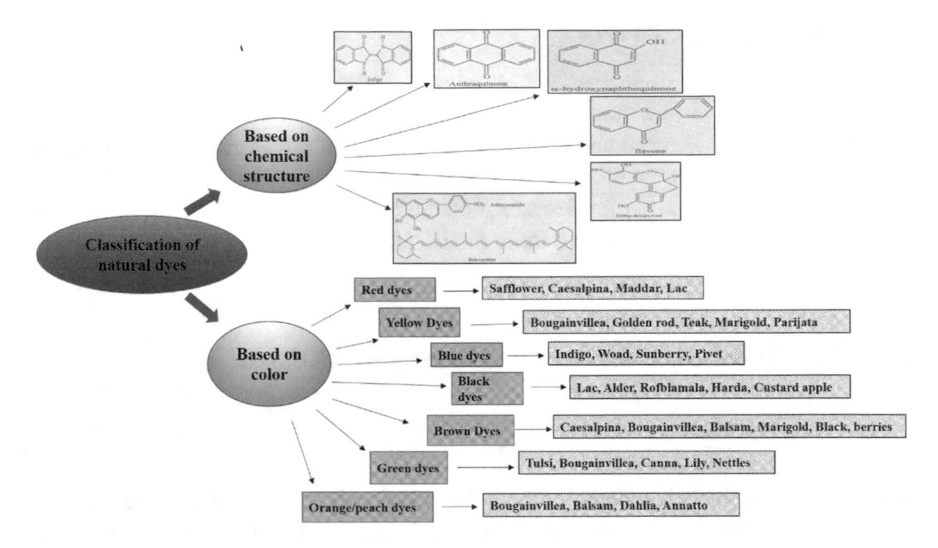

Fig. 2 Classification of natural dyes based on chemical structure and colour [101]

2.1.2 Synthetic Dyes

Synthetic dyes are man-made dyes generated from organic and inorganic chemicals, and they are usually created from petroleum by-products or earth minerals as synthetic resources. Synthetic dyes are used in textile dyeing, pharmaceutical, food, paper printing, cosmetic, colour photography, and leather industries [17, 53]. The ease of application, low cost, and wide range of pigments rendered these dyes popular. Synthetic dyes, on the other hand, can be toxic, mutagenic, carcinogenic, and clastogenic to living organisms, resulting in huge contamination of water and soil [4]. Synthetic dyes are divided into different classes as following.

2.1.3 Direct Dyes

Direct dyes are anionic dyes that are water soluble. These dyes are referred to as direct dyes since they do not require any sort of fixing. Protein fibres and synthetic fibres such as nylon and rayon are dyed with direct dyes and they bind to the fibre due to Van der Waals forces. These dyes are applied in an electrolyte and ionic salts under aqueous bath [84]. Direct dyes are azo dyes having multiple azo bonds, as well as stilbene, phthalocyanine, and oxazine compounds. Cotton, leather, paper, wool, silk, and nylon are all dyed with direct dyes.

2.1.4 Basic Dyes

Basic dyes are cationic water-soluble dyes that are commonly used on acrylic fibres [39]. They have cationic functional groups i.e. $-NR3^+$ or $= NR2^+$. In 1856, H. W. Perkin developed mauveine guilty basic dye, the world's first synthetic dye. Acetic acid is commonly added to the dyebath to enhance dye uptake onto the fibre, and the dyes connect to the fibres by forming ionic bonds with the anionic groups of the fibres. Paper is dyed with the basic dyes as well.

2.1.5 Acidic Dyes

Acid dyes are water-soluble anionic dyes that are used to colour fabrics such as silk, nylon, wool, and modified acrylic fibres in neutral to acid dyebaths. At least half of the dye's attachment to the fibre is linked to salt formation between anionic groups in the dyes and cationic groups in the fibre [61] (Ansari and Seyghali 2013). This category includes the majority of synthetic food colours.

2.1.6 Reactive Dyes

Reactive dyes are most commonly used on cellulosic fibres, and they make a covalent link with the right textile functionality. Certain functional groups in the dye molecule can react with the –OH, –SH, and –NH2 groups found in textile fibres [73]. Reactive dyes have around 29 percent of the global textile dyes market. Their emergence to the market in the mid-1950s gave rise to a new class of cellulose dyes to replace sulphur, vats, azoics.

2.1.7 Vat Dyes

Cotton and other cellulosic materials are dyed with Vat Dyes, which are soluble leuco salts. They are also employed in the production of rayon and wool dyes. The main chemical classes of these water-insoluble colours are anthraquinon (including polycyclic quinones) and indigoids [39].

2.1.8 Disperse Dyes

Disperse dyes are water insoluble dyes that were originally developed to colour cellulose acetate. The dyes are finely ground and marketed as a spray-dried and marketed as a powder in the presence of a dispersing agent [39]. They are most generally used to colour polyester, although they can also cellulose triacetate, colour nylon and acrylic fibres. The small particle size allows for a wide surface area, which aids in dissolving and fibre uptake and the dispersion agent employed during the grinding process might have a big impact on the dyeing rate.

2.1.9 Sulphur Dyes

Sulphur dyes are water insoluble in nature and used to colour cotton [13]. Dyeing is done by submerging the cloth in an organic compound solution, which interrelates with the sulphide source to produce dark colours that stick to the fabric [66].

2.1.10 Azoic Dyes

Azoic dye contains a functional group R–N = N–R', with R and R' typically being aryl groups. Azoic dyeing is a process that involves applying an insoluble azo dye directly to or within the fibre. This is accomplished by combining diazoic and coupling components in a fibre. However, because the chemicals employed are harmful, this procedure is not employed when colouring cottons.

2.2 Classification Based on Ionic Nature

Dyes are categorized as cationic (all basic dyes), anionic (direct, acid, and reactive dyes), or non-ionic (all those other dyes) based on particle charge upon dispersion in aqueous application medium (dispersed dyes).

2.2.1 Cationic Dyes

Cationic dyes are positively charged, water soluble, and produce colourful cations when mixed in solution. Cationic dyes are also known as basic dyes and are used in the dyeing of acrylic, silk, nylon, silk, and wool. This set of dyes encompasses a wide range of chemical structures, most of which are based on substituted aromatic groups [24]. These dyes act as a toxin and has been linked to skin irritation, allergic dermatitis, mutations, and cancer [25]. Cationic functionality can be found in a variety of dyes, most notably methane dyes and cationic azo dyes, as well as anthraquinone, di- and tri-arylcarbenium, phthalocyanine dyes, and a variety of polycarbocyclic and solvent colours. Anthraquinone dyes are costly and weak, whereas azo dyes are strong and cost effective.

2.2.2 Anionic Dyes

Anionic dyes encompass a wide range of compounds from many dye classes that have structural variances (e.g., azoic, triphenylmethan, anthraquinone, and nitro dyes), but all include water-solubilizing, ionic substituents as a common property. These can be used to dye protein fabrics like wool, silk, and nylon. To improve solubility, acid dyes have a complicated structure with big aromatic molecules, a sulfonyl group, and an amino group [86].

2.3 Classification Based on Chemical Structures

2.3.1 Azo Dyes

Azo dyes are the most vital group among the half of all marketable dyes [39, 90]. Azo dyes have been explored extensively than any other class of dyes, and currently a large amount of data is available. Azo dyes contain one nitrogen-nitrogen (N = N) double bond, but they can have a variety of structures [35]. Diazonium salts, phenols or aromatic/naphthyl amines are used to prepare azo dyes by the coupling reactions. These dyes are the most widely used in the textile, leather, pharmaceutical, paper, plastic, and printing industries, owing to their moderately simple synthesis.

2.3.2 Anthraquinone Dyes

Anthraquinone dyes are the second most common type of dye after azo dyes. This class have anthraquinone chromophore groups and are widely utilised in the textile industry. Anthraquinone dyes have an intricate and stable structure than azo dyes, and they are more hazardous to microorganisms and human cells [70]. Anthraquinone dyes exhibited a number of advantages, including higher illumination and virtuous fastness [34]. Anthraquinones are essential natural compounds found in bacteria, fungi, lichens, and plants, unlike azo dyes, which have no natural analogues [43]. The nucleus of substituted anthraquinones and quinizarin (1,4-dihydroxyanthraquinone) is usually prepared by synthesis of the nucleus using phthalic anhydride and a benzene derivative. Most anthraquinone dyes exhibit prominent absorption bands and large molar extinction coefficients in the long-wavelength range. Anthraquinone dyes have notable photo stability, as a result, they have been employed in dye-sensitized solar cells [55].

2.3.3 Nitro and Nitroso Dyes

Nitro and nitroso dyes are of modest commercial value currently, but their minuscule molecular structures are intriguing. Nitro or nitroso groups are conjugated with an electron donating group through an aromatic system. Nitrous acid reacts with phenols and naphthols to generate hydroxynitroso compounds. The nitroso compounds do not dye themselves, but they can produce metal complexes that are either pigments or acid dyes [34].

2.3.4 Triarylmethane Dyes

Triarylmethane dyes are prepared by the condensation of ketone, aldehydes, acid anhydride or acid chlorides with aromatic amines or phenols (naphthols). Triarylmethane dyes are employed as textile dyes for silk, cotton, wool, and as well as in the manufacture of inks, dyeing of paper, colourants in foods, pharmaceuticals and cosmetics. They are also used as biological stains, and as anti-infective, antibacterial, and anthelmintic agents [22, 57, 64]. They have also been used to clean blood in vitro of flagellate parasites like Trypanosoma cruzi, as well as dye-assisted laser inactivation of enzymes [80].

2.3.5 Indigoid Dyes

Indigoid dyes are the one of the oldest organic dyes belongs to vat dyes. These are blue-coloured chemical compound contains carbonyl groups. Indigo dye has been used for textile dyeing for over 5000 years, when it was initially derived from plants. Indigo is nearly entirely used to colour denim pants and jackets. People like it because of its blue colour and the fact that it fades over time to reveal softer blue colours. In the late nineteenth century these dyes were derived from natural sources, but it became one of the first natural chemicals to be synthesised with the development of the modern chemical industry. Dyers favoured synthetically made indigo over indigo from plants because it was of higher quality. Almost all imported indigo is now synthetically generated, and most branded commercial manufacturing is based on versions of the Pflegers technique [32].

3 Common Treatment Strategies in Dye Removal Process

There are three most common treatment strategies of textile effluent physical methods (membrane filtration processes and sorption techniques), chemical methods (coagulation or flocculation and conventional oxidation processes), and biological methods [83, 87].

3.1 *Physical Methods*

These treatment methods are mostly used for small-scale dye removal. Physical methods for dye removal are costly and influenced by other effluent components and also have disposal problems. Major physical treatment processes are given below.

3.1.1 Adsorption

The adsorption technique is the most popular physical method. Despite a lower utilisation rate, this procedure has recently acquired interest because to its high dye removal effectiveness when compared to traditional procedures. Physical adsorption is caused by weak bonding interactions between the adsorbate and the adsorbent, such as hydrogen bonding, dipole–dipole, and van der wall interactions [38]. Due to its high adsorption property activated carbon is the best-known adsorbent that can reduce COD and 92% of color from the textile wastewater [30]. Other adsorbents that can be used to remove colours from wastewater include rice husks, sawdust, pine sawdust, alkaline white mud, bauxite, silica, and chitin, [111]. Some disadvantages of this procedure include the need for high-cost chemicals for precipitation and pH changes, as well as problems with disposal and dewatering of generated sludge following treatment.

3.1.2 Membrane Filtration

Another major approach for extracting colours from an aqueous solution is membrane filtration. The dye liquid is filtered through a micropore membrane of a specific size. The saturate can be reused after these membrane filters separate the dye from the effluent [10]. However, it acts as a physical barrier for the filtration of dyes, chemicals, and processed water for reuse and does not degrade or decolorize the dye [59]. After primary and secondary treatment of dye effluents, it is utilised as a principal post-treatment technique such as reverse osmosis, ultrafiltration, nanofiltration, and microfiltration. Solutes can be retained by membranes with varying pore sizes according to their different molecular weight cut-offs (MWCO) and are classified as: reverse osmosis (<1000 MWCO), nanofiltration (500–15,000 MWCO), and ultrafiltration membranes (1000–100,000 MWCO). The method effectively eliminates all types of dyes, although it generates a lot of sludge. High costs of labor and membrane replacement, since membranes are prone to clogging and fouling are major drawbacks of this method.

3.1.3 Irradiation

Irradiation is a useful approach for degrading dyes that are resistant to chemical oxidation /reduction. The amount of radiation and the availability of oxygen both influence the rate of deterioration [106]. For organic molecules to be successfully broken down by radiation, dissolved oxygen is required. It is an effective technique for the removal of reactive, acid, and disperse dyes as well as toxic organic compounds. The photocatalytic oxidation processes such as UV/H_2O_2, UV/TiO_2, UV/Fenton, and UV/O_3 form free radicals due to UV irradiations [45]. The method's main shortcomings are such high chemical cost and low UV light penetration in highly coloured effluent [113].

3.1.4 Coagulation/Flocculation

The use of a chemical reagent, a coagulant, disrupts the electrostatic connections that exist between the molecules of reactive hydrolyzed dyes (or auxiliaries) and water [5]. Coagulation is employed in conjunction with flocculation or sedimentation, and its effectiveness is determined by the medium's pH and the type of flocculant utilised. [5]. The usage of a high amount of chemicals and the formation of a large amount of sludge are the main disadvantages of this method [74].

3.2 Chemical Methods

3.2.1 Fenton's Oxidation Method

The Fenton oxidation process generates radicals (-OH-) from Fenton's reagent (Fe^{2+}/H_2O_2) when the Fe^{2+} ion is oxidised by H_2O_2 [98]. Both soluble and insoluble dyes can be efficiently removed from effluent. The procedure is generally inexpensive and effective in treating wastewater, resistant to biological treatment. It also has a high COD removal efficiency [88]. The main disadvantage of Fenton's approach is that it is difficult to treat effluents with an alkaline pH, as well as the precipitation of ferric ion salts, which results in a huge volume of sludge and the formation of radicals, which slows down the reaction rate [1].

3.2.2 Ozonation

Ozone was first used in the early 1970s because of its high instability, which made it a good oxidising agent (oxidation potential, 2.07) when compared to chlorine (1.36), and H_20_2 (1.78). Ozonation is a chemical process in which ozone due to its oxidizing property decolorizes the dye at faster rates. In industrial processes, ozone breaks the conjugated bonds of chromophore groups in dye structure resulting in their decolorization with no sludge generation of toxic metabolites formation. Ozonation decolorizes water-soluble dyestuffs at a high rate as compared to non-soluble dyestuffs. One of the advantages is that it is applied in its gaseous state, so it does not affect the overall volume of wastewater. The disadvantage of the ozonation method is the short half-life (20 min) of ozone. Dye materials, pH, temperature, and the presence of effluent salts all affect the half-life. In addition, the cost of operation is higher than with electrochemical treatment. Another drawback of the ozonation process is the formation of secondary products such as aldehydes and dicarboxylic acids which are known to be more toxic than parent molecules [72].

3.2.3 Electrochemical Destruction

In the mid-1990s, the electrochemical destruction technique was developed. An electrochemical reduction/oxidation, electrocoagulation, and electroflotation reaction occur when an electric current is applied to the effluent using the electrodes (anode and cathode). Electrochemical oxidation of pre-treated textile effluent using boron-doped diamond (BDD) anode system resulting in 80% reduction in the COD after 12 h. A novel bioelectrochemical system (BES) based on electrodes; dissolved oxygen and bacterial biofilm formed on electrode had been reported by [109] for treatment of Methyl Orange (MO) that leads to the complete decolorization of the dye (0.15 mM) after 5.5 h. The main advantage of this method is no or little consumption of chemicals and no sludge formation. The major drawback of this technique is high electricity cost, foaming, the short life span of electrodes, poor decolorization of some dyes, and generation of unwanted products [16, 77].

3.2.4 Photochemical Oxidation

This method degrades dye molecules to CO_2 and H_2O by UV treatment in the presence of H_2O_2. High concentrations of hydroxyl radicals are produced during the reaction which results in the degradation of the dye molecules. These hydroxyl radicals broke down the chromophore group of unsaturated dye molecules. Advantages of this method are no sludge formation, a great reduction in foul odors. However, the dye removal rate is influenced by the intensity of the UV radiation, pH, dye structure, and the dye bath composition [27, 94].

3.3 Biological Process of Dye Removal

Dyes, organic pollutants, are well known for their harmful effects on human health and the environment. They are used in a wide variety of industries such as textile, pharmaceutics, cosmetics, food, leather, printer ink, or leather producing industries. The effluents released from these industries contain a wide variety of dyes [31]. Due to complex chemical structure and high molecular weight, the biodegradability of these stable dyes is more difficult [42, 104] and hence, direct release of such potent pollutants causes serious environmental impacts [31]. The discharge of these dyes has persistent colour and a high BOD (Biological Oxygen Demand) load. The toxic carcinogenic dyes obstruct the sunlight penetration into the water, inhibiting photosynthesis and growth of aquatic biota, and interferes with gas solubility [9]. Hence, dye removal from industrial effluent is a necessary process to meet environmental regulations. There are three different methods of dye removal—physical, chemical, and biological method. Physicochemical methods (adsorption, chemical and electrochemical oxidation, ion exchange, ozonation, coagulation/flocculation, membrane process, irradiation, sonication, and filtration with coagulation) are easy to use and some are cost-effective but there are some drawbacks associated with these methods [76]. High electricity consumption [18, 19], non-reusable by-products and sludge, multistage processing with long retention time, use of costly chemicals for pH modification (precipitation and coagulation), and dewatering are some of the prominent challenges associated with the physicochemical methods [49, 100].

Biological methods involve the conversion of synthetic dyes into less toxic inorganic compounds with the help of microorganisms (Table 1). These methods are considered to be cost-efficient, environmentally friendly, and results in reduced sludge production when compared with other techniques. The process uses aerobic or anaerobic organisms such as bacteria, fungi, algae and plants, and the enzymatic system.

3.3.1 Biological Removal of Dye Using Bacteria:

Bacteria can degrade organic pollutants by using them as an energy source and can also oxidize sulphur containing dyes into sulphuric acid. Azo dyes are most commonly treated with the two-stage bacterial process [56]. The first stage involves direct or indirect reduction of azo bond anaerobically and to facilitate this reaction auxiliary substrate must be present to generate reduction equivalents. The breakdown is brought by the azoreductase enzyme resulting in the colourless solution of aromatic amines that can be a mutagenic, toxic and potent carcinogen. The metabolites so formed are further catabolized by the aerobic method in the second stage.

3.3.2 Biological Removal of Dye Using Fungi:

Ligninolytic fungi has shown the most promising results of effluent treatment. It produces laccase, lignin peroxidase, and manganese peroxidase enzymes that are proven to oxidize soluble, insoluble, phenolic, and non-phenolic dyes. Lignin peroxidase oxidizes aromatic ring to cationic radical. Manganese peroxidase oxidizes Mn (II) to Mn (III) which further oxidize phenolic compounds to phenoxyl radicals and Laccase also oxidizes phenols to phenoxyl radicals. Fungal cells are first immobilized either by entrapment (microorganisms are captured in porous or fibrous material or are restrained in a solid or porous matrix-like natural polymeric gels and synthetic polymeric gels [47] or attachment (adherence to the surface like polyurethane foam or nylon foam or to other organisms). When immobilized fungi along with the effluent are kept in the bioreactor, decolourization takes place due to enzymatic action [56].

Table 1 Summary of various biological processes for dye removal

Method	Example	References
Biodegradation of dye using bacteria	*Pseudomonas aeruginosa* has been used to remove Remazol Red dye, *Pseudomonas putida* to remove Orange II dye, *Lactobacillus delbrueckii* can remove Reactive Black 5 and Reactive Orange 16, and *Streptomyces microflavus* CKS6g can remove Crystal Violet and Safranin T, etc	Nur Hazirah et al. (2014), Buntić et al. (2017), [76]
Biodegradation of dye using fungi	Extracellular enzymes like Lignin peroxidase, laccase, and Manganese peroxidase secreted by *Phanerochaete chrysosporium* are used to treat the effluent discharged from textile, paper, or pulp industries. *Aspergillus oryzae* has been used to degrade azo dyes, Laccase enzyme produced by *Trametes polyzona* is used to degrade bisphenol-A and anthraquinone dye, etc. A biodegradation-ozonation hybrid model has been successfully used to treat dyes and tannins. *Penicillium* spp. and *Aspergillus* spp. are the fungi that have been extensively studied for this hybrid model	(Singh et al. 2010; Singh, et al. 2012; Senthilkumar et al. 2014; [12]
Biodegradation of dye using algae	Methylene blue degraded by *Chlorella pyrenoidosa,* acid red 247 degraded by *Spirogyra Rhizopus,* Azo dyes degraded by *Nostoc muscorum, Cosmarium* spp., *Pithophora* spp., *Ulva lactuca, Desmodesmus* spp., *Sargassum,* etc	(Omar et al. 2008; Thirumagal et al. 2016; Pathak et al. 2015)

(continued)

Table 1 (continued)

Method	Example	References
Biodegradation of dye using plants	Hydroponically *Nasturtium officinale* can remove Acid Blue 92, *A. filiculoides* remove Basic Red 46, *Hibiscus furcellatus* remove Poly R 478. Using plant tissue culture technique, many plants were grown in vitro and have shown decolorization results like *Portulaca grandiflora, Petunia grandiflora, Nopalea cochenilifera,* etc. grown in vitro degrade Brilliant Blue R, Reactive Red 196, Green HE4B, Direct Red 5B, and Reactive Blue 160. An enzyme such as peroxidase produced by *Saccharum spontaneum* can decolorize Supranol Green up, Navy Blue HER, and Brilliant Blue H-7G and *Ipomea palmate* can decolorize Methyl Orange, Supranol Green, Brilliant Green, and Chrysoidine. Strategic action of plant *Bouteloua dactyloides* and bacterial species namely *Pseudomonas aeruginosa, Ochrobactrum* spp. and *Providencia vermicola* has been implemented to treat Industrial effluents	(Zheng et al. 2000; Shaffiqu et al. 2002; Davies et al. 2005; Patil et al. 2012; Khandare et al. 2015)

3.3.3 Biological Removal of Dye Using Algae:

Algae are potent biosorbents as they are having high surface area and high binding ability. They are found in both marine and freshwater. The decolourization by algae takes place by three different mechanisms—utilizing chromophores for harvesting algal biomass, carbon dioxide, and water, converting chromophore to non-chromophoric material, and finally, the resulting product is attached to algal biomass. It has been reported that algae may produce azoreductase dye enzyme against azo dyes resulting in decolourization [12]. The product obtained by enzymatic action is further degraded into organic compounds or carbon dioxide. Some algae can even use azo dyes as the sole source of carbon and nitrogen [9]. Dead algal biomass is more efficient than living algal biomass because of the nutritional requirement of living cells and hence, algal biomass can be stored and use for longer periods.

3.3.4 Biological Removal of Dye Using Plants:

Plants can also be used for the bioremoval of dyes from industrial effluents. Plants are autotrophic systems having large biomass and having little nutrient cost. They are accepted because of their easy handling, environmental sustainability, and aesthetic demand [36, 48]. Hydroponic system, plant tissue culture (hairy root, whole plant, callus, cell suspension), purified plant enzymes, and synergistic action of plants and microbes have been implemented/ explored for the biodegradation of dyes.

4 Importance of Anaerobic Process in Dye Removal

Due to their environmental safety and low input requirements, anaerobic treatment has been deemed the most promising technology for wastewater treatment in dyeing industry [107]. Various anaerobic treatment methods in the degradation of a wide range of synthetic dyes have been proved efficient many times [20]. Anaerobic decolourization uses a hydrogen-based oxidation–reduction reaction rather than free molecular oxygen in an aerobic environment to decolorize azo and other water-soluble pigments [83]. Anaerobic reduction of azo dyes can be a cost-effective and efficient way to remove colour from textile wastewater [26]. Chemical reduction by sulphide is partially responsible for the anaerobic transformations of Acid Orange 7, according to certain experiments. The experimental results were mathematically evaluated, and it was discovered that autocatalysis was crucial, as 1-amino-2-naphthol hastened the chemical reduction of the azo link. Under anaerobic conditions, decolourization of reactive water-soluble azo dyes was performed using glucose as a carbon source [14]. The addition of tapioca starch improved the efficacy of colour removal from synthetic blue effluent. Under anaerobic circumstances, methanogenic granular sludge was used to decrease and decolourize Mordant Orange 1 and Azodisalicylate [81]. In a traditional sewage treatment system, Reactive Red 141 was similarly decolorized under anaerobic surroundings. The chemical identification of dye degradation products revealed that decolourization occurred via a reduction mechanism.

Tartrazine, a synthetic dye, was revealed to be easily decolorized in an anaerobic baffled reactor (Bell et al. 2000; Plumb et al. 2001). In anoxic sediment–water systems, Disperse Blue 79 was likewise decreased, with the principal breakdown products being N, Ndisubstituted 1,4 azobenzene and 3-bromo-6-nitro-1,2-diaminobenzene (Weber and Adams 1995). The impact of various modern technologies on the decomposition rate of dyes, as well as the effect of the presence of other chemicals in the media, has received a lot of attention. The creation of high-rate systems, in which hydraulic retention times are uncoupled from solids retention times, has recently been discovered to facilitate the removal of dyes from textile industry wastewaters (Lier van et al. 2001). Another work (Zee van der et al. 2001) demonstrated the viability of using anaerobic granular sludge for total decolourization of 20 azo dyes. It was also shown that using the redox mediator anthraquinone-2,6-disulfonic acid greatly accelerates up the breakdown of azo dyes (Zee van der et al. 2001b). Under anaerobic conditions, the effect of salts (nitrate and sulphate) on the breakdown rate of the azo dye Reactive Red 141 was investigated. The findings revealed that nitrate slows the commencement of breakdown, whereas sulphate had no effect on the biodegradation process (Carliell et al. 1998).

Several high-rate anaerobic reactors, such as the upflow anaerobic sludge blanket (UASB) and the anaerobic baffled reactor (ABR), have been used in textile wastewater treatment in various investigations (Sen et al. 2003). The ability of a thermophilic UASB anaerobic system to discolour synthetic textile dye wastewater as a unit operation clearly shows that it has a considerable advantage over a comparable mesophilic system and can efficiently decolourize such wastewater with a greater efficiency (Willetts et al. 2000). Under anaerobic conditions, using a sequencing batch biofilter, discolouration of commercially relevant azo dyes Orange II, Black 2 HN exhibited > 99 percent colour removal up to a dye concentration of 400 mg/l for both colours [67]. The anaerobic filter and the UASB reactor outperformed the other reactor types evaluated in terms of colour removal efficiency.

5 Integration of Aerobic/Anaerobic Systems to Work as a Hybrid System

The co-existence of anaerobic and aerobic microbes in a single biofilm is the foundation of this system (Zitomer et al. 1998). An integrated anaerobic–aerobic system can be maintained by supplying oxygen to an oxygen-tolerant anaerobic consortium (Tan et al. 1999). Initially, to remove certain dyes such as Mordant Yellow 10 and 4-Phenylazophenol, Expanded Granular Sludge Bed (EGSB), which is an integrated method with oxygenation of recovered effluent was used (Tan et al. 1999; Tan et al. 2001). Reactive Red 5 was removed using a baffled reactor with anaerobic and aerobic compartments, demonstrating excellent colour removal efficiency (Gottlieb et al. 2002). Van der Zee and Villaverdee (2005) in their review article mentioned the degradation of biological azo dye in two stages where the first step is the reductive cleavage of the azo bond, leading to colour reduction which is more efficient in the anaerobic digestion process but also results in the production of aromatic amines which can, in turn, have a high toxicity. It is more appropriate to have a subsequent degradation of the produced aromatic amines, in aerobic environments, which are a very efficient method for the removal of aromatic amines. Due to limitations of aerobic digestion leading to low colour removal and anaerobic digestion resulting in the generation of aromatic amines and limitation of its mineralization, many authors like [23] and [60] have suggested using both the aerobic—anaerobic process for treating effluents with different types of azo dyes.

Despite the effectiveness of anaerobic reduction in the removal of azo dyes the intermediate products such as carcinogenic aromatic amines need to be degraded by an aerobic process. In a study where this type of hybrid system was involved, decolourization rates were found to be 20%, 72%, and 78% for Acid Yellow 17, Basic Blue 3, and Basic Red 2, respectively [2]. This method has successfully used for the degradation of bisazo vinylsulphonyl, anthraquinone vinylsulphonyl and anthraquinone monochlootriazine reactive dyes (Panswad and Luangdilok, 2000). A study has also shown the removal of Basic Red Dye using a sequential anaerobic/aerobic filter system [11]. In a similar study, further two-stage anaerobic/aerobic system has shown the successful decomposition of sulfonated azo dyes such as Acid Orange 10, Acid Red 14, and Acid Red 18 (FitzGerald and Bishop, 1995). In a study by [91], it was further demonstrated that an anaerobic/aerobic treatment is highly useful for the breakage of the azo bond in various azo dyes. The investigation comprising of the removal of reactive diazo Remazol Black B dye by aerobic and anoxic along with anaerobic/aerobic sequencing batch reactor (SBR) activated sludge processes showed that longer anoxic and anaerobic period-promoted decolourization [75].

5.1 *Typical Effluent Treatment Process*

The typical effluent treatment process can be categorized into the primary, secondary, and tertiary treatment processes. Since most of the dye using industries like textile, paper printing, leather, plastic, cosmetics, etc. consume large quantities of water at every stage of their process, wastewater coming from such industries is the main effluent laden with dye and other wastes. An overview provided by [95] of typical primary, secondary and tertiary treatment technologies. The typical primary treatment process consists of screening, sedimentation, homogenization, neutralization, mechanical flocculation, and or chemical coagulation. Secondary treatment can include any or combination of aerobic and or anaerobic treatment which can be sequential, integrated or staggered, aerated lagoons, activated sludge process, trickling filtration, oxidation ditch, and pond. Tertiary treatment which is also considered as the end to pipe or final treatment stage includes technologies individually or in combination like membrane filtration (modified cellulose membrane), adsorption (biochar, agricultural modified waste, etc.) oxidation technique, electrolytic precipitation, foam fractionation, electrochemical processes, ion exchange method, photocatalytic degradation and or thermal evaporation. Below we have tried to list some of the key novel technologies which can be used for the removal of dye residues or by-products generated because of its treatment.

5.2 Hybrid Model (Anaerobic–Aerobic Digestion) with Photocatalysis

[44] demonstrated the use of photocatalysis (fixed titanium dioxide) for the treatment of azo, anthraquinone, and phthalocyanine textile dyes. Photocatalysis was applied on treated product generated post-biological (anaerobic–aerobic) treatment of azo dyes. The product generated post photocatalysis showed non-toxicity to methanogenic bacteria.

5.3 Hybrid Model with Partial Ozonisation

[52] combined biological and chemical treatment into a sequential batch reactor (SBR) for the treatment of residual dyehouse liquors, which has a high concentration of reactive azo dye and recalcitrant compounds. SBR had anaerobic and aerobic phases where water-soluble reactive dyes were reductively cleaved and decolorized by a facultative anaerobic bacterial mixed culture under anaerobic conditions. Further mineralization of the cleaved products produced during anaerobic degradation was subjected to the mixed bacterial population in aerobic conditions. For mineralization of recalcitrant products present initial, ozone was used for partial oxidation in the aerobic phase. Ozone supported mineralization by increasing selectivity of reaction and better biological mineralization. At the end of the process, complete decolourization was achieved along with overall 90% degradation of dissolved organic carbon (DOC).

5.4 Adsorption with Anaerobic Digestion

[37] demonstrated a combination of the adsorption process with anaerobic biological treatment. Anaerobic treatment is the most used method to treat complex wastes present in textile effluent. Sawdust was used as adsorption material for sludge generated post anaerobic treatment. This resulted in ease of sludge collection and disposal.

When compared to other sequential systems, the advantage of these systems is the small-scale requirement. One thing to keep in mind is that these systems were largely used in lab settings using synthetic colours. To determine the usefulness of such a system in an industrial textile mill, a full-scale test should be conducted.

6 Conclusion

Dye decolourization and degradation is now one of the most common problems in textile plants as a result of the current stringent standards. The aesthetic and environmental difficulties of discharging textile effluents in a water body should be considered while designing an effluent treatment facility. The activated sludge technique is the most widely utilized treatment in textile wastewater treatment plants around the world. According to published research, such a technique is unable to entirely remove colour from effluent. Several applications are using lengthy course of actions to increase the bio-adsorption of dyes into bio-sludge using just aerobic processes. This type of process operation may meet local rules for effluent needs in some countries; however, it is ineffective for dye degradation since dyes remain in the sludge. Anaerobic reduction, particularly with azo dyes, can be a cost-effective and efficient way to remove colour from textile effluent. Another significant benefit is that the biogas produced by anaerobic degradation can be used in a cogeneration system to generate both heat and energy. The best treatment plant structure generally incorporates spatially consecutive anaerobic and aerobic treatments. The tertiary treatment procedure should be focused on for effective removal of dye residues remaining after secondary treatment, as this is where the main load of leftover residues like cleaved products post dye degradation needs to be addressed. However, it can be concluded that a combination of different technologies like hybrid model involving anaerobic–aerobic treatment and or usage of adsorption, filtration, ozonation, photocatalysis, etc. needs to be selected appropriately to cater in their removal and thus no toxic by-products are left to enter the environment which can further cause damage.

References

1. Abou-Gamra ZM (2014) Kinetic and thermodynamic study for fenton-like oxidation of amaranth red dye. Advances in chemical engineering and science (2014)
2. An H, Qian Y, Gu X, Tang WZ (1996) Biological treatment of dye wastewaters using an anaerobic-oxic system. Chemosphere 33(12):2533–2542
3. Ajaz M, Shakeel S, Rehman A (2020) Microbial use for azo dye degradation—a strategy for dye bioremediation. Int Microbiol 23(2):149–159. https://doi.org/10.1007/s10123-019-00103-2
4. Ali H (2010) Biodegradation of synthetic dyes - a review. Water Air Soil Pollut 213(1–4):251–273. https://doi.org/10.1007/s11270-010-0382-4
5. Allegre C, Maisseu M, Charbit F, Moulin P (2004) Coagulation–flocculation–decantation of dye house effluents: concentrated effluents. J Hazard Mater 116(1–2):57–64
6. Amani-Ghadim AR, Aber S, Olad A, Ashassi-Sorkhabi H (2013) Optimization of electrocoagulation process for removal of an azo dye using response surface methodology and investigation on the occurrence of destructive side reactions. Chem Eng Process 64:68–78. https://doi.org/10.1016/j.cep.2012.10.012
7. Ayed L, Ksibi IE, Charef A, Mzoughi RE (2021) Hybrid coagulation-flocculation and anaerobic-aerobic biological treatment for industrial textile wastewater: pilot case study. J Textile Inst 112(2):200–206

8. Baêta BEL, Lima DRS, Queiroz Silva S, Aquino SF (2016) Influence of the applied organic load (OLR) on textile wastewater treatment using submerged anaerobic membrane bioreactors (SAMBR) in the presence of redox mediator and powdered activated carbon (PAC). Braz J Chem Eng 33(4):817–825. https://doi.org/10.1590/0104-6632.20160334s20150031
9. Banat IM, Nigam P, Singh D, Marchant R (1996) Microbial decolorization of textile-dyecontaining effluents: a review. Biores Technol 58(3):217–227
10. Barredo-Damas S, Alcaina-Miranda MI, Iborra-Clar MI, Mendoza-Roca JA (2012) Application of tubular ceramic ultrafiltration membranes for the treatment of integrated textile wastewaters. Chem Eng J 192:211–218
11. Basibuyuk MESUT, Forster CF (1997) The use of sequential anaerobic/aerobic processes for the biotreatment of a simulated dyeing wastewater. Environ Technol 18(8):843–848
12. Bhatia D, Sharma NR, Singh J, Kanwar RS (2017) Biological methods for textile dye removal from wastewater: a review. Crit Rev Environ Sci Technol 47(19):1836–1876. https://doi.org/10.1080/10643389.2017.1393263
13. Božič M, Kokol V (2008) Ecological alternatives to the reduction and oxidation processes in dyeing with vat and sulphur dyes. Dyes Pigm 76(2):299–309. https://doi.org/10.1016/j.dyepig.2006.05.041
14. Carliell CM, Barclay SJ, Buckley CA (1996) Treatment of exhausted reactive dyebath effluent using anaerobic digestion: laboratory and full-scale trials. Water SA 22(3):225–233
15. Chinwetkitvanich S, Tuntoolvest M, Panswad T (2000) Anaerobic decolorization of reactive dyebath effluents by a two-stage UASB system with tapioca as a co-substrate. Water Res 34(8):2223–2232
16. Cui L, Wu J, Li J, Ge Y, Ju H (2014) Electrochemical detection of Cu2+ through Ag nanoparticle assembly regulated by copper-catalyzed oxidation of cysteamine. Biosens Bioelectron 55:272–277
17. Couto SR (2009) Dye removal by immobilised fungi. Biotechnol Adv 27(3):227–235
18. Demirbas A (2009) Agricultural based activated carbons for the removal of dyes from aqueous solutions: a review. J Hazard Mater 167(1–3):1–9
19. De Gisi S, Lofrano G, Grassi M, Notarnicola M (2016) Characteristics and adsorption capacities of low-cost sorbents for wastewater treatment: a review. Sustain Mater Technol 9:10–40
20. Delee W, O'Neill C, Hawkes FR, Pinheiro HM (1998) Anaerobic treatment of textile effluents: a review. J Chem Technol Biotechnol: Int Res Process, Environ Clean Technol 73(4):323–335
21. dos Santos AB, Cervantes FJ, van Lier JB (2007) Review paper on current technologies for decolourisation of textile wastewaters: perspectives for anaerobic biotechnology. Biores Technol 98(12):2369–2385. https://doi.org/10.1016/j.biortech.2006.11.013
22. Duxbury DF (1993) The photochemistry and photophysics of triphenylmethane dyes in solid and liquid media. Chem Rev 93(1):381–433
23. Koupaie EH, Moghaddam MR, Hashemi SH (2011) Post-treatment of anaerobically degraded azo dye acid red 18 using aerobic moving bed biofilm process: enhanced removal of aromatic amines. J Hazard Mater 195:147–154
24. Eren E, Afsin B (2008) Investigation of a basic dye adsorption from aqueous solution onto raw and pre-treated bentonite surfaces. Dyes Pigm 76(1):220–225. https://doi.org/10.1016/j.dyepig.2006.08.019
25. Eren E (2009) Investigation of a basic dye removal from aqueous solution onto chemically modified Unye bentonite. J Hazard Mater 166(1):88–93. https://doi.org/10.1016/j.jhazmat.2008.11.011
26. Erkurt HA (2010) Biodegradation of azo dyes. Springer
27. Forgacs E, Cserhati T, Oros G (2004) Removal of synthetic dyes from wastewaters: a review. Environ Int 30(7):953–971
28. Fernandes NC, Brito LB, Costa GG, Taveira SF, Cunha-Filho MSS, Oliveira GAR, Marreto RN (2018) Removal of azo dye using Fenton and Fenton-like processes: evaluation of process factors by Box-Behnken design and ecotoxicity tests. Chem Biol Interact 291:47–54. https://doi.org/10.1016/j.cbi.2018.06.003

29. Gaydardzhiev S, Karthikeyan J, Ay P (2006) Colour removal from model solutions by coagulation - surface charge and floc characterisation aspects. Environ Technol 27(2):193–199. https://doi.org/10.1080/09593332708618633
30. Ghasemian E, Palizban Z (2016) Comparisons of azo dye adsorptions onto activated carbon and silicon carbide nanoparticles loaded on activated carbon. Int J Environ Sci Technol 13(2):501–512
31. Gholami M, Nasseri S, Alizadehfard MR, Mesdaghinia A (2003) Textile dye removal by membrane technology and biological oxidation. Water Quality Res J 38(2):379–391
32. Głowacki ED, Voss G, Leonat L, Irimia-Vladu M, Bauer S, Sarıçiftçi NS (2012) Indigo and Tyrian purple—from ancient natural dyes to modern organic semiconductors. Isr J Chem 52:1–12
33. Gottlieb A, Shaw C, Smith A, Wheatley A, Forsythe S (2003) The toxicity of textile reactive azo dyes after hydrolysis and decolourisation. J Biotechnol 101(1):49–56
34. Gregory P (1990) Classification of dyes by chemical structure. In: Waring DR, Hallas G (eds) The chemistry and application of dyes. Plenum Press, New York
35. Growther L, Meenakshi M (2011) Biotechnological approaches to combat textile effluents. ChemInform 42(29)
36. Ghodake GS, Talke AA, Jadhav JP, Govindwar SP (2009) Potential of Brassica juncea in order to treat textile—effluent—contaminated sites. Int J Phytorem 11(4):297–312
37. Gobinath R, Saravanan SP, Kovendiran S, Hema S, Anas SM, Thangaraj M, Sangeetha M (2015) Treatment of textile wastewater by adsorption process combined with anaerobic digestion. Int J Appl Eng Res 10(53):101–106
38. Guo X, Yao Y, Yin G, Kang Y, Luo Y, Zhuo L (2008) Preparation of decolorizing ceramsites for printing and dyeing wastewater with acid and base treated clay. Appl Clay Sci 40(1–4):20–26
39. Gupta VK, Suhas (2009) Application of low-cost adsorbents for dye removal - a review. J Environ Manage 90(8):2313–2342. https://doi.org/10.1016/j.jenvman.2008.11.017
40. Hai FI, Yamamoto K, Fukushi K (2007) Hybrid treatment systems for dye wastewater. Crit Rev Environ Sci Technol 37(4):315–377. https://doi.org/10.1080/10643380601174723
41. Haapala R, Linko S (1993) Production of phanerochaete chrysosporium lignin peroxidase under various culture conditions. Appl Microbiol Biotechnol 40(4):494–498
42. Hao OJ, Kim H, Chiang P-C (2000) Decolorization of wastewater. Crit Rev Environ Sci Technol 30(4):449–505
43. Han YH, Van der Robert H, Verpoorte R (2001) Biosynthesis of anthraquinones in cell cultures of the Rubiaceae. Plant Cell Tiss Org 67:201–220
44. Harrelkas F, Paulo A, Alves MM, El Khadir L, Zahraa O, Pons MN, Van Der Zee FP (2008) Photocatalytic and combined anaerobic–photocatalytic treatment of textile dyes. Chemosphere 72(11):1816–1822
45. Isaev AB, Aliev ZM, Adamadzieva NK, Alieva NA, Magomedova GA (2009) The photocatalytic oxidation of azo dyes on Fe2 O3 nanoparticles under oxygen pressure. Nanotechnol Russ 4(7):475–479
46. Jiang H, Bishop PL (1994) Aerobic biodegradation of azo dyes in biofilms. Water Sci Technol 29(10–11):525
47. Katzbauer B, Narodoslawsky M, Moser A Classification system for immobilization techniques." Bioprocess Engineering 12, no. 4 (1995): 173–179.
48. Kagalkar AN, Jagtap UB, Jadhav JP, Bapat VA, Govindwar SP (2009) Biotechnological strategies for phytoremediation of the sulfonated azo dye Direct Red 5B using Blumea malcolmii hook. Biores Technol 100(18):4104–4110
49. Kim S-H, Kim T-W, Cho D-L, Lee D-H, Kim J-C, Moon H (2002) Application of characterization procedure in water and wastewater treatment by adsorption. Korean J Chem Eng 19(5):895–902
50. Konsowa AH, Abd El-Rahman HB, Moustafa MA (2011) Removal of azo dye acid orange 7 using aerobic membrane bioreactor. Alex Eng J 50(1):117–125. https://doi.org/10.1016/j.aej.2011.01.014

51. Khandare RV, Govindwar SP (2015) Phytoremediation of textile dyes and effluents: current scenario and future prospects. Biotechnol Adv 33(8):1697–1714
52. Krull R, Hemmi M, Otto P, Hempel DC (1998) Combined biological and chemical treatment of highly concentrated residual dyehouse liquors. Water Sci Technol 38(4–5):339–346
53. Kuhad RC, Sood N, Tripathi KK, Singh A, Ward OP (2004) Developments in microbial methods for the treatment of dye effluents. Adv Appl Microbiol 56:185–213. https://doi.org/10.1016/S0065-2164(04)56006-9
54. Kusumawati N, Santoso AB, Sianita MM, Muslim S (2017) Extraction, characterization, and application of natural dyes from the fresh mangosteen (Garcinia mangostana L.) peel. Int J Adv Sci, Eng Inf Technol 7(3):878–884. https://doi.org/10.18517/ijaseit.7.3.1014
55. Li C, Yang X, Chen R, Pan J, Tian H, Zhu H, Wang X, Hagfeldt A, Sun L (2007) Anthraquinone dyes as photosensitizers for dye-sensitized solar cells. Sol Energ Mat Sol C 91:1863–1871
56. Libra JA, Borchert M, Vigelahn L, Storm T (2004) Two stage biological treatment of a diazo reactive textile dye and the fate of the dye metabolites. Chemosphere 56(2):167–180
57. Lillie RD, Conn HJ (1991) HJ Conn's biological stains. Sigma Chemical Co
58. Luangdilok W, Panswad T (2000) Effect of chemical structures of reactive dyes on color removal by an anaerobic-aerobic process. Water Sci Technol 42(3–4):377–382
59. Lu Y, Shi W, Qin J, Lin B (2010) Fabrication and characterization of paper-based microfluidics prepared in nitrocellulose membrane by wax printing. Anal Chem 82(1):329–335
60. Silva MER, Firmino PIM, Sousa MR, Dos Santos AB (2012) Sequential anaerobic/ aerobic treatment of dye-containing wastewaters: colour and cod removals, and ecotoxicity tests. Appl Biochem Biotechnol 166:1057–1069
61. Mani S, Bharagava RN (2010) 3 Textile industry wastewater environmental and health hazards and treatment approaches, pp 47–69
62. Meerbergen K, Willems KA, Dewil R, Van Impe J, Appels L, Lievens B (2018) Isolation and screening of bacterial isolates from wastewater treatment plants to decolorize azo dyes. J Biosci Bioeng 125(4):448–456. https://doi.org/10.1016/j.jbiosc.2017.11.008
63. Melgoza AM, Cruz A, Buitrón G (2004) Anaerobic/aerobic treatment of colorants present in textile effluents. Water Sci Technol 50(2):149–155. https://doi.org/10.2166/wst.2004.0111
64. Marmion DM (1991) Handbook of US colorants: foods, drugs, cosmetics, and medical devices. Wiley
65. Mitrović J, Radović M, Bojić D, Anđelković T, Purenović M, Bojić A (2012) Decolorization of textile azo dye reactive orange 16 with UV/H2O2 process. J Serb Chem Soc 77(4):465–481
66. Mani S, Bharagava RN (2018) Textile industry wastewater: environmental and health hazards and treatment approaches. In: Recent advances in environmental management, pp 47–69. CRC Press
67. Manu B, Chaudhari S (2002) Anaerobic decolorisation of simulated textile wastewater containing azo dyes. Biores Technol 82(3):225–231
68. Miao Y, Ouyang L, Zhou S, Xu L, Yang Z, Xiao M, Ouyang R (2014) Electrocatalysis and electroanalysis of nickel, its oxides, hydroxides and oxyhydroxides toward small molecules. Biosens Bioelectron 53:428–439
69. Nakamura Y, Sawada T, Sungusia MG, Kobayashi F, Kuwahara M, Ito H (1997) Lignin peroxidase production by Phanerochaete chrysosporium immobilized on polyurethane foam. J Chem Eng Jpn 30(1):1–6
70. Novotný Č, Dias N, Kapanen A et al (2006) Comparative use of bacterial, algal, and protozoan tests to study the toxicity of azo- and anthraquinone dyes. Chemosphere 63(9):1436–1442
71. Omar HH (2008) Algal decolorization and degradation of monoazo and diazo dyes. Pak J Biol Sci 11(10):1310–1316
72. Patil NN, Shukla SR (2015) Decolorization of reactive blue 171 dye using ozonation and UV/H2O2 and elucidation of the degradation mechanism. Environ Prog Sustain Energy 34(6):1652–1661
73. Panda R, Panda H, Panda A (2009) Reactive dyes. Colourage 56(6):44–49. https://doi.org/10.1201/b21336-17

74. Pandey R, Patel S, Pandit P, Nachimuthu S, Jose S (2018) Colouration of textiles using roasted peanut skin-an agro processing residue. J Clean Prod 172:1319–1326
75. Panswad T, Techovanich A, Anotai J (2001) Comparison of dye wastewater treatment by normal and anoxic+ anaerobic/aerobic SBR activated sludge processes. Water Sci Technol 43(2):355–362
76. Piaskowski K, Świderska-Dąbrowska R, Zarzycki PK (2018) Dye removal from water and wastewater using various physical, chemical, and biological processes. J AOAC Int 101(5):1371–1384
77. Pelegrini R, Peralta-Zamora P, de Andrade AR, Reyes J, Durán N (1999) Electrochemically assisted photocatalytic degradation of reactive dyes. Appl Catal B 22(2):83–90
78. Purkait MK, DasGupta S, De S (2005) Adsorption of eosin dye on activated carbon and its surfactant based desorption. J Environ Manage 76(2):135–142. https://doi.org/10.1016/j.jenvman.2005.01.012
79. Rai HS, Bhattacharyya MS, Singh J, Bansal TK, Vats P, Banerjee UC (2005) Removal of dyes from the effluent of textile and dyestuff manufacturing industry: a review of emerging techniques with reference to biological treatment. Crit Rev Environ Sci Technol 35(3):219–238. https://doi.org/10.1080/10643380590917932
80. Ramirez LE, Lages-Silva E, Pianetti GM, Rabelo RMC, Bordin JO, Moraes-Souza H (1995) Prevention of transfusion-associated Chagas' disease by sterilization of Trypanosoma cruzi-infected blood with gentian violet, ascorbic acid, and light. Transfusion 35(3):226–230
81. Razo-Flores E, Luijten M, Donlon B, Lettinga G, Field J (1997) Biodegradation of selected azo dyes under methanogenic conditions. Water Sci Technol 36(6–7):65–72
82. Rodríguez Couto S (2009) Dye removal by immobilised fungi. Biotechnol Adv 27(3):227–235. https://doi.org/10.1016/j.biotechadv.2008.12.001
83. Robinson T, Marchant R, Nigam P (2001) Remediation of dyes in textile effluent: a critical review on current treatment technologies with a proposed alternative. Biores Technol 77:247–255
84. Royer B, Cardoso NF, Lima EC, Vaghetti JCP, Simon NM, Calvete T, Veses RC (2009) Applications of Brazilian pine-fruit shell in natural and carbonized forms as adsorbents to removal of methylene blue from aqueous solutions-Kinetic and equilibrium study. J Hazard Mater 164(2–3):1213–1222. https://doi.org/10.1016/j.jhazmat.2008.09.028
85. Salem SS, Mohamed AA, Gl-Gamal MS, Talat M, Fouda A (2019) Biological decolorization and degradation of azo dyes from textile wastewater effluent by Aspergillus niger. Egyptian J Chem 62(10):1799–1813. https://doi.org/10.21608/EJCHEM.2019.11720.1747
86. Salleh MAM, Mahmoud DK, Karim WAWA, Idris A (2011) Cationic and anionic dye adsorption by agricultural solid wastes: a comprehensive review. Desalination 280(1–3):1–13. https://doi.org/10.1016/j.desal.2011.07.019
87. Saratale RG, Saratale GD, Chang JS, Govindwar SP (2011) Bacterial decolorization and degradation of azo dyes: a review. J Taiwan Inst Chem Eng 42(1):138–157
88. Saratale RG, Sivapathan S, Saratale GD, Banu JR, Kim DS (2019) Hydroxamic acid mediated heterogeneous Fenton-like catalysts for the efficient removal of Acid Red 88, textile wastewater and their phytotoxicity studies. Ecotoxicol Environ Saf 167:385–395
89. Sarkar S, Banerjee A, Halder U, Biswas R, Bandopadhyay R (2017) Degradation of synthetic azo dyes of textile industry: a sustainable approach using microbial enzymes. Water Conserv Sci Eng 2(4):121–131. https://doi.org/10.1007/s41101-017-0031-5
90. Shah M (2014) Effective treatment systems for azo dye degradation: a joint venture between physico-chemical and microbiological process. Int J Environ Bioremed Biodegradat 2(5):231–242. https://doi.org/10.12691/ijebb-2-5-4
91. Seshadri S, Bishop PL, Agha AM (1994) Anaerobic/aerobic treatment of selected azo dyes in wastewater. Waste Manage 14(2):127–137
92. Şen S, Demirer GN (2003) Anaerobic treatment of real textile wastewater with a fluidized bed reactor. Water Res 37(8):1868–1878
93. Singh PK, Singh RL (2017) Bio-removal of azo dyes: a review. Int J Appl Sci Biotechnol 5(2):108–126 (2017)

94. Slokar YM, Le Marechal AM (1998) Methods of decoloration of textile wastewaters. Dyes Pigm 37(4):335–356
95. Srebrenkoska V, Zhezhova S, Risteski S, Golomeova S (2014) Methods for waste waters treatment in textile industry. In: International scientific conference UNITECH, pp 248 – 252
96. Soares OSGP, Órfão JJM, Portela D, Vieira A, Pereira MFR (2006) Ozonation of textile effluents and dye solutions under continuous operation: Influence of operating parameters. J Hazard Mater 137(3):1664–1673. https://doi.org/10.1016/j.jhazmat.2006.05.006
97. Tan NC, Borger A, Slenders P, Svitelskaya A, Lettinga G, Field JA (2000) Degradation of azo dye Mordant Yellow 10 in a sequential anaerobic and bioaugmented aerobic bioreactor. Water Sci Technol 42(5–6):337–344
98. Tantak NP, Chaudhari S (2006) Degradation of azo dyes by sequential Fenton's oxidation and aerobic biological treatment. J Hazard Mater 136(3):698–705
99. Thirumagal J, Panneerselvam A (2016) Isolation of azoreductase enzyme in its various forms from Chlorella Pyrenoidosa and its immobilization efficiency for treatment of water. Int J Sci Res 5:2133–2138
100. Ukiwe LN, Ibeneme SI, Duru CE, Okolue BN, Onyedika GO, Nweze CA (2014) Chemical and electro-coagulation techniques in coagulation-flocculation in water and wastewater treatment-a review. J Adv Chem 9(3):2321–807
101. Vankar PS (2000) Chemistry of natural dyes. Resonance 5(10):73–80. https://doi.org/10.1007/bf02836844
102. Van der Zee FP, Lettinga G, Field JA (2000) The role of (auto) catalysis in the mechanism of an anaerobic azo reduction. Water Sci Technol 42(5–6):301–308
103. Van der Zee FP, Villaverde S (2005) Combined anaerobic–aerobic treatment of azo dyes—a short review of bioreactor studies. Water Res 39(8):1425–1440
104. Wang S, Li H (2007) Kinetic modeling and mechanism of dye adsorption on unburned carbon. Dyes Pigm 72(3):308–314
105. Willetts JRM, Ashbolt NJ (2000) Understanding anaerobic decolourisation of textile dye wastewater: mechanism and kinetics. Water Sci Technol 42(1–2):409–415
106. Wojnarovits L, Takacs E (2008) Irradiation treatment of Azo dye containing wastewater: an overview. Radiat Phys Chem 77(3):225–244
107. Xu H, Yang B, Liu Y, Li F, Shen C, Ma C, Sand W (2018) Recent advances in anaerobic biological processes for textile printing and dyeing wastewater treatment: a mini-review. World J Microbiol Biotechnol 34(11):1-9
108. Yagub MT, Sen TK, Afroze S, Ang HM (2014) Dye and its removal from aqueous solution by adsorption: a review. Adv Coll Interface Sci 209:172–184. https://doi.org/10.1016/j.cis.2014.04.002
109. Yang Q, Wang J, Wang H, Chen X, Ren S, Li X, Li X (2012) Evolution of the microbial community in a full-scale printing and dyeing wastewater treatment system. Bioresour Technol 117:155–163
110. Yurtsever A, Sahinkaya E, Aktaş Ö, Uçar D, Çinar Ö, Wang Z (2015) Performances of anaerobic and aerobic membrane bioreactors for the treatment of synthetic textile wastewater. Biores Technol 192:564–573. https://doi.org/10.1016/j.biortech.2015.06.024
111. Zazouli MA, Balarak D, Mahdavi Y (2013) Application of Azolla for 2-chlorophenol and 4-chrorophenol removal from aqueous solutions. Iranian J Health Sci 1(2):43–55
112. Zitomer DH, Speece RE (1993) Sequential environments for enhanced biotransformation of aqueous contaminants. Environ Sci Technol 27(2):226–244
113. Zyoud A, Zu'bi A, Helal MH, Park D, Campet G, Hilal HS (2015) Optimizing photo-mineralization of aqueous methyl orange by nano-ZnO catalyst under simulated natural conditions. J Environ Health Sci Eng 13(1):1–10

Activated Sludge: Conventional Dye Treatment Technique

Rudy Laksmono Widajatno, Edwan Kardena, Nur Novilina Arifianingsih, and Qomarudin Helmy

Abstract Bio-decolorization is a biological method intended to remove dyestuffs contained in textile wastewater. The biological method has three mechanisms that can be carried out at the decolorization stage, namely biodegradation, bioaccumulation, and biosorption. These three methods have been shown to have the potential to remove dyes from textile wastewater. Biodegradation is an environmentally friendly waste treatment at a lower cost than physical and chemical methods. The use of microorganisms in degrading synthetic dyes is to break the cyclic chain or the double bond of its chemical structure. Microorganisms used to degrade bacterial dyes will produce enzymes and change the chemical structure of the pollutant to be simpler so that the level of toxicity is low. The enzymes produced by the bacteria will then be used to degrade the dye. The biodegradation process carried out on the dye will also change the chemical structure of the chromophore or ausochrome groups. The reductive enzymes produced by several specific bacterial strains function to break the $N = N$ double bond which will be replaced by two molecules of NH2. The product produced in the reduction process is two aromatic aminos which will cause

R. L. Widajatno
Energy Security Study Program, Peace and Security Center, Indonesian Defence University, Sentul, West Java, Indonesia
e-mail: rudy.laksmono@idu.ac.id

E. Kardena · Q. Helmy (✉)
Water and Wastewater Engineering Research Group, Faculty of Civil and Environmental Engineering, Institute of Technology, Ganesa St. 10, Bandung, West Java 40132, Indonesia
e-mail: helmy@tl.itb.ac.id

E. Kardena
e-mail: kardena@pusat.itb.ac.id

N. N. Arifianingsih
Graduate Institute of Environmental Engineering, National Taiwan University, No.1, Sec.4, Roosevelt Rd, Taipei 106, Taiwan
e-mail: d09541008@ntu.edu.tw

Q. Helmy
Bioscience and Biotechnology Research Center, Institute of Technology Bandung, Ganesa St. 10, Bandung 40132, Indonesia

A. Khadir et al. (eds.), *Biological Approaches in Dye-Containing Wastewater*, Sustainable Textiles: Production, Processing, Manufacturing & Chemistry,
https://doi.org/10.1007/978-981-19-0545-2_5

no absorption of light in the visible spectrum which indicates that the reduction process and azo dye decolorization process occur.

1 Introduction

1.1 Textile Industry Process Overview

The textile dyeing and refinement industry is a water-intensive industry, one that uses large amounts of water to meet process needs. Specific water requirements vary from 25 to 180 L/kg-product. The use of water in the cotton dyeing process is more than that of synthetic fabrics. In the wet dyeing and refinement facility, about 72% of the total water used is process water which will then become wastewater. Details of water use in wet dyeing facilities are shown in Fig. 1.

Wastewater produced by the textile industry is a mixture of various compounds, consisting of fibers, biodegradable organic compounds, surfactants, salts such as sodium chloride and sulfate, alkalis that contribute to high pH, oils and fats, hydrocarbons, harmful heavy metals, recalcitrant compounds (difficult to degrade chemically) and persistent compounds (difficult to degrade biologically) from aromatic and heterocyclic compounds, and volatile compounds [51].

The World Bank estimates that around 17–20% of industrial wastewater comes from dyeing and fabric finishing facilities. The typical components of textile wastewater are shown in Fig. 2.

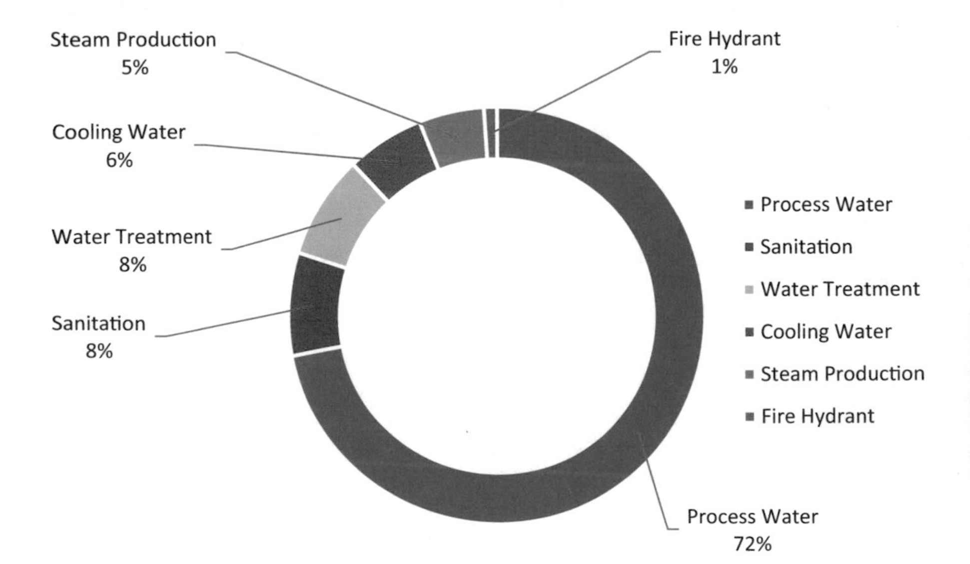

Fig. 1 Typical use of water in textile industry wet dyeing facilities

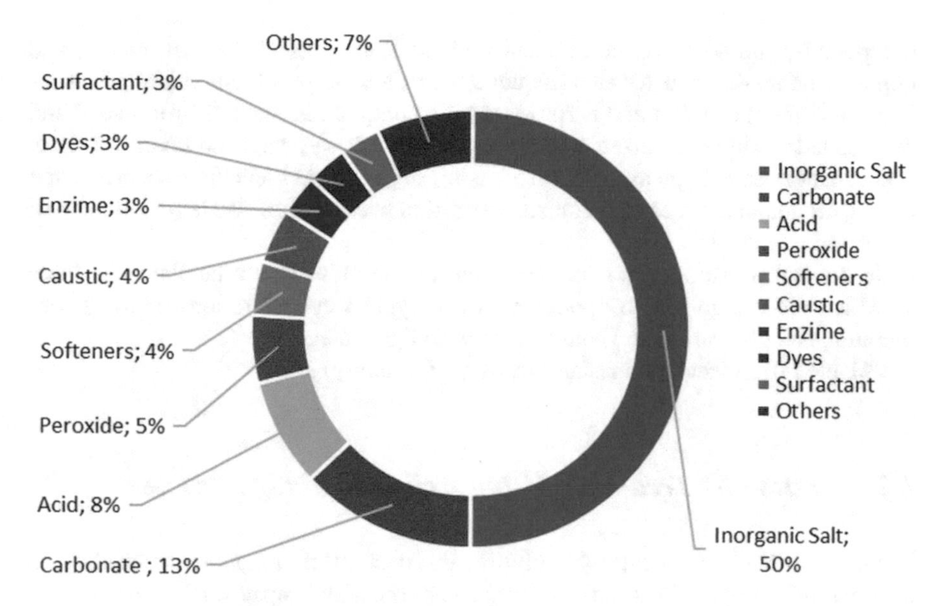

Fig. 2 Typical types of contaminants in textile wastewater

The main pollutants in the textile industry wastewater are classified into three types, floating materials, suspended solids, and dissolved solids, as shown in Fig. 3.

Textile wastewater is generally alkaline with a high content of organic compounds (700–2,000 mg/L BOD). Textile wastewater contains suspended solids, surfactants, and other organic compounds, including phenols and halogenated organic

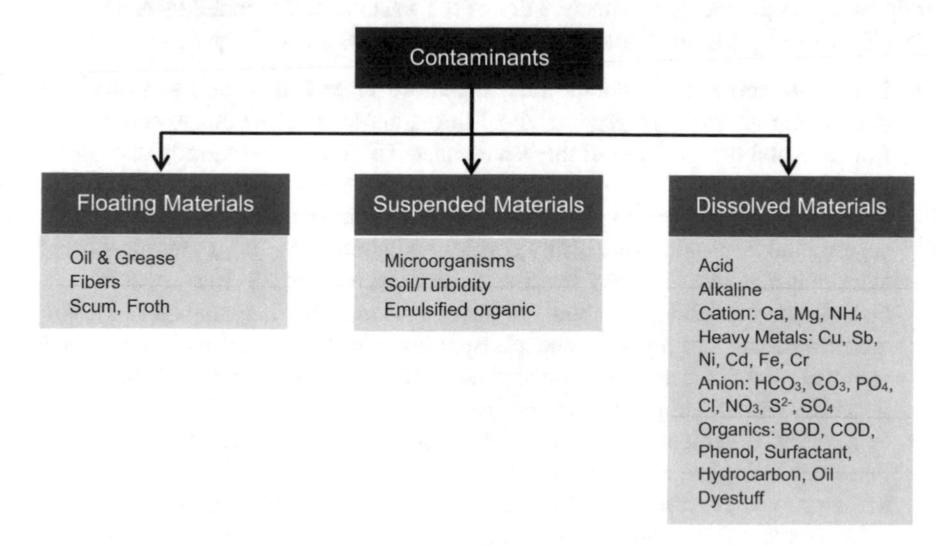

Fig. 3 Typical classification of major pollutant types in textile industry wastewater

compounds, and some contain metals such as lead, zinc, chromium, nickel, and copper. The textile industry also frequently uses a variety of flame retardant chemicals and other persistent and harmful organic compounds, such as brominated and chlorinated compounds, arsenic, and mercury, which can enter the textile industry wastewater stream. Typically, final processing steps such as bleaching, scouring, and dyeing are major sources of pollutant generation because they use large amounts of water and chemicals.

In general, wastewater sources from the textile industry are tabulated in Table 1, which are categorized by pollutant level (typical concentration), namely low, medium, and high strenght. Liquid waste with light category.

Adapted from IMI (Indonesian Ministry of Industry) 2020.

1.2 Wastewater Treatment Using Activated Sludge Process

Nature has the ability to neutralize pollution that occurs if the amount is small enough, but if it is in large quantities and loads it can have a negative impact on the environment because it can lead to changes in the environmental balance so that the waste is said to have polluted the environment. This can be prevented by treating industrial waste before being discharged into water bodies. There are many choices of technologies and techniques in wastewater management with certain characteristics. Determining the right technology for an industry is influenced by many factors such as waste characteristics, quantity, available technology, environmental impact, available land, available human resources, financial availability, ease of installation, operation and maintenance, and so on. The levels and stages of wastewater treatment can vary, depending on the specific characteristics of the wastewater generated by an industry [8]. Typical stages of the wastewater treatment process are as follows:

- Pre-Treatment. Pre-Treatment aims to remove large solids such as gravel and debris left over from the process. In addition, this stage also aims to equalize the flow rate and composition of the wastewater. This stage is generally carried out physically.
- Primary Treatment. Primary processing aims to remove settleable materials, both organic and inorganic materials by gravity, and remove floating materials through skimming methods. Primary treatment can reduce about 25–30% organics, 30–50% suspended solids, and about 60–80% of the total fat/oil contained in wastewater. Some organic nitrogen and phosphorus content as well as heavy metals attached to solids can be separated during this primary treatment. This stage is generally carried out physically and chemically.
- Secondary Treatment. Secondary treatment generally involves the removal of dissolved biological material and colloidal organic matter using an aerobic biological treatment process. The combination of secondary and primary treatment is able to remove BOD and SS levels up to 80–90% of the initial level as well as some heavy metals. This stage is generally carried out biologically and chemically.

Table 1 Identification of potential sources of textile industry wastewater generation

Potential sources	Characteristics
Low strenght	
Storm water	If the housekeeping of the drainage is not proper, there is the potential for small amounts of suspended solids, oil, and grease to go along with the water flow. Rainwater should be harvested for industrial process water
Leaks in heat exchangers, boilers, cooling towers, and steam traps	The amount of dissolved solids is low, in contrast to the condensate which sometimes has a high pH. Ideally, wastewater is separated prior to treatment
Medium strenght	
Machinary clean-up and maintenance	Oil & grease, surfactant
Boiler blowdown and maintenance	High temperature, High dissolved solid
Cooling tower blowdown	High suspended and dissolved solid
Filter and softeners backwash	High suspended and dissolved solid, salts
High strenght	
Sizing: Process for removing fine fibers that arise on the surface of the fabric so that the fabric becomes smooth, even, and clean. Sizing is a very important process for textile materials and must be carried out before mercerization, dyeing, and printing processes	High in organics content (COD/BOD) from excess starch, cellulose, wax, polyvinyl alcohol, and wetting agents. Typical water requirements are 0.5–8,5 L/kg with an average of 4.5 L/kg product
Desizing: Process for removing sizing agents which are mainly starch-based and can inhibit further textile processing	High in organics (COD/BOD), suspended and dissolved solids. Wastewater contains starch, cellulose, polyvinyl alcohol, surfactants, persulfate, and complexing agents. Typical water requirements are 2.5–21 L/kg with an average of 11.5 L/kg product
Scouring: Process for removing various impurities and contaminants from fibers using hot water and detergent	High in organics (COD/BOD) and pH. Wastewater contains surfactants, complexing agents, metals ion, and detergents. Typical water requirements are 20–45 L/kg with an average of 32.5 L/kg product
Bleaching: Process for bleaching textile materials generally uses chlorine-based bleach (sodium hypochlorite and sodium chlorite) or hydrogen peroxide	High in suspended and dissolved solids, fibers, and alkalinity, but low in organics contents. Wastewater contains chlorine, peroxide, sulfite, phosphate, enzime, fibers, and surfactants. Typical water requirements are 24–48 L/kg with an average of 35 L/kg product

(continued)

Table 1 (continued)

Potential sources	Characteristics
Mercerizing: Process for treating cellulose fibers using a caustic alkali solution (or ammonia solution) to increase the strength and affinity of the material for dyes	Low in organics (COD/BOD), dissolved solid, and oil&grease, but high in pH. Wastewater contains caustic, ammonia, wax, sulfate, antifoam, and complexing agents. Typical water requirements are 17–32 L/kg with an average of 24.5 L/kg
Dyeing: Process for giving color to textiles. In addition to dyes, various chemical additives are often used to increase the efficiency of the process. Liquor ratio (LR) is the mass ratio between the total dry matter and the total liquid/solution used. This value varies from 3:1 to 50:1 (for dyes with low affinity and low process efficiency)	High in organics, color, and dissolved solid, but low in suspended solid. Wastewater contains dyes, reducing agents, wetting agents, urea, solvents, phenolic compounds, aromatic hydrocarbons, sulfonic acid, formaldehyde, metals ion
Padding: Process where dyes and auxiliaries can penetrate into the textile material with the help of padder rollers	Wastewater contains polyacrilamide, foaming agents, polyacrylate, polymers, and sulfate
Printing: Process for giving a motif to textile using dyes or other materials, usually in the form of paste or ink	High in organics, color, oil and grease, suspended solids, and slightly alkaline. Low volume of wastewater that contains dyes, starch, binder, thickeners, reducing agents, polymers, polyacrylate, mineral oils, isopropanol, melamine derivative, and volatile organics compounds
Finishing: Process for producing a textile product to suit its intended purpose or end use. At this stage, the quality of the fabric will be increased through one or various stages of the refinement process	Low volume of wastewater that contains organics, and solvents
Coating/repellent: The process of finishing textile fabrics with various coating materials	Low volume of wastewater that contains resins, fluorocarbon, polysiloxane, aluminum, zinc, and chromium

- Tertiary Treatment. The tertiary/advanced process aims to remove suspended solids or dissolved solids remaining after going through primary and secondary processing. Advanced processing independently or separately is carried out to remove nutrients such as nitrogen, phosphorus, and organic materials that are difficult to handle (refractory organics), and heavy metals. This stage is generally carried out physically, chemically, and biologically.

If necessary, before being discharged into the environment, the effluent is disinfected first. The disinfection process aims to reduce or kill pathogenic microorganisms present in wastewater so that it does not pollute the environment. In addition, the

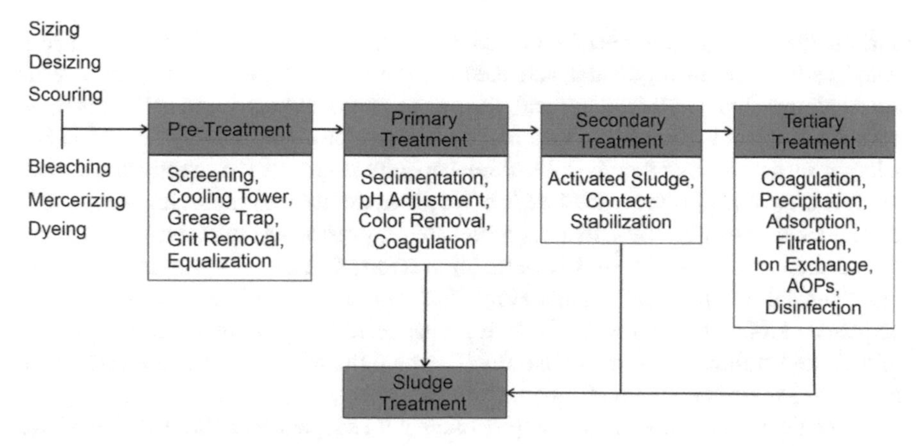

Fig. 4 Process technology options flow chart for textile wastewater treatment

wastewater treatment process generally produces sludge. The level of pollutants in the sludge will be removed or lowered to the quality standard so that it can be utilized or discharged into the environment. In general, the various technology options that can be used to treat textile wastewater are shown in Fig. 4.

Textile industry wastewater contains two elements that have the potential to pollute the environment, namely a relatively strong color and other organic carbon. In particular, the color content of textile wastewater is difficult to remove given the complex chemical structure of dyestuffs. The process of dyeing fabrics in the textile industry generally uses synthetic dyes so that wastewater discharged into the environment still contains residual dye because not all of it is absorbed by the fabric. In the textile dyeing process, about 10–15% of the dye used will be wasted into the environment as liquid waste, so the textile industry wastewater containing color must be treated first before being discharged into the environment. The textile industry wastewater treatment technology that is widely applied to date is a combination of chemical and biological processes. The chemical process commonly used is the coagulation–flocculation process using ferrous sulfate (FeSO4) and lime Ca(OH)2. With the addition of these two chemicals, a CaSO4 precipitate will be formed which will adsorb the dye. In addition, in the process of flocculation and deposition, this precipitate will also adsorb and trap other organic substances. The remaining organic substances that are still contained in the wastewater will be further processed using biological processes. In this case, the biological process commonly used is the activated sludge process or trickling filter (biofiltration). These chemical and biological treatment methods require high operating costs, especially for the procurement of chemicals and the generation of chemical sludge which is a new problem. Another treatment method that has been tried is to do physical processing using an adsorption process with powdered or granular activated carbon to remove color and organic matter at once, but this process also requires expensive costs because it has to replace saturated activated carbon. Activated carbon as an adsorbent has a limited operational time because it will experience saturation. In the adsorption process with activated carbon, the use of granular activated carbon is more economical than powdered activated carbon, because granular activated carbon can be regenerated to restore its adsorption ability. On the other hand, the use of powdered activated carbon is more competitive than granular activated carbon because powdered activated carbon has a larger surface area so that the adsorption capacity is greater [65]. Activated carbon regeneration is an effort to extend the life of activated carbon by removing compounds that have been absorbed by activated carbon, thereby restoring the adsorption capacity of activated carbon to be reused. Regeneration efforts that are often carried out to date are chemical and physical regeneration. However, in a combined process between activated carbon and microorganisms, it is suspected that a bioregeneration process of activated carbon can occur so that it will extend the life of the activated carbon in addition to increasing the performance of this system.

Biological wastewater treatment processes with suspended culture systems have been widely used worldwide for domestic and industrial wastewater treatment. This process is principally an aerobic process in which organic compounds are oxidized to carbon dioxide, water, and new cell biomass. Oxygen supply is usually by blowing air mechanically through a blower or aerator. The most widely used suspended culture wastewater treatment system is the activated sludge system. Activated sludge is a complex biological mass produced when organic waste is treated aerobically. Sludge will contain a wide variety of heterotrophic microorganisms including bacteria, protozoa, and higher life forms. In other words, activated sludge is a mixture of sludge and microorganisms that have the ability to treat waste. Since the activated sludge system was first demonstrated by Edward Arden and William T. Lockett at the Davyhulme Sewage Works in Manchester, United Kingdom in 1914 [28], various modifications of the activated sludge system have been developed. But basically, it has two basic concepts, namely the biochemical stage in the aeration tank and the physical stage in the settling tank. The suspension liquid in the aeration tank in the wastewater treatment process with an activated sludge system is referred to as mixed liquor suspended solids (MLSS), which is a mixture of wastewater with microbial biomass and other suspended solids. MLSS is the total amount of suspended solids in the form of organic and mineral materials, including microorganisms.

Wastewater treatment with conventional activated sludge process generally consists of a primary settling basin, aeration basin, and a secondary settling basin, followed by a chlorination tank to kill pathogenic bacteria. In general, the treatment process is as follows, wastewater originating from pollutant sources is accommodated in an equalization tank. This equalization tank serves to regulate the discharge of wastewater which is equipped with a coarse screen to separate large impurities. The wastewater is then pumped to the primary settling tank. This primary settling basin serves to reduce suspended solids by about 30–50%, and organic content by about 25–30%. The runoff water from the primary settling basin is channeled into the aeration tank by gravity. In this aeration tank, the wastewater is aerated with air so that the existing microorganisms will decompose the organic substances present in the wastewater. The energy obtained from the decomposition of organic substances is used by microorganisms for their growth process. Thus, in the aeration tank, biomass will grow and develop in large enough quantities. This biomass or microorganism

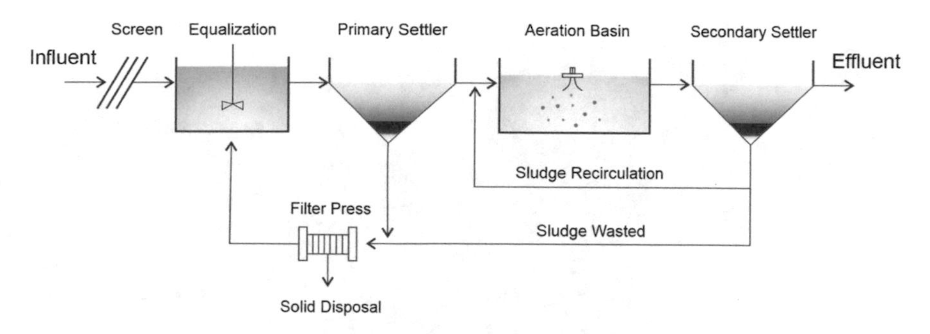

Fig. 5 Typical scheme of conventional activated sludge system

will continuously decompose the pollutant compounds present in the wastewater. From the aeration tank, the water flows into the secondary settling basin. In this tank, the activated sludge, which is a mass of microorganisms, is deposited and pumped back to the inlet of the aeration tank by a mud circulation pump. Overflow from the secondary settling basin is channeled to the chlorination bath. In this chlorine contactor tank, wastewater is contacted with chlorine compounds to kill pathogenic microorganisms. The suspended solids that settle in the secondary settling basin will be recirculated back to the aeration basin, and some of the sludge at a certain time will be disposed of as wasted sludge to control the age of the sludge in the activated sludge system (Tchobanouglous et al. 2004). Schematic of wastewater treatment process with standard or conventional activated sludge system can be seen in Fig. 5.

The biomass is separated in the secondary sedimentation tank so that it flocculates and settles. This causes bacteria, protozoa, filamentous microbe, and other microorganisms to form macroscopic flocs which will eventually settle. The attachment of these microorganisms is aided by the polysaccharide matrix produced by these microbes. An illustration of the formation of activated sludge floc can be seen in Fig. 6.

1.3 Operational Variables in Activated Sludge Process

The operational variables commonly used in wastewater treatment processes with activated sludge systems are as follows:

Hydraulic Retention time (HRT): The hydraulic retention time is the average time required for wastewater to enter the aeration tank or tank [40]. For the activated sludge process, the value is inversely proportional to the dilution rate (D).

$$HRT = \frac{V}{Q} = \frac{1}{D}$$

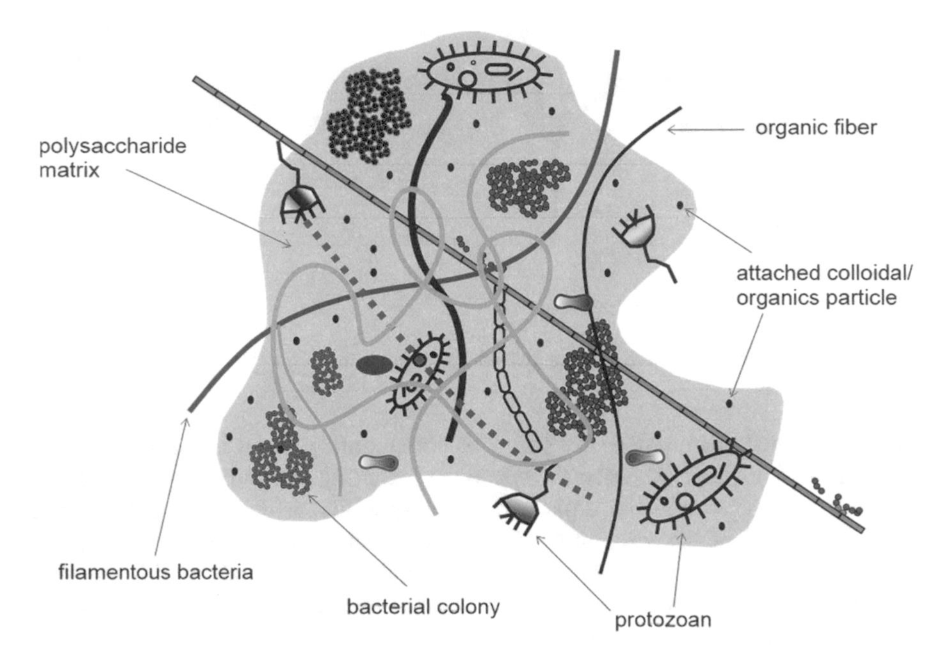

Fig. 6 Illustration of floc forming structure in activated sludge

where V is the volume of the reactor or aeration tank (m^3), Q is the discharge of wastewater entering the aeration tank (m^3/hour), and D is the dilution rate ($hour^{-1}$).

BOD Loading Rate or Volumetric Loading Rate: BOD load is the total mass of BOD in the influent wastewater divided by the reactor volume.

$$BOD\,Load = \frac{Q \times S_0}{V}\frac{kg}{m^3}.day$$

where Q is the incoming wastewater discharge (m^3/day), S_0 is the BOD concentration in the incoming wastewater (kg/m^3), and V is the reactor volume (m^3). For the conventional activated sludge process, the BOD load is generally in the range of 0.3–0.8 $^{kg}/m^3$.day, while for the extended aeration activated sludge process the BOD load is generally in the range of 0.15–0.25 $^{kg}/m^3$.day.

Microbial Degrader Concentration: The liquid suspension in the aeration tank in the wastewater treatment process with an activated sludge system are referred to as mixed liquor, which is a mixture of wastewater with microbial biomass and other suspended solids. Mixed liquor suspended solid (MLSS) is the total amount of suspended solids in the form of organic and mineral materials, including microorganisms. The MLSS was determined by filtering the mixed liquor with filter paper, then the filter was dried at a temperature of 105 °C, and the weight of the solids in the sample was weighed. The portion of organic matter in MLSS is represented by Mixed Liquor Volatile Suspended Solid (MLVSS), which contains both living and dead microbes, non-microbial organic matter, and cell debris. MLVSS is measured by continuously heating a dried filter sample of MLSS to a temperature of 600 °C, and the value is close to 65–75% of the MLSS.

***Food-to-Microorganism (F/M) Ratio**:* This parameter indicates the amount of organic matter (BOD) to be removed divided by the total mass of microorganisms in the aeration tank. The value of the F/M ratio is generally expressed in kilogram of BOD per kilogram of MLSS per day [44]. F/M can be calculated using the following formula:

$$\frac{F}{M} = \frac{Q(S_0 - S)}{MLSS \times V}$$

where Q is the effluent flow rate m^3 per day, S_0 is the BOD concentration in the wastewater entering the reactor area (kg/m^3), S is the BOD concentration in the effluent (kg/m^3), MLSS in (kg/m^3), and V is the volume of the reactor or aeration tank (m^3). The F/M ratio can be controlled by adjusting the circulation rate of activated sludge from the circulating final settling basin to the aeration basin and/or adjusting the wasted sludge. The higher the activated sludge circulation rate, tipically the higher the F/M ratio. For wastewater treatment with conventional activated sludge systems, the F/M ratio is 0.2–0.5 kg BOD per kg MLSS per day, but can be as high as 1.5 if pure oxygen is used (Hammer and [23, 26]. A low F/M ratio indicates that the number of microorganisms in the aeration tank is too much compared to organic compounds as food. This condition can cause starvation of healthy forming floc microbe, which is followed by several issues such as the formation of bulking, pin floc which triggers an increase in suspended solids content in the effluent.

***Sludge Age**:* Sludge age is often called the mean cell residence time. This parameter indicates the average residence time of microorganisms in the activated sludge system. If HRT requires retention in hours, the residence time of microbial cells in an aeration tank can be days. This parameter is inversely proportional to the rate of microbial growth [67]. The age of the sludge can be calculated by the following formula:

$$Sludge\ Age = \frac{MLSS \times V}{SS_e \times Q_e + SS_w \times Q_w}$$

where V is the volume of the aeration basin (L), SS_e is the suspended solids in the effluent (mg/l), SS_w is the suspended solids in the sewage sludge (mg/l), Q_e is the effluent rate of waste (m^3/day), and Q_w is the rate of influent waste (m^3/day). Sludge age can vary between 5 and 15 days for conventional activated sludge systems. In winter, it can be longer than in summer. Important parameters controlling activated sludge operation are organic load, oxygen supply, and control and operation of the final settling basin. This final settling basin has two functions, for clarification and thickening of the sludge [7, 14, 58].

Sludge Volume Index (SVI): The conventional way to observe the settleability of a sludge is to determine the Sludge Volume Index. To determined SVI, take a mixture of sludge and wastewater (mixed liquor) from the aeration tank and put in a 1 L conical cylinder and left for 30 min, and record the volume of the sludge formed.

SVI is an indication of the volume occupied by 1 g of sludge. SVI can be calculated using the following formula:

$$SVI = \frac{SV}{MLSS} \times \frac{1000\,mg}{gram}$$

where SV is the volume of sludge deposited in a conical cylinder after 30 min of settling (ml), MLSS is mixed liquor suspended solid (mg/l). In wastewater treatment units with conventional activated sludge systems with MLSS < 3500 mg/l, typical SVI values are in the range of 50–150 ml/gr [54]. Considering that there are many operational parameters in the activated sludge process that must be controlled, the wastewater treatment process using the activated sludge process is quite complicated and requires sufficient operator expertise.

1.4 Modification of Activated Sludge Process

The development of the industrial sector will have positive and negative impacts, where one of the negative impacts that arise is the increase in the amount of waste produced. This waste will disrupt the flow process in the industry so that the waste must be removed from the industrial process and generally the waste will be discharged into the environment. However, before being disposed of, the waste must be treated properly first so that it does not have bad consequences such as damage to the surrounding environmental ecosystem due to pollution or disturbance to public welfare. The problem is that the waste generated from the production of an item will have different treatment methods and technology so that the problem of treating this waste is a complex problem. In this sub-chapter, several modifications to the process and operation unit of the activated sludge WWTP will be discussed. The wastewater treatment process that contains color pollutants that are widely used in Indonesia currently is the activated sludge process. The problem faced in the current WWTPs is that the processed water often does not meet the quality standards to comply with the new regulations in 2019 [30]. The parameter that often exceeds the quality standard is the color parameter. Several factors that are often encountered include too short in hydraulic residence time, very large fluctuations in waste discharge, poor aeration process, and operational errors due to inadequate knowledge of the operator about the process. To overcome the problems mentioned above, technological innovation is needed to improve the efficiency of wastewater treatment, especially the conventional activated sludge process. One example to improve the activated sludge process is by adding activated carbon to the aeration tank. By adding activated carbon to the aeration basin, microorganisms will multiply on the surface of the activated carbon thus increasing the density of microorganisms that decompose organic matter and dye in the aeration basin.

Granular Activated Sludge (GAS): In biological wastewater treatment systems, microorganisms can exist as biofilm and biogranule. A biofilm consists of microorganisms attached to an inert substrate, while a biogranule is a microorganism that is completely self-immobilized [36]. Biogranules cannot occur in the natural environment. Strong selective pressure is the requirement to trigger biogranulation in reactors [37]. Observations of this phenomenon were first made in the sludge of anaerobic sewage treatment systems in 1980 [35]. The first granular sludge to be detected in an aerobic system was found in 1997 in the Sequencing Batch Reactor (SBR) [45]. About two decades after the discovery, aerobic granular sludge has been studied for applications such as treatment of domestic and high organic load wastewaters, bioremediation/biotransformation of toxic aromatic pollutants (including phenol, toluene, pyridine), treatment of industrial effluents (textile, dairy, brewery), adsorption of heavy metals, recovery of high value-added products, and others [53]. In comparison to conventional activated sludge systems, GAS has several advantages, including good settling ability, high biomass retention, tolerance to toxicity, high tolerability, and higher extracellular polymer substances (EPS) production [1, 46, 53, 63]. EPS is known as bacterium secreted sticky material containing proteins, polysaccharides, humic acids, and lipids that initiate the granulation process [55]. Aerobic granulation is affected by several operational parameters, including substrate composition, organic loading rate (OLR), hydrodynamic shear force, feast-famine regime, feeding strategy, dissolved oxygen concentration, reactor configuration, solids retention time, sequential batch reactor (SBR) cycle time, settling time, and volume exchange ratio [53].

The formation mechanism of aerobic particles consists of four stages [48]. Firstly, intercellular contact with microorganisms. Secondly, under the pressure of external disturbances, the microbes react similar to Quorum sensing and gather with each other to form initial aggregates. Thirdly, microorganisms unite with each other and enhance adhesion through the production of EPS. This EPS is an extracellular polymer, which is one of the important physiological capabilities of microorganisms. It can break down molecules in the environment (in vitro) into small molecules and then take them into the body. Finally, gradually form particles through hydrodynamic shearing force (by aeration bubbles), and switch between aerobic and anaerobic cycles in a single reactor tank, as the oxygen concentration gradient decreases with the depth of the particles, so obvious stratification is gradually formed. Generally, the hydrodynamic shear caused due to bubble aeration induces extracellular polymeric substances (EPS) production, cell surface hydrophobicity, and trigger cell–cell interactions contributing to initiation of granule formation. The gel forming exopolysaccharide components of the EPS can play a role in the structural stability. Segregated distribution of microorganisms in the granular activated sludge illustrated in Fig. 7.

After the mature particles are formed, the reactions in different layers are not the same. The reaction that can be achieved by the internal anaerobic and external aerobic composition is the removal of carbon, nitrogen, and phosphorus. The outermost aerobic part is decomposed by heterogeneous bacteria for COD and nitrifying bacteria for digestion and phosphorus uptake. The excessive anoxic layer in the middle can carry out the denitrification reaction The innermost anaerobic

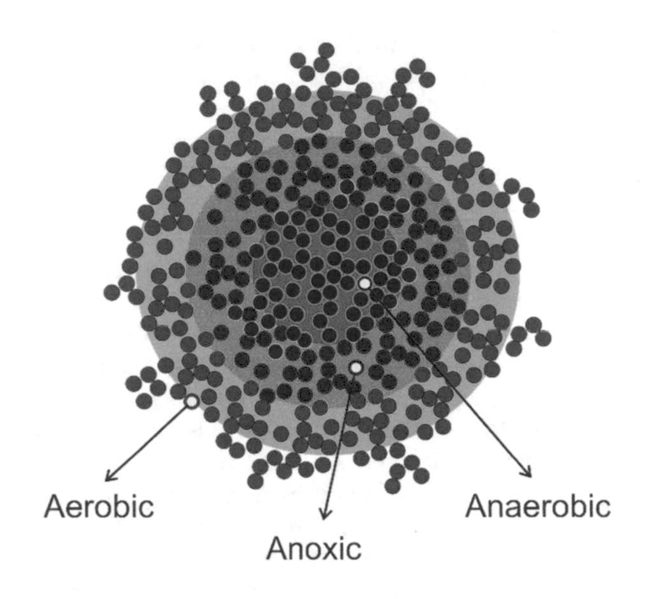

Fig. 7 Graphical representation on segregated distribution of microorganism in the individual granular activated sludge. Dyestuff in the wastewater was biotransformed to amines in the anaerobic zone, which were subsequently degraded in the anoxic–aerobic zone [9, 11, 41, 48, 71]

layer releases phosphorus. Anaerobic phosphorus release aerobic phosphorus uptake. Aerobic granular sludge (AGS) is a novel microbial community which allows simultaneous removal of carbon, nitrogen, phosphorus, and other pollutants in a single sludge system [18]. The biological dye removal process consists of 3 steps, firstly, anaerobic reduction occurs to cleave the azo dye molecule on its chromophore (-N = N-) to form colorless aromatic amine intermediates. The second is the cleavage of the aromatic ring of amine compounds into simpler aliphatic ones, and the third is mineralization in the next aerobic phase to form carbon dioxide, water, and ammonia. The granular activated sludge is then settled in the clarifier tank and due to its heavier characteristics than conventional flocs, and has a much faster settling speed (>12 m/hour), the sludge deposition process can be carried out in a smaller clarifier tank. It also allows for higher concentrations of biomass in the aeration tank, leading to more efficient degradation of wastewater [71].

Biological Activated Carbon (BAC): What is meant by biologically activated carbon is the biological process of activated sludge with a modification of the process in the form of adding a certain amount of activated carbon into a biological reactor containing microorganism culture. This combination treatment has been successfully developed by DUPONT as a PACT™ (Powder Activated Carbon Treatment) process [17, 27, 56]. This is done with the aim of improving the quality of the WWTP effluent, considering that the addition of activated carbon into the biological reactor will improve the system's performance and is more economical than the physical process with activated carbon adsorption which is carried out separately after the biological process [70]. The combined activated sludge and activated carbon system has better performance due to the following phenomena (as shown in Figs. 8, 9,10).

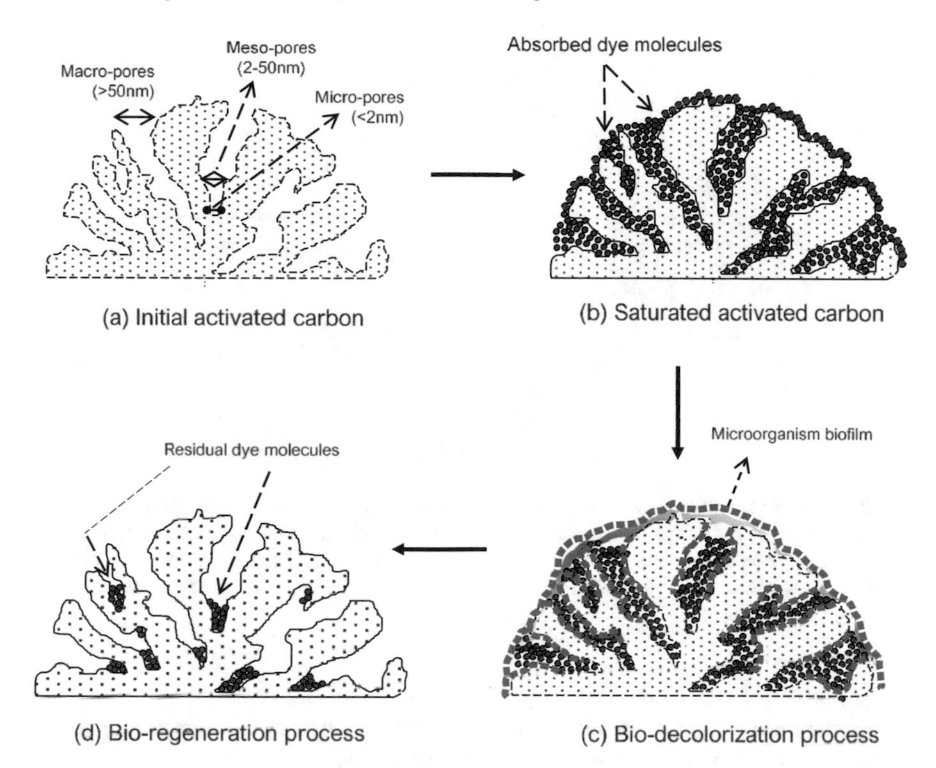

Fig. 8 The artistic scheme of the combined activated sludge and activated carbon system in the biodecolorization process of dye compounds, **a** initial activated carbon, **b** saturated activated carbon after absorbing dyes molecules, **c** breakdown of dyes molecules by microorganisms activity, and **d** bioregeneration process of activated carbon. [68, 69] Adapted from

a. Activated Carbon Adsorption. Activated carbon adsorbs dissolved substances in wastewater, including non-biodegradable substances. These non-biodegradable molecules are easier to adsorb, so their levels in wastewater are reduced.
b. Interaction Effect. Activated carbon increases microbial activity, because the surface of activated carbon becomes a matrix for microbial growth so that the rate of substrate removal increases. In addition, the absorption of the substrate on the surface of the activated carbon causes the substrate concentration to decrease. By decreasing the concentration of this substrate, the process of substrate biodegradation by microbes becomes faster.
c. Contact Optimization. Substrates that are adsorbed by activated carbon and the presence of microbes attached to the surface of activated carbon will increase the contact between the substrate and its degrading microbes.
d. Biological Regeneration Process. The presence of a substrate that is absorbed by activated carbon will be utilized by microorganisms, as a result, activated carbon becomes active again for further adsorption processes.

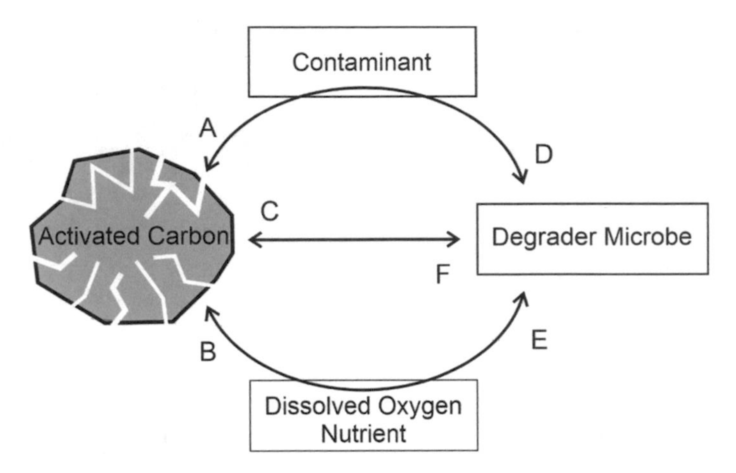

Fig. 9 Simplified relationship of four factors in biological activated carbon. Activated carbon adsorbs both dissolved oxygen and contaminants (A-B), degrader microbe utilizing contaminant and nutrient to grow on the surface of activated carbon (C-D-E), and by the decomposition of contaminants initially adsorbed by activated carbon, promote the regeneration process of the saturated activated carbon (F)

e. Microbial Flocs Core. Carbon particles have a certain density so that they can function as ballast elements and as a core for flocculation of biomass, as a result, the deposition of activated sludge flocs increases.

It can be explained that the transport of microorganisms during the bioregeneration process enters the pore structure through the macropores, mesopores, and micropores, it is possible because the size of microorganisms is smaller than the macro- and meso-pore diameters, but in the micropore structure, microorganisms cannot enter because of their size. larger microorganisms. From the test results of activated carbon that has been regenerated, the efficiency of color removal, COD, and absorption ability tends to decrease over time from the initial activated carbon [68]. The process of using activated carbon coated with a biofilm of microorganisms for water and wastewater treatment processes was developed in the 1970s. [52] examining the bioregeneration of activated carbon used to purify textile wastewater. For this case, it was operated for 11 months and it was stated that the results were good, but the mathematical bioregeneration process was not reported and the mechanism was not mentioned. Little attention has been paid to the development of biologically activated carbon regeneration research as an alternative to traditional methods of activated carbon regeneration. The experimental practice was carried out with three different treatments, namely activated carbon which was added to the biological system, the second was fine sand which was added to the activated sludge system, and the third was control in the form of activated sludge without adding other materials. In these three treatments, the growth of microorganisms was observed with the results in the first treatment, the number of microorganisms in activated sludge reached 15×10^5 colonies/cc, the second treatment resulted in 25×10^2

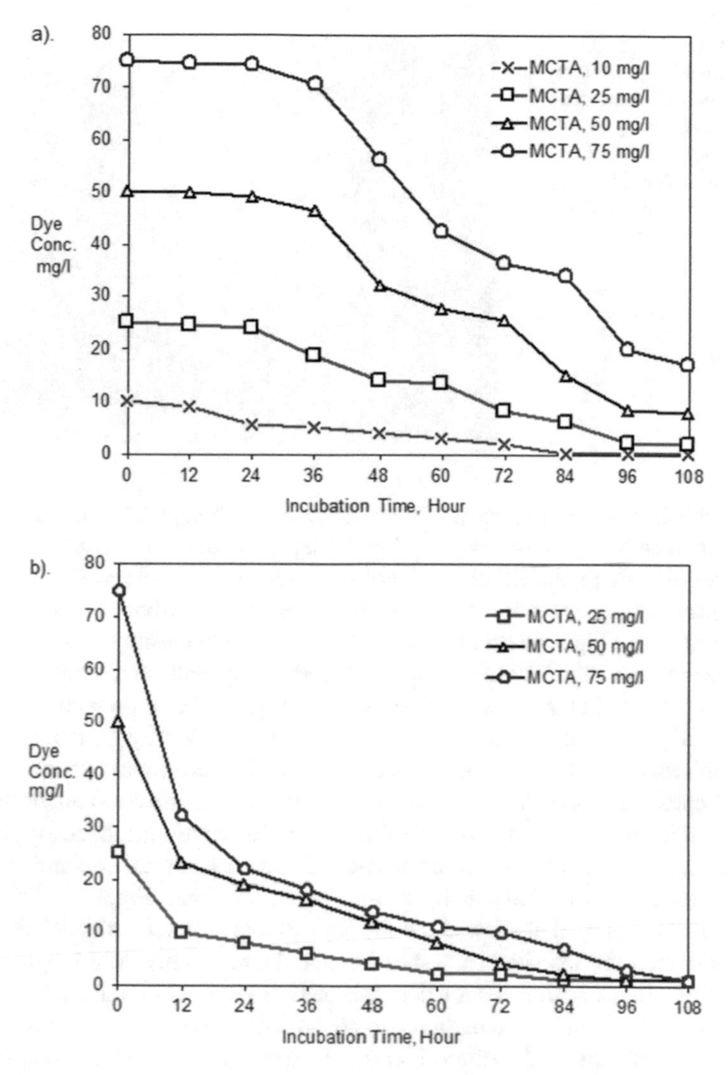

Fig. 10 Biodegradation assay of monochlorotriazine reactive red (MCTA-RR) dye for 108-h incubation time using Pseudomonas rudinensis. The presence of microorganism activity in the system caused a decrease in the concentration of MCTA dye (a), while a faster decolorization was observed in the BAC system where the color removal process was caused by both adsorption of activated carbon and due to biodegradation action of microorganisms. Adapted from [68]

colonies/cc, and the third resulted in 5×10^2 colonies/cc. Biodegradation of organic pollutants occurring in the BAC column overgrown with a large amount of aerobic biomass will result in a longer operating time than unadulterated carbon and this makes for a low cost treatment. Although the BAC process has been widely used, but the mechanism is not widely known especially the relationship between biodegradation and carbon sequestration, these two can be promoted individually or they occur

Fig. 11 Visual image of biodecolorization process of 25 mg/l monochlorotriazine reactive red dye using *Pseudomonas rudinensis* (left: control reactor; right: after 108-h incubation time)

simultaneously in the biological column of activated carbon [49, 50, 66]. The biological activated carbon process was developed based on the activated carbon process, which uses the synergistic effect of the pollutant biodegradation process by attached growth microbes and the adsorption process on activated carbon. Activated carbon is known to have a high specific surface area and a highly developed pore structure, so it is characterized by its great effect in absorbing dyestuff and organic matter in textile wastewater. In the process of Biological Activated Carbon technology, both granular and powdered activated carbon is used as a carrier, by growing or immobilizing microorganisms under proper temperature and nutritional conditions on the activated carbon surface and finally forming BAC, which can exert adsorption and biodegradable roles simultaneously. Biological activated carbon technology consists of the interaction of activated carbon particles, microorganisms, contaminants, and dissolved oxygen, in the mixed liquor in the aeration tank. Figure 9 illustrates a simplified relationship that shows how the four factors interact with each other. The relationship between activated carbon and contaminants is only the adsorption effect of activated carbon, and the reaction depends on the nature of the activated carbon and contaminants. While activated carbon can adsorb DO and microorganisms adsorbed on the surface of activated carbon, DO administration will degrade contaminants. In summary, with the interaction of these four factors, the goal of removing both dyes and contaminants from textile wastewater can be achieved using a biologically activated carbon process.

Generally, the widely adopted textile wastewater treatment is a sequential anaerobic–aerobic biological treatment process. This treatment method has proven to be the most successful yet economical in decolorizing the dye content in textile wastewater due to the breakdown of azo bonds (—N = N—) in an anaerobic process into intermediate compounds, before entering the aerobic process which then continues the mineralization of the intermediate compound into carbon dioxide and water. Currently, biological treatment for textile wastewater using microorganisms has been widely studied compared to physical and chemical treatment due to its environmental friendliness, reproducible efficiency treatment and economical implementation. BAC can be categorized as a biofilm which is best described as a community of microorganisms attached to the activated carbon surface as a carrier. BAC can be developed by single or multispecies microorganisms which have the ability to form on both living and nonliving surfaces. The growth of microorganisms in the form of biofilms protects them from adverse environmental conditions and acts as a shield against environmental stresses, enabling communication and exchange of genetic material and nutrient availability and persistence in different metabolic states. Biofilms can have very long biomass residence times when treatment requires slow-growing organisms with poor biomass yield or when wastewater concentrations are too low to sustain activated sludge floc growth [47]. The decolorization process in this biofilm system can be similar to the decolorization mechanism in a granular activated sludge system.

2 Biological Dye Removal Process

The biological textile wastewater treatment process can be carried out under two conditions, namely aerobic conditions and anaerobic conditions or a combination of anaerobic and aerobic [72]. Aerobic biological processes are usually used for textile wastewater treatment with a low Biological Oxygen Demand (BOD) load, while anaerobic biological processes are used for wastewater treatment with very high BOD loads [32, 39]. Generally, biological wastewater treatment can be divided into three, namely biological processes with suspended culture, biological processes with attached culture and treatment processes using a lagoon or pond system. Biological process with suspended culture is a treatment system using the activity of microorganisms to decompose pollutant compounds present in the water and the microorganisms used are cultured in suspension in a reactor. Biological process with attached culture is a treatment system which the microorganisms used are cultured and attached the surface of the media. This process is also known as the microbiological film process or biofilm process. Examples of wastewater treatment technologies in this way include: trickling filters, submerged biofilters, rotary biological contact reactors, and others. The process of biological wastewater treatment with a lagoon or pond is to accommodate wastewater in a large pond with a long residence time so that with the activity of microorganisms that grow naturally, pollutant compounds present in the water will decompose [62]. The dye molecule is a combination of unsaturated organic substances with chromophores as color carriers and auxochromes as color binders with fibers. Unsaturated organic substances found in the formation of dyestuffs are aromatic compounds, including aromatic hydrocarbons and their derivatives, phenols and their derivatives, and nitrogen-containing hydrocarbon compounds. The chromophore group is the group that gives the molecule its color. Based on the source origin, dyes are divided into two, namely natural dyes and synthetic dyes. Natural dyes are natural dyes or dyes derived from plants, while synthetic dyes are aromatic hydrocarbon derivatives such as benzene, toluene, and naphthalene. The names and chemical structures of the chromophore are shown in Table 2. Based on the coloring

Table 2 Chemical structure of dye chromophore

Chemical bonding group	Chemical structure
Azo group	—N = N—
Ethylene Group	—C = C—
Carbonyl Group	—C = O
Carbon Group – Nitrogen	—C = NH; C = N—
Sulfur Carbon Group	—C = S; —C–S–S–C—
Nitroso	NO; —N – OH
Nitro	NO_2; NN–OOH

process, the dyestuffs are classified as reactive dyes, dispersion dyes, direct dyes, vat dyes, sulfur dyes, basic dyes, acid dyes, and solvent dyes.

Many studies have been carried out on the ability of microorganisms to degrade textile dyes because the dye contains organic nutrients that can be used for bacterial growth. Reports on the use of synthetic dye-degrading bacteria have been identified including: Alcaligenes eutrophus, Bacillus subtilis, Klebsiella pneumonia, Pseudomonas stutzeri, and sphingomonas sp. [61]. This dye decolorization research began with the discovery of the metabolism of mammals fed a mixture of azo dyes. The azo dyes that enter the digestive tract of these animals are reduced by the microflora in the digestive tract under anaerobic conditions. This reduced azo bond produces side products or intermediate compounds, namely amino azo benzene derivatives which are known to be carcinogens. Azo dyes reduction catalyzed by azo reductase enzymes in the liver is the same as azo reduction by microorganisms present in contamination under anaerobic conditions. From the results of these studies, further research has been developed for anaerobic digestion of dyestuffs. Furthermore, the biodegradation of dyes under anaerobic conditions is quite potential to remove textile dyes. The use of microorganisms in degrading synthetic dyes is to break the cyclic chain or the double bond of the chromophore compound. Microorganisms used to degrade bacterial dyes will produce enzymes and change the chemical structure of the pollutant to be simpler so that the level of toxicity is low. The enzymes produced by bacteria will then used to degrade dyes. The biodegradation process carried out on the dye will change the chemical structure of the chromophore or auxochrome groups through the process of reducing the N = N double bonds found in azo dyes replaced by HN-HN from NH_2.

The removal of the dyes from the textile and dyestuff manufacturing industry biologically can be broadly classified into three categories: aerobic treatment, anaerobic treatment, and a combination of both. The color removal process that occurs in textile wastewater treatment is the process of breaking the double bond in the chemical structure of the dye. The result of breaking the dye chain is several chemical compounds as intermediate products including aromatic amine compounds. Aromatic amines are compounds that are very harmful to the environment because of their very toxic nature, so a series of intermediate product degradation processes are needed to become the final product in the form of minerals that are not harmful to the environment. The results of the research on breaking the chemical structure of color in both aerobic and anaerobic activated sludge systems are summarized in Table 3.

The textile wastewater treatment plant which is widely applied by the textile industry in Indonesia is the conventional activated sludge system. In this installation, the aerobic systems that dominate are contact aeration, step aeration, and contact stabilization. Biological treatment, aerobically or anaerobically, is generally considered the most cost- and technically effective way to remove major pollutants from complex or high-strength organic textile wastewater such as BOD, COD, and TSS. On the other hand, only the biological process treatment of the textile wastewater, may not be sufficient to meet the quality standards which over time and by increasing in environmental awareness demands tightening both the quality and quantity of wastewater to be discharged into the environment. As a case study that occurred in Indonesia, regulations regarding color parameters were not regulated before 2019, but starting in 2019, the nationally applied color parameter quality standard regulations state that the entire textile industry is required to treat color parameters with a value of no more than 200 platinum cobalt units (pt–co). Regulation of the Indonesian Ministry of Environment and Forestry Number 16, 2019 which regulates textile industry wastewater quality standards, as shown in Table 4 depicted the spirit in regulating, limiting, and improving environmental quality by tightening the concentration value of wastewater parameters.

To cope with the newly more stringent regulatory thresholds value, many textile industries have to modify or even rebuild their existing WWTPs by adapting the latest wastewater treatment technologies. The most widely applied WWTP modification is to add a decolorization unit with a coagulation flocculation process using ferrous sulfate ($FeSO_4$) and lime $Ca(OH)_2$. With the addition of these two chemicals, a $CaSO_4$ precipitate will be formed which will adsorb the dye. In addition, in the process of flocculation and deposition, this precipitate will also adsorb and trap other organic substances before being biodegraded further in the activated sludge unit. Another process modification that is generally carried out in existing WWTPs is to add an anaerobic decolorization unit since the chemical structure of the dye is a double bond in the chromophore group. Many studies suggest that the cleavage of the dye double bond will occur optimally under anaerobic conditions. However, dyestuffs can be biodegraded anaerobically although not completely, but only part of their chemical groups will be degraded. Therefore, adding an anaerobic unit will increase the efficiency of decolorization with further intermediate compound destruction taking place

Table 3 Dyes bio-decolorization in activated sludge system

Dyes	Biodecolorization treatment	Results	References
Aerobic			
Reactive Red (RR) (monochlorotriazine)	Degrader culture: Pseudomonas rudinensis	After 108 h of incubation periods, P. rudinensis able to completely decolorized RR with initial concentration of 10 mg dye/l; 92%, 84%, 77.3% color removal efficiency with initial concentration of 25, 50, 75 mg dye/l, respectively. Using biological activated carbon system with the same culture, higher color removal efficiency was observed with 96%, 98%, and 98.7% color removal with initial concentration of 25, 50, 75 mg dye/l, respectively	[68]
Acid Dyes (Acid Red-119)	Degrader culture: Bacillus thuringiensis SRDD	Bacillus thuringiensis exhibited 50–60% decolorization of 5000 ppm Acid Red-119 in 7 days of incubation. B. thuringiensis was also able to decolorized more than 98%, 92%, 95%, and 95% of C.I. Acid brown 14, C.I. Acid black 210, C.I. Acid violet 90, and C.I. Acid yellow 42 azo dyes at 100 ppm concentration in 24 h, respectively. When the developed isolate of B. thuringiensis was examined for bioremediation of actual azo dye contaminated waste it removed 70% color from the waste in 24 h	[13]
Reactive Red BS C.I. 111	Degrader culture: Pseudomonas aeruginosa NGKCTS	P. aeruginosa culture exhibited 91% decolorization of 300 ppm Reactive Red BS dye within 5.5 h over a wide variation of pH ranging from 5.0 to 10.5 and temperature from 30 to 40 ^{0}C under static conditions in the presence of either glucose, peptone, or yeast extract. The addition of 300 ppm of Reactive Red BS, in each step feeding, gave more than 90% decolorization within 2 h corresponding to 136 mg per liter per hour dye removal rate	[57]
Reactive Red 195 (Sulfonated Azo Dyes)	Degrader culture: Bacillus cereus M1 and M6	Bacillus cereus M1 and M6 were proved to decolorizing sulfonated azo dyes (Reactive Red 195) under aerobic conditions for more than 97% after 72 h of incubation. Carbon and nitrogen source used in this study is maltose and peptone	[43]

(continued)

Table 3 (continued)

Dyes	Biodecolorization treatment	Results	References
Reactive Red 198 (RR198)	Degrader culture: Aspergillus flavus	Results showed that bioremoval of RR198 by Aspergillus flavus under optimized conditions (initial dye concentration 50 ppm, pH 4, and 12 ml culture inoculum volume) increased to over 84.96% with increasing time until equilibrium was reached after a period of 24 h. A low pH was the most effective, which can be advantageous when applied to real wastewater which is generally acidic	[15]
Reactive Red 198 (RR198)	Degrader culture: Enterococcus faecalis and Klebsiella variicola	The removal efficiency of RR198 dye at an initial concentration of 10–25 mg/L was more than 98% within 72 h incubation period; however, removal efficiency was reduced to 55.62%, 25.82%, and 15.42% with initial concentrations of 50, 75, and 100 mg/L, respectively. The highest removal efficiency occurred at pH 8.0, reaching 99.26% after 72 h of incubation, occured after increasing the incubation temperature from 25 °C to 37 °C	[16]
Acid Orange 7	Degrader culture: Bacillus cereus (MTCC 9777) RMLAU1	The maximum decolorization of 68.5% from initial 100 mg dye/l was achieved at optimum pH 8.0 and 33 °C under static culture conditions during the 96-h incubation period. When using real textile wastewater, the maximum decolorization of 52.5% occurred when the system was supplemented with optimized exogenous carbon and nitrogen sources along with B. cereus augmentation. Sulfanific acid was identified as acid orange 7 dye degradation product, and other metabolic products indicated the presence of amino and hydroxyl functional groups. Researchers suggest that this strain may be suitable to be employed for in situ decolorization of textile industrial effluent under wide environmental conditions	[20]

(continued)

Table 3 (continued)

Dyes	Biodecolorization treatment	Results	References
Coomassie brilliant blue, Bromcresol purple, Congo red, and Sarranine	Degrader culture: Bacillus amyloliquefaciens W36	The optimum decolorization conditions were that the strain Bacillus amyloliquefaciens W36 was grown in a medium using 1 g/L maltose as carbon source, and 1 g/L (NH4)2SO4 as nitrogen source, supplemented with 100 mg/L different dyes at pH 6.0, incubated at 30 °C, and stirred at 200 rpm for 96 h. The bacteria could aerobically decolorize dyes, such as Coomassie brilliant blue, Bromcresol purple, Congo red, and Sarranine with an efficiency of 96 h. 95.42%, 93.34%, 72.37%, and 61.7%, respectively	[38]
Triphenylmethane Brilliant Green, Fluorone Erythrosine, Triphenylmethane Crystal Violet, Azo Evans Blue, Fluorone Bengal Rose, Azo Congo Red	Degrader culture: Mixed culture of total 31 bacteria, isolated from WWTP aeration tank	Using Kimura medium, mixed bacterial culture was multiplied for 48 h and after that fed with various dyes with an initial concentration of 100 mg dye/l and incubated at temperature 26 ^{0}C for 144 h incubation period. Color removal efficiency was achieved for triphenylmethane brilliant green (85.7%), fluorone erythrosine (78.9%), triphenylmethane crystal violet (65.5%), azo Evans blue (64.4%), fluorone Bengal rose (61.0%), and azo Congo red (57.4%), respectively	[21]
Azo dye C.I. Procion Red H-3B	Degrader culture: Pseudomonas stutzeri	Maximum decolorization of Procion Red H-3B achieved at 96%, when P.stutzeri was grown in a medium contains 50 mg dye/l, 1% fructose, and 0.5% peptone as carbon and nitrogen source with pH 7.5 and 30 ^{0}C temperature for 24 h of incubation period	[4]
Disperse Red 54	Degrader culture: Brevibacillus laterosporus	Under optimized conditions (pH 7, 40 °C), Brevibacillus laterosporus led to 100% decolorization of DR54 (at 50 mg L(-1)) within 48 h. Yeast extract and peptone, supplemented in medium enhanced the decolorization efficiency of the bacterium. Researchers identified the final biodegradation product was N-(1λ(3)-chlorinin-2-yl)acetamide, and suggested that the metabolites obtained after biodegradation of DR54 were non-toxic as compared to the untreated dye	[34]

(continued)

Table 3 (continued)

Dyes	Biodecolorization treatment	Results	References
Methanil Yellow G (MY-G)	Degrader culture: Mixed bacterial consortium ZW1 (Halomonas, 49.8%; Marinobacter, 30.7%; and Clostridiisalibacter, 19.2%)	Mixed bacterial consortium ZW1 was enriched under saline (10% salinity), alkaline (pH 8.0), and temperature (40 °C) conditions to decolorize Methanil Yellow G. Addition of yeast extract in the medium led to 93.3% decolorization of 100 mg/L MY-G within 16 h of incubation period (compared with 1.12% for control)	[24]
Acid Blue 113 (di-Azo Dye)	Degrader culture: Pseudomonas stutzeri AK6	Biodecolorization of Acid Blue 113 dye, a commonly used textile di-azo dye, has been conducted using Pseudomonas stutzeri AK6 strain. The initial dye concentration of 300 ppm was decolorized up to 86.2% within 96 h of the incubation period	[33]
Anaerobic			
Acid Red 14	Degrader culture: Oerskovia paurometabola	Decolorization batch tests with 20–100 mg/l AR14 in a synthetic textile wastewater supplemented with yeast extract indicated that Oerskovia paurometabola has a high color removal capacity for a significant range of AR14 concentrations (91% after 24 h in static anaerobic culture). Further analysis confirmed that decolorization occurred through azo bond (-N = N-) cleavage under anaerobic conditions, the azo dye being completely reduced after 24 h of anaerobic incubation for the range of concentrations tested. Another interesting research finding, partial (up to 63%) removal of one of the resulting aromatic amines (4-amino-naphthalene-1-sulfonic acid) was occurred when subsequently subjected to aerobic conditions	[19]
Alizarin Yellow R (AYR)	Degrader culture: Activated sludge microflora	AYR decolorization under anaerobic–aerobic–anoxic SBR shown the optimum removal efficiency of 85.7% and 66.8% at initial AYR concentrations of 50 and 200 mg/l, conversely higher AYR concentration of 400 mg/l indicates inactivation of the activated sludge due to the insufficient support of electron donors in the anaerobic process. Further decolorized by-products p-phenylenediamine and 5-aminosalicylic were completely decomposed in the aerobic stage of the treatment applied for 50 and 200 mg/l of initial dye	[10]

(continued)

Table 3 (continued)

Dyes	Biodecolorization treatment	Results	References
Alizarin Yellow R (AYR)	Degrader culture: Activated sludge microflora	A four-compartment anaerobic baffled reactor (ABR) incorporated with membraneless biocatalyzed electrolysis system (BES) was examined for the treatment of azo dye AYR with initial concentration of 200 mg/l. The decolorization efficiency in the ABR-BES (8 h HRT, 0.5 V) was found higher than that in ABR-BES without electrolysis (95.1 ± 1.5% compared to 86.9 ± 6.3%), while higher power supply (0.7 V) give higher efficiency up to 96.4 ± 1.8% and VFAs removal	[12]
Acid Orange 7 (AO7)	Degrader culture: Activated sludge microflora	Two-stage anaerobic system (acidogenic and methanogenic phase) were used to investigate the removal of an azo dye AO7 (2.14 mM initial concentration), with starch (1 g/l) as the primary co-substrate. Research results discovered that under 5 days HRT, the methanogenic phase accounted for about 90% of the entire AO7 removal, and the obtained removal rate constant for AO7 was 2.93-fold higher of that in the acidogenic phase. Effluent from the acidogenic phase containing readily available electron donors was fed as the influent for the methanogenic phase, in which AO7 was preferred to be reduced	[29]
Acid Orange 7 (AO7)	Degrader culture: Non-adapted methanogenic granular sludge microflora	Fed-batch and continuous reactor under anaerobic conditions was used to investigated the decolorization of AO7 with loading rate of 1.7 mM/day (590 mg/l.day), and fed with glucose (2 g/l) as co-substrate able to remove 92% AO7. It was noticed that when the co-substrate was reduce (AO7 0.3 mM and glucose 0.25 g/l, AO7 removal efficiency was decreased significantly to 78%	[42]
Acid Orange 6 (AO6) Acid Orange 7 (AO7)	Degrader culture: Anaerobic sludge from a full scale UASB plant microflora	Sequential fixed-film anaerobic batch reactor (SFABR) was used to investigated decolorization of two azo dyes (AO6 and AO7), with varying initial dye concentration and co-substrate (0.5 g/l glucose). At 300 mg/l AO6 and AO7, more than 90% efficiency was achieved with removal rates were 168 mg/l.day and 176 mg/l.day, respectively. Degradation by-products identified as 4-aminobenzenesulfonate and Aminoresorcinol were discovered to be resistant to further degradation under anaerobic environment	[59]

(continued)

Table 3 (continued)

Dyes	Biodecolorization treatment	Results	References
Acid Orange 7 (AO7) Direct Red 254 (DR254)	Degrader culture: Mixed anaerobic bacterial consortia	The anaerobic treatment of a monoazo dye (AO7) and a diazo dye (DR254), was investigated in a methanogenic laboratory-scale Upflow Anaerobic Sludge Blanket (UASB), fed with acetate as primary carbon source. A color removal achieved more than 88% for both dyes at a HRT of 24 h, while at a HRT of 8 h, a more extensive reductive decolorization was observed for DR254 (82%) compare to AO7 (56%). Research results suggested that methanogenic cultures predominantly to perform azo bond cleavage	[6]
Reactive Orange 16 (RO16)	Degrader culture: Granular sludge from a full scale UASB plant microflora	Submerged anaerobic membrane bioreactors (SAMBRs) was used to investigated the decolorization of azo dyes containing textile wastewater and operated for almost four months with increasing RO16 concentration from 0.060 to 3.2 g/l. The results indicated that high removal efficiency of 99% was achieved by SAMBRs even fed with high concentration (3.2 g/l) of dye	[60]
Yellow Gold Remazol (YGR)	Degrader culture: Anaerobic sludge from a pilot scale UASB reactor microflora	Two continuous bench-scale treatment were used to investigated decolorization of Yellow Gold Remazol. Upflow Anaerobic Sludge Blanket followed by an activated sludge (UASB/AS) system and UASB followed by a shallow polishing pond (UASB/PP) were fed with 50 mg/l YGR and 350 mg/l pretreated residual yeast, as co-substrate. Results indicated that the UASB/PP system achieved the highest removal of YGR and chemical oxygen demand with efficiency of 23% and 85%, respectively	[2]
Reactive Red 2 (RR2)	Degrader culture: Mixed culture containing Desulfovibrio aminophilus, Thermoanaerobacter, Lactococcus raffinolactis, Ruminiclostridium and Rhodopirellula	An integrated hydrolysis/acidification (HA) and multiple anoxic/aerobic (AO) process was used to investigated the removal of RR2 and nitrogen from azo dye containing wastewater. The RR2 and nitrogen removal percentages of the HA and AO process with HRT of 12 h, treating initial concentration of 30 mg/l RR2 and 114 mg/l NH4Cl were 89.4% and 54.0%, respectively	[22]

(continued)

Table 3 (continued)

Dyes	Biodecolorization treatment	Results	References
Methyl Orange (MO)	Degrader culture: Denitrifying anaerobic methane oxidator microflora (Methanomethylovorans, Moranbacteria)	A methane-based hollow fiber membrane bioreactor (HfMBR) inoculated with an enriched anaerobic methane oxidation (AOM) culture was used to investigated the decolorization of MO at various initial concentration of 400–800 mg/l at temperature 35 °C, and HRT ranging from 12 to 48 h. Results showed the MO decolorization achieved ranging from 88 to 100% with maximum decolorization rates of 883 mg/l.day	[3]
Remazol Brilliant Blue R (RBBR)	Degrader culture: Mixed culture containing Proteobacteria, Spirochaetae, Aminicenantes, Bacteroidetes, Thermotogae, and Chloroflexi	An anaerobic dynamic membrane bioreactor (AnDMBR) was used to investigated the RBBR decolorization operated for a total of 120 days in three stages at mesophilic environment of 37 °C. Initial concentration of RBBR ranging from 0.5 g/l (stages I–II) to 1 g/l (stage III), with addition of 4.5 g/l glucose as co-substrate (stages I–II). The AnDMBR was run under hydraulic retention time (HRT) of 5 and 2.5 days, while organic loading rates in stages I–III were kept at 5.0, 5.0, and 1.0 g COD/l.day, respectively. Results showed satisfactory for soluble COD and RBBR removal efficiency of 98.5% and 97.5%, respectively	[5]
Reactive Yellow 15 (RY15)	Degrader culture: Mixed culture collected from Activated Sludge WWTP containing Bacillus, Pseudomonas, and E-Coli	Immobilized mixed cells using biocarrier of sodium alginate (SA), starch (St), and Gelatin (Ge) cross-linking with polyvinyl alcohol (PVA) were used to investigated decolorization of RY15 via a sequential anaerobic–aerobic process. Results showed complete decolorization (100%) of RY15 was occurred, and COD removal were 92% ± 6.8, 96% ± 3.5, and 100%, using PVA-SA, PVA-St, and PVA-Ge at RY15 initial concentrations of 10 mg/l, under the overloading rate (OLR) and Hydraulic retention time (HRT) of the aerobic bioreactor are 24.5 mg/l.hour and 41.37 h, respectively	[25]

Table 4 Textile industry wastewater quality standards in Indonesia [30]

Parameters	Unit	Threshold value		
Flowrate	m^3/day	≤ 100	100 < F < 1000	≥ 1000
BOD	mg/l	60	45	35
COD	mg/l	150	125	115
TSS	mg/l	50	40	30
Total phenol	mg/l	0.5	0.5	0.5
Total chromium	mg/l	1	1	1
Total ammonia	mg/l	8	8	8
Sulfide	mg/l	0.3	0.3	0.3
Oil & grease	mg/l	3	3	3
pH	–	6–9	6–9	6–9
Color	Pt–Co	200	200	200
Temperature	°C	Dev. 2°	Dev. 2°	Dev. 2°
Maximum flowrate	m^3/ton product	100	100	100

in the existing aerobic unit. Schematic process flow diagram of several suggested WWTPs modification is described in Figs. 12, 13.

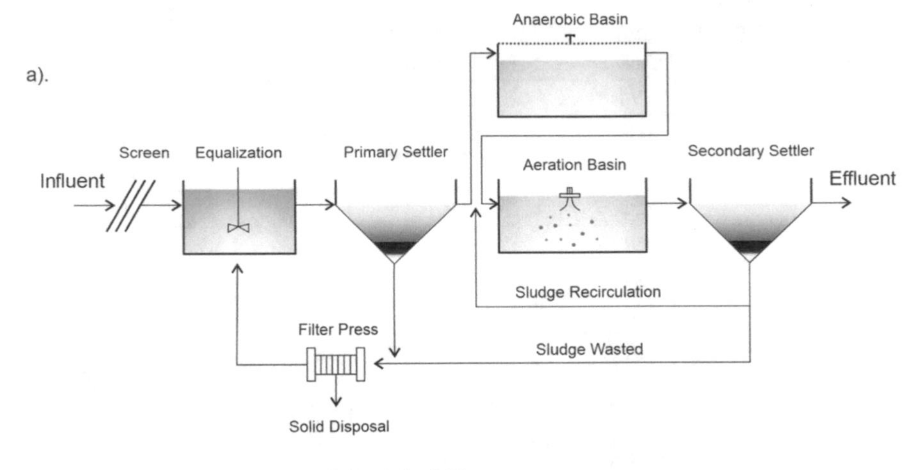

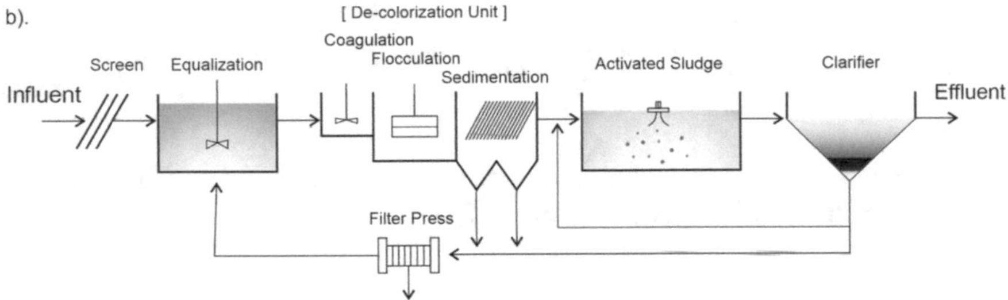

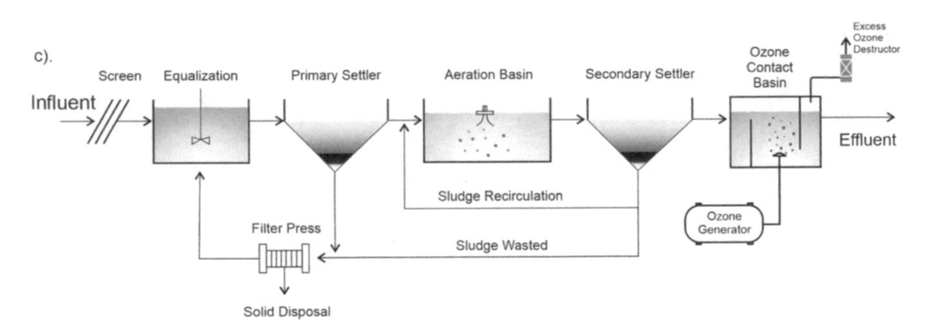

Fig. 12 Schematic process flow diagram of commonly practiced in the modification of existing WWTP to cope with newly stringent regulatory threshold value in textile wastewater by simply adding anaerobic unit before aeration basin (**a**), installing decolorization unit using coagulation-flocculation process (**b**), and installing ozone unit for color destruction as pre- and/or post-treatment (**c**)

Fig. 13 WWTP Cisirung (aerial view, upper), an integrated communal wastewater treatment plant located in Bandung, West Java, Indonesia, processes up to 175 L per second wastewater collected from 22 nearby textile industries. To cope with newly stringent regulatory threshold value in textile wastewater, WWTP operator company made modification by simply adding anaerobic unit (lower right) before aeration basin (lower left)

References

1. Adav SS, Lee DJ, Show KY, Tay JH (2008) Aerobic granular sludge: Recent advances. Biotechnol Adv 26(5):411–423. https://doi.org/org/10.1016/j.biotechadv.2008.05.002
2. Bahia M, Passos F, Adarme OFH, Aquino SF, Silva SQ (2018) Anaerobic-Aerobic Combined System for the Biological Treatment of Azo Dye Solution using Residual Yeast. Water Environ Res 90(8):729–737. https://doi.org/10.2175/106143017X15131012153167
3. Bai YN, Wang XN, Zhang F, Wu J, Zhang W, Lu YZ, Fu L, Lau TC, and Zeng RJ. 2020. High-rate anaerobic decolorization of methyl orange from synthetic azo dye wastewater in a methane-based hollow fiber membrane bioreactor. J Hazard Mater 388: 121753. doi: https://doi.org/10.1016/j.jhazmat.2019.121753
4. Bera SP and Tank SK,. 2021. Screening and identification of newly isolated Pseudomonas sp. for biodegrading the textile azo dye C.I. Procion Red H-3B. J Appl Microbiol., 130(6):

1949–1959. doi: https://doi.org/10.1111/hour.14920.
5. Berkessa YW, Yan B, Li T, Jegatheesan V, and Zhang Y,. 2020. Treatment of anthraquinone dye textile wastewater using anaerobic dynamic membrane bioreactor: Performance and microbial dynamics. Chemosphere. 238:124539. doi:https://doi.org/10.1016/j.chemosphere.2019.124539
6. Brás R, Gomes A, Ferra MI, Pinheiro HM, Gonçalves IC (2005) Monoazo and diazo dye decolourisation studies in a methanogenic UASB reactor. J Biotechnol 115(1):57–66. https://doi.org/10.1016/j.jbiotec.2004.08.001
7. Cakici A, Bayramoglu M (1995) An Approach to Controlling Sludge Age in the Activated Sludge Process. War Res 29(4):1093–1097
8. Chan YJ, Chong MF, Law CL, Hassell DG (2009) A review on anaerobic-aerobic treatment of industrial and municipal wastewater. Chem Eng J 155:1–18
9. Chen H (2006) Recent Advances in Azo Dye Degrading Enzyme Research. Curr Protein Pept Sci 7(2):101–111
10. Chen A, Yang B, Zhou Y, Sun Y, Ding C (2018) Effects of azo dye on simultaneous biological removal of azo dye and nutrients in wastewater. R Soc Open Sci 5(8):180795. https://doi.org/10.1098/rsos.180795
11. Cruz A, Buitron G (2001) Biodegradation of dispersion blue 79 using sequenced anaerobic/aerobic biofilters. Water Science Technol. 44(4):159–166
12. Cui D, Guo YQ, Lee HS, Wu WM, Liang B, Wang AJ, Cheng HY (2014) Enhanced decolorization of azo dye in a small pilot-scale anaerobic baffled reactor coupled with biocatalyzed electrolysis system (ABR-BES): a design suitable for scaling-up. Bioresour Technol 163:254–261. https://doi.org/10.1016/j.biortech.2014.03.165
13. Dave SR, Dave RH (2009) Isolation and characterization of Bacillus thuringiensis for acid red 119 dye decolourisation. Bioresource Technol 100(1):249–253. https://doi.org/10.1016/j.biortech.2008.05.019
14. Ekama G (2010) The role and control of sludge age in biological nutrient removal activated sludge systems. Water Sci Technol 61(7):1645–1652. https://doi.org/10.2166/wst.2010.972
15. Esmaeili A, Kalantari M (2012) Bioremoval of an azo textile dye, Reactive Red 198, by Aspergillus flavus. World J Microbiol Biotechnol 28(3):1125–1131. https://doi.org/10.1007/s11274-011-0913-1
16. Eslami H, Shariatifar A, Rafiee E, Shiranian M, Salehi F, Hosseini SS, Eslami G, Ghanbari R, Ebrahimi AA (2019) Decolorization and biodegradation of reactive Red 198 Azo dye by a new Enterococcus faecalis-Klebsiella variicola bacterial consortium isolated from textile wastewater sludge. World J Microbiol Biotechnol 35(3):38. https://doi.org/10.1007/s11274-019-2608-y
17. Flynn BP, Stadnik JG (1979) Start-Up of a Powdered Activated Carbon-Activated Sludge Treatment System. Journal (Water Pollution Control Federation) 51(2):358–369
18. Franca RDG, Vieira A, Mata AM, Carvalho GS, Pinheiro HM, Lourenço ND (2015) Effect of an azo dye on the performance of an aerobic granular sludge sequencing batch reactor treating a simulated textile wastewater. Water Res 15(85):327–336. https://doi.org/10.1016/j.watres.2015.08.043
19. Franca RDG, Vieira A, Carvalho G, Oehmen A, Pinheiro HM, Crespo MTB, and Lourenco ND,. 2020. Oerskovia paurometabola can efficiently decolorize azo dye Acid Red 14 and remove its recalcitrant metabolite. Ecotoxicol Environ Saf., 191:110007. doi: https://doi.org/10.1016/j.ecoenv.2019.110007
20. Garg SK, Tripathi M (2013) Process parameters for decolorization and biodegradation of orange II (Acid Orange 7) in dye-simulated minimal salt medium and subsequent textile effluent treatment by Bacillus cereus (MTCC 9777) RMLAU1. Environ Monitor Assess 185(11):8909–8923. https://doi.org/10.1007/s10661-013-3223-2
21. Godlewska EZ, Przystaś W, Sota EG (2018) Possibilities of Obtaining from Highly Polluted Environments: New Bacterial Strains with a Significant Decolorization Potential of Different Synthetic Dyes. Water Air Soil Pollut 229(6):176. https://doi.org/10.1007/s11270-018-3829-7

22. Gu M, Yin Q, Wang Z, He K, Wu G (2018) Color and nitrogen removal from synthetic dye wastewater in an integrated mesophilic hydrolysis/acidification and multiple anoxic/aerobic process. Chemosphere 212:881–889. https://doi.org/10.1016/j.chemosphere.2018.08.162
23. Guo N, Zhang J, Xie HJ, Tan LR, Luo, JN, Tao ZY, and Wang SG. 2017. Effects of the Food-to-Microorganism (F/M) Ratio on N_2O Emissions in Aerobic Granular Sludge Sequencing Batch Airlift Reactors. Waters, 9, 477; doi:https://doi.org/10.3390/w9070477
24. Guo G, Hao J, Tian F, Liu C, Ding K, Xu J, Zhou W, and Guan Z. 2020. Decolorization and detoxification of azo dye by halo-alkaliphilic bacterial consortium: Systematic investigations of performance, pathway and metagenome. Ecotoxicol Environ Saf., 204:111073. doi: https://doi.org/10.1016/j.ecoenv.2020.111073
25. Hameed BB, Ismail ZZ (2021) Biodegradation of reactive yellow dye using mixed cells immobilized in different biocarriers by sequential anaerobic/aerobic biotreatment: experimental and modeling study. Environ Technol 42(19):2991–3010. https://doi.org/10.1080/09593330.2020.1720306
26. Hammer MJ Sr, and Hammer MJ Jr. 2012. Water and Wastewater Technology-7th Edition. Pearson New International Edition
27. Heath HW Jr. 1981. Full-scale Demonstration Of Industrial Wastewater Treatment Utilizing Du Ponts Pact Process. USEPA Research and Develpoment, EPA-600/S2–81–159, USA
28. Herschy RW. 1998. Activated sludge process: Historical. In: Encyclopedia of Hydrology and Lakes. Encyclopedia of Earth Sciences. Springer, Dordrecht. Doi:org/https://doi.org/10.1007/1-4020-4497-6_7
29. Huang J, Shi B, Yin Z, Guo K, Fu C, Tang J (2020) Two-stage anaerobic process benefits removal for azo dye orange II with starch as primary co-substrate. Water Science Technol. 81(11):2401–2409. https://doi.org/10.2166/wst.2020.294
30. IMEF. 2019. Wastewater Quality Standards for Textile Industry Activities. Indonesia Ministry of Environment and Forestry, Regulation No. P.16/Menlhk/ Setjen/ Kum.1/ 4/2019
31. IMI. 2020. Wastewater Treatment Plant Design Guidelines for Small and Medium Textile Industries. Ministry of Industry, Indonesia
32. Isik M, Sponza T (2004) Monitoring of toxicity and intermediates of C.I. Direct Black 38 azo dye through decolorization in an anaerobic/aerobic sequencing reactor system. J Hazard Mater B114:29–39
33. Joshi AU, Hinsu AT, Kotadiya RJ, Rank JK, Andharia KN, and Kothari RK,. 2020. Decolorization and biodegradation of textile di-azo dye Acid Blue 113 by Pseudomonas stutzeri AK6. 3 Biotech., 10(5): 214. doi: https://doi.org/10.1007/s13205-020-02205-5.
34. Kurade MB, Waghmode TR, Khandare RV, Jeon BH, Govindwar SP (2016) Biodegradation and detoxification of textile dye Disperse Red 54 by Brevibacillus laterosporus and determination of its metabolic fate. J Biosci Bioeng 121(4):442–449. https://doi.org/10.1016/j.jbiosc.2015.08.014
35. Lettinga G, F.M., van V., S.W., H., W.J., de Z., & A., K. (1980) Use of the upflow sludge blanket (USB) reactor concept for biological wastewater treatment. Biotechnol Bioeng 22:699–734
36. Liu YQ, and Tay JH. (2007). Influence of cycle time on kinetic behaviors of steady-state aerobic granules in sequencing batch reactors. In Enzyme and Microbial Technology (Vol. 41, Issue 4, pp. 516–522). https://doi.org/10.1016/j.enzmictec.2007.04.005
37. Liu Y, Tay JH (2004) State of the art of biogranulation technology for wastewater treatment. Biotechnol Adv 22(7):533–563. https://doi.org/10.1016/j.biotechadv.2004.05.001
38. Liu Y, Shao Z, Reng X, Zhou J, Qin W (2021) Dye-decolorization of a newly isolated strain of Bacillus amyloliquefaciens W36. World J Microbiol Biotechnol 37(1):8. https://doi.org/10.1007/s11274-020-02974-4
39. Luangdilok W, Panswad T (2000) Effect of chemical structure of reactive dyes on color removal by anaerobic-aerobic process. Water Science. Technol. 42:377–382
40. Manu B, Chaudhari S (2003) Decolorization of indigo and azo dyes in semicontinuous reactors with long hydraulic retention time. Process Biochem 38:1213–1221
41. Melgoza RM, Cruz A, Buitron G (2004) Anaerobic/aerobic treatment of colorants present in textile effluents. Water Science Technol. 50(2):149–155

42. Mendez-Paz D, Omil F, and Lema JM. Anaerobic treatment of azo dye Acid Orange 7 under fed-batch and continuous conditions. Water Res., 39(5): 771–8. https://doi.org/10.1016/j.watres.2004.11.022
43. Modi HA, Rajput G, Ambasana C (2010) Decolorization of water soluble azo dyes by bacterial cultures, isolated from dye house effluent. Bioresour Technol 101(16):6580–6583. https://doi.org/10.1016/j.biortech.2010.03.067
44. Moreno G, Cruz A, Buitron G (1999) Influence of So/Xo ratio on anaerobic activity test. Water Science. Technol. 40:9–15
45. Morgenroth E, Sherden T, Van Loosdrecht MCM, Heijnen JJ, Wilderer PA (1997) Aerobic granular sludge in a sequencing batch reactor. Water Res 31(12):3191–3194. https://doi.org/10.1016/S0043-1354(97)00216-9
46. Moy BYP, Tay JH, Toh SK, Liu Y, Tay STL (2002) High organic loading influences the physical characteristics of aerobic sludge granules. Lett Appl Microbiol 34(6):407–412. https://doi.org/10.1046/j.1472-765X.2002.01108.x
47. Mubarak MFM, Nor MHM, Sabaruddin MF, Bay HN, Lim CK, Aris A, Ibrahim Z (2018) Development and performance of BAC-ZS bacterial consortium as biofilm onto macrocomposites for raw textile wastewater treatment. Malaysian Journal of Fundamental and Applied Sciences 14(1):257–262
48. Nancharaiah YV, Reddy GKK (2018) Aerobic granular sludge technology: Mechanisms of granulation and biotechnological applications. Biores Technol 247:1128–1143. https://doi.org/10.1016/j.biortech.2017.09.131
49. Ong SA, Toorisaka E, Hirata M, Hano T (2008) Granular activated carbon-biofilm configured sequencing batch reactor treatment of C.I. acid orange 7. Dyes Pigments 76:142–146
50. Pasukphun N, Vinitnantharat S, Gheewala S (2010) Investigation of Decolorization of Textile Wastewater in an Anaerobic/Aerobic Biological Activated Carbon System (A/A BAC). Pak J Biol Sci 13:316–324
51. Queiroz MTA, Queiroz CA, Alvim LB, Sabara MG, Leao MMD, and de Amorim CC. 2019. Restructuring in the flow of textile wastewater treatment and its relationship with water quality in Doce River, MG, Brazil. Gestão & Produção, 26(1), e1149. https://doi.org/10.1590/0104-530X1149-19
52. Rodman CA, and Shunney EL, 1971. Bio-regenerated activated carbon treatment of textile dye wastewater, Environmental Protection Agency, EP 2.10:12090DWM01/71, United State.
53. Rollemberg SLS, Barros ARM, Firmino PIM, dos Santos AB (2018) Aerobic granular sludge: Cultivation parameters and removal mechanisms. Biores Technol 270(July):678–688. https://doi.org/10.1016/j.biortech.2018.08.130
54. Said NI, Utomo K (2007) Domestic Wastewater Treatment with Activated Sludge Process Filled with Bioball Media. Indonesian Water Journal 3(2):160–174
55. Schmidt JEE, Ahring BK (1994) Extracellular polymers in granular sludge from different upflow anaerobic sludge blanket (UASB) reactors. Appl Microbiol Biotechnol 42(2–3):457–462. https://doi.org/10.1007/BF00902757
56. Schultz JR, Keinath TM (1984) Powdered Activated Carbon Treatment Process Mechanisms. Journal (Water Pollution Control Federation) 56(2):143–151
57. Seth NT and Dave SR, 2009. Optimization for enhanced decolourization and degradation of Reactive Red BS C.I. 111 by Pseudomonas aeruginosa NGKCTS. Biodegradation, 20(6): 827–36. doi:https://doi.org/10.1007/s10532-009-9270-2.
58. Seviour RJ, 2010, Microbial Ecology of Activated Sludge. Water Intelligence Online, 9, doi: https://doi.org/10.2166/9781780401645
59. Singh P, Sanghi R, Pandey A, Lyengar L (2007) Decolorization and partial degradation of monoazo dyes in sequential fixed-film anaerobic batch reactor (SFABR). Bioresour Technol 98(10):2053–2056. https://doi.org/10.1016/j.biortech.2006.08.004
60. Spagni A, Casu S, Grilli S (2012) Decolourisation of textile wastewater in a submerged anaerobic membrane bioreactor. Bioresour Technol 117:180–185. https://doi.org/10.1016/j.biortech.2012.04.074

61. Stolz A (2001) Basic and applied aspect in the microbial degradation of azo dyes. Appl Microbiol Biotechnol 56:69–80
62. Supaka N, Juntongjin K, Damronglerd S, Delia ML, Strehaiano P (2004) Microbial decolorization of reactive azo dyes in a sequential anaerobic-aerobic system. Chem eng J 99:169–176
63. Tay J, Liu Q, Liu Y (2001) Microscopic observation of aerobic granulation in sequential aerobic sludge blanket reactor. 639798:168–175
64. Tchobanoglous G, Stensel H, Tsuchihashi R, Burton F (2014) Wastewater Engineering: Treatment and Resource Recovery, 5th edn. McGraw-Hill, New York
65. USACE. 2001. Engineering and Design Adsorption Design Guide. Department of the Army, U.S. Army Corps of Engineers. DG No. 1110–1–2
66. Walker GM, Weatherley LR (1999) Biological activated carbon treatment of industrial water in stirred tank reactors. Chem eng J 75:201–206
67. Wang LK, Wu Z, and Shammas NK. (2009). Activated Sludge Processes. In LK. Wang et al. (Eds.) Handbook of Environmental Engineering, Volume 8: Biological Treatment Processes, 207–281. Humana Press, Totowa, NJ, doi:https://doi.org/10.1007/978-1-60327-156-1_6
68. Widajatno RL, Wahyudi B, Wijaya DGO (2009) Biodegradation of monochlorotriazinyl reactive red by pseudomonas rudinensis and pseudomonas diminuta. LPPM UPN Veteran Press, Surabaya, Indonesia, Monograph
69. Zdravkov BD, Cermak JJ, Sefara M, Janku J (2007) Pore classification in the characterization of porous materials: A perspective. Cent Eur J Chem 5(2):385–395
70. Zhao X, Hickey RF, Voice TC (1999) Long-term evaluation of adsorption capacity in a biological activated carbon fluidized bed reactor system. Water Res 33:2983–2991
71. Yan LKQ, Fung KY, Ng KM (2018) Aerobic sludge granulation for simultaneous anaerobic decolorization and aerobic aromatic amines mineralization for azo dye wastewater treatment. Environ Technol 39(11):1368–1375. https://doi.org/10.1080/09593330.2017.1329354
72. Yoo ES, Libra J, Adrian L (2001) Mechanism of decolorization of azo dyes in anaerobic mixed culture. J Environ Eng 127:844–849

Role of Moving Bed Bioreactor (MBBR) in Dye Removal

Roumi Bhattacharya

Abstract Dye containing effluents disposed from various industries include several toxic chemicals and have adverse health effects on human as well as derogatory impacts on ecosystem. Among various dyes, chromophores with azo bonds are most abundant and are investigated to be a potential carcinogen and mutagen. Dyes in wastewater are difficult to degrade as they are made to be stable under the action of external factors including temperature, microorganisms and chemicals including bleaches. Analysing the strengths and weaknesses of various treatment technologies available, biodecolourization is often economically and environmentally favoured. Moving Bed Bioreactors (MBBRs) are one of the advanced biological systems that allow degradation of a wide range of recalcitrant compounds with notable advantages over other treatments and have been modified as well as coupled with several other technologies to obtain complete mineralisation of the dyes leaving non-toxic by-products. The present work reviews the aspects of dye removal from effluents in MBBR with critical outlook on various investigations undertaken in the reactor. The major points include the following (1) Dye degradation takes place mostly by azo bond cleavage (bacterial biomass) or enzymatic action (fungi), along with biosorption (10–50% at most) and bioaccumulation. (2) Under identical experimental conditions, dye removal in MBBR is comparatively higher (51.6%) than in activated sludge process (26.8%). (3) Anaerobic–aerobic MBBR results in 85% colour removal whereas complete decolourization was recorded in MB-SBBR and SBR-MBBR combination. (4) Around 95% colour removal was obtained in case of MBBR combined with coagulation, ozonation and membrane filtration. (5) All these experimental results are highly influenced by a number of parameters and efficiency also includes economical aspect of every process. Generally, initial dye concentration, pH, HRT and biocarrier concentration has a threshold value which yields maximum removal, above or below which removal efficiency decreases.

Keywords Moving Bed Bioreactor (MBBR) · Dye removal · Anaerobic · Aerobic · Combined physico-chemical and biological processes · Biocarriers · Biofilm

---

R. Bhattacharya (✉)
Department of Civil Engineering, Indian Institute of Engineering, Science and Technology, J-308 Paharpur road, Garden Reach, Kolkata 700024, India

A. Khadir et al. (eds.), *Biological Approaches in Dye-Containing Wastewater*, Sustainable Textiles: Production, Processing, Manufacturing & Chemistry,
https://doi.org/10.1007/978-981-19-0545-2_6

1 Introduction

At present, dyeing process comprises of more than 8000 chemical compounds with structural, compositional and behavioural variations that produce up to 1,00,000 different dyes [15, 21, 67, 77]. These dyes are majorly used in textile industries along with substantial application in paper, rubber, electroplating, food, printing, leather tanning, cosmetic and pharmaceutical sectors [2, 21, 30, 52, 132]. The contribution of major industrial sectors towards the production of dye containing effluents is shown in Fig. 1 [41, 134]. For more stability of colour on different products manufactured in these industries, dyes are constantly upgraded and new compounds are combined that results in more resistance to degradation by water, sunlight, temperature, detergents or any washing substances [5, 163]. Dye wastewater is considered one of the most toxic industrial effluents [39]. Around 17–20% of pollution caused by textile industry is due to various dyeing mechanisms [67] which contains a total of 72 toxic chemicals, only 30 of which can be treated by conventional treatment processes [31]. Dye wastewater often exhibits highly fluctuating pH, generally towards the higher range [159], high temperature and high COD concentration along with hazardous and xenobiotic compounds [74, 144].

Based on origin, dyes can be organic or synthetic. Synthetic dyes have more complex structure and are made to be more resistant to chemical action and fading thus making them less susceptible to biodegradation [141]. Synthetic dyes include acidic, reactive, basic, disperse, azo, diazo, anthraquinone-based and metal complex dyes [15], among which, the most toxic group of dyes includes basic and direct diazo dyes [136]. Azo dyes, which constitute more than 50% of the dyes [91], are characterized by –N = N– (azo) bonds and are often xenobiotic in nature [141]. Azo groups are present as chromophores in anionic and non-ionic dyes which makes them the most abundant type of dyes in wastewater. Dyes can be broadly classified as anionic, cationic and non-ionic dyes [89, 102]. According to Wang et al. [161] and

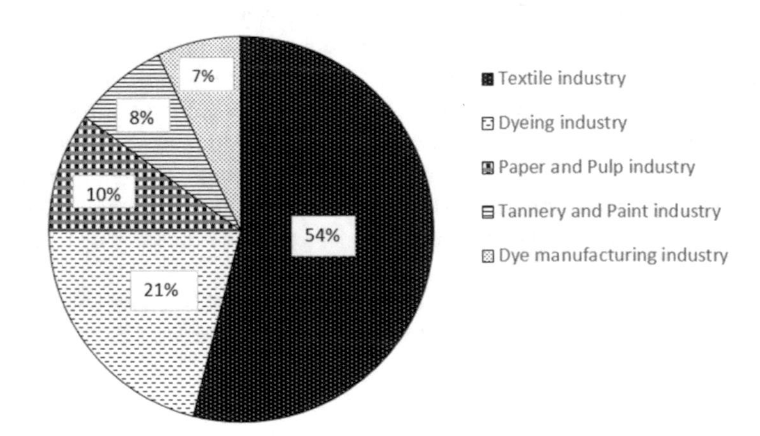

Fig. 1 Contribution of different industrial sectors towards production dye wastewater [41, 134]

Katheresan et al. [77], fibre dyes can be produced by one or combination of more than one of the following groups:

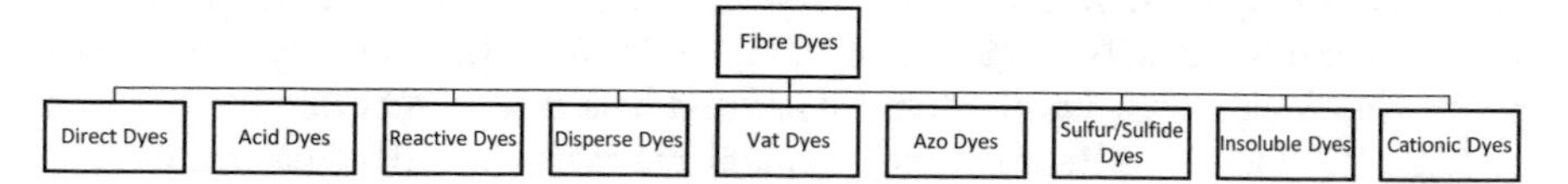

Considering water quality, one of the major concerns of untreated dye wastewater is its elevated chemical oxygen demand (COD) along with the presence of organics that are not easily biodegradable. Dye effluents are often dark coloured and thus block sunlight when disposed into any water body affecting the aquatic ecosystem of that place [12, 57, 159]. Dye effluents having a high pH can alter the pH in disposed of waterbody along with its dissolved oxygen [72]. Synthetic dyes often react with other chemicals in the environment to form more recalcitrant compounds [21]. Presence of organic chemical-based dye fixing agents such as formaldehyde, softeners having hydrocarbons, chlorinated stain removers also found in dyeing wastewater have carcinogenic effects [3]. Several dyes have been seen to cause bioaccumulation in biotic species due to its stable nature and resistant to biodegradation. Metal-based dyes when eliminated into water system may release the metals (like chromium) which have adverse health effects on animals in the surrounding ecosystem [15]. Adverse human health effects associated with residual dye effluents include irritation, permanent eye injuries when in contact with eyes, respiratory problems, reproductive failures, effects on immune system along with genetic mutation and carcinogenic effects [72, 136].

Besides aesthetic concerns, one of the major concerns of adverse health effect on humans is the potential carcinogenicity of several dyes [99, 149]. Among the diversified categories of dyes, the azo groups constitute the largest, contributing to about 70% by weight of total dyes in the effluent [34], being most common and most toxic commercial dyes [37, 139]. Azo dyes are reduced to toxic, mutagenic and/or carcinogenic intermediate by-products [60] in anaerobic conditions of sediments [112, 132, 163] and intestines of humans [35, 141]. As many as 46 different strains of gut bacteria have been isolated that can reduce a number of azo dyes [36]. Human skin bacteria are also reported to reduce the dye Direct Blue 14 to amines that are carcinogenic [126]. These dyes are directly linked with bladder cancer, mutation of chromosomes and splenic sarcomas [133]. It has been observed that the faecal anaerobic bacteria in human digestive tract degrade tetrazine, an azo dye to form amines like benzidine and 4-aminoaniline which are carcinogenic in nature [25, 112]. Benzidine based azo dyes are shown to produce various aromatic amines including N-acetylated derivative that is tumorigenic in the urine of sample mammals [131]. Even after treatment of azo dye containing wastewater, the effluents are often found to have toxic effects along with several azo dyes being completely unaltered [110].

Treating wastewater economically and effectively is often considered as one of the challenging issues due to the presence of harmful recalcitrant substances [124]. At present, effluents released from dye utilizing industries must comply with the standards as prescribed by The International Dye Industry Wastewater Discharge Quality Standards [46]. The methods developed for treatment of dye wastewater are not individually employed because of incomplete and inadequate treatment along with associate disadvantage for each treatment method [76, 119]. Generally, a biological process is selected along with a chemical pretreatment process for effective treatment of dyestuff effluent [53]. Physical and chemical treatment methods often do not necessarily eliminate complex contaminants of dye wastewater [21]. Effective physico-chemical processes are often coupled with high cost treatment plants and operational expenses, intensive energy consumption, excess amount of chemical requirement and thus sludge production, which necessitates the cost of sludge handling [50, 53, 71]. These treatment methods are also sensitive to a variable wastewater input, which is often experienced in industries [15, 144]. Chemical coagulation alone uses a lot of coagulants, producing a large amount of sludge with comparative low efficiency in treatment [80]. Electrochemical oxidation produces a number of intermediary pollutants which needs to be treated in associative treatment systems, thus increasing plant and operational costs [80].

Microbial degradation of dyes or biodecolourization is often considered as a cost effective and ecologically safe method for treatment of dye wastewater [15, 52, 169]. Dyes are often toxic to microbial species and may lead to failure of conventional biological treatment system [109] although it has been observed that microorganisms growing in vicinity of areas where there is regular disposal of dye effluent can utilize the dyes as their nutrient sources [72]. Dye wastewater is generally characterized by very low BOD to COD ratio, even around 0.1 [40, 147] which often necessitates the use of a pretreatment process to make it more biodegradable [53]. The effectiveness of using multi-staged reactor comprising of anaerobic process followed by aerobic treatment system is thoroughly studied throughout the past years [105]. The mechanism for biodecolourization of azo dyes by bacteria may involve both degradation of azo bond cleavage by reductase enzymes as well as cell adsorption [112]. The cleavage of azo bond in absence of oxygen is brought about by non-specific enzymes [172] which is the main reason behind easy anaerobic degradation compared to aerobic degradation where the enzymes are dye specific [177]. Decolourization by fungi is brought about by non-specific enzymes as well as adsorption by dead fungal cells [54].

MBBR is often regarded as one of the most effective and promising water treatment technology due to its capability to degrade a wide range of wastewater [144]. It comprises a homogenously mixed reactor vessel where biomass remains attached to carriers in fluidized condition [108]. In aerobic condition, the carriers remain suspended with the help of aeration supplied by diffusers at the bottom of the tank whereas in anaerobic condition, it is done using mechanical stirrers. A schematic representation of MBBR is shown in Fig. 2 which illustrates the reactor operating in aerobic and anaerobic mode. Biocarriers are characterized by density close to that of water which helps to remain in suspended condition and integrity when kept in water for a long period [23]. MBBR has proven to treat a large volume of wastewater at once thus reducing reactor volume [169]. Sustainability of more biomass

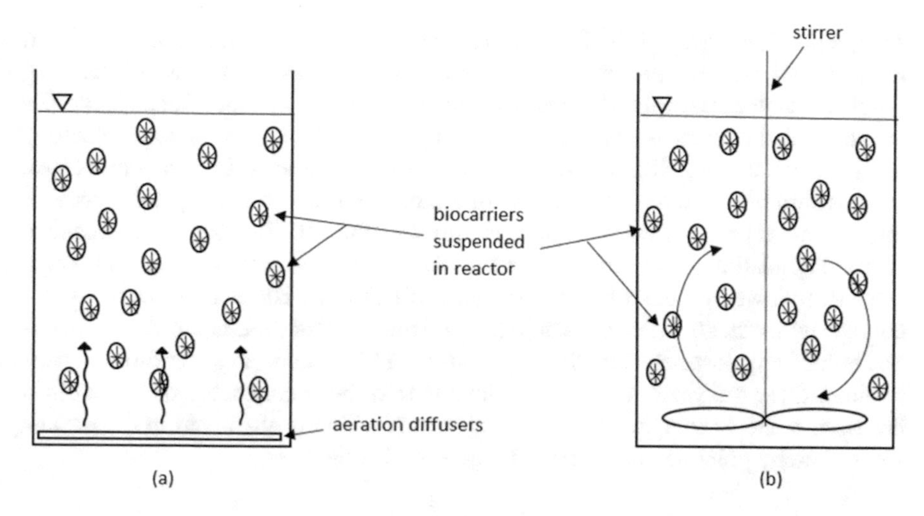

Fig. 2 Schematic representation of moving bed bioreactor in **a** aerobic condition and **b** anaerobic condition. Note that both homogenous mixing of water as well as fluidization of carriers in aerobic reactor is caused by upward aeration whereas that in anaerobic reactor is caused by mechanical stirrer

in the reactor along with maintaining a stable concentration of microorganism is one of the basic reasons behind the increasing acceptance of MBBR as a versatile treatment technology [108]. It provides comparatively larger solid retention times (SRT) where older cells with diminished supplementary nutrient supply exhibit better degradability [166]. Fluidization of biocarriers increases the contact opportunity of wastewater with the biomass thereby removing the desired pollutants more effectively [121]. Since there is no need for separate sedimentation tank, the constructional and operational cost as well as reactor footprint is drastically reduced [80].

Being attached to a support medium, biofilm cells are observed to be more toxicity resistant than suspended biomass [49] due to the extrapolymeric substances (EPS) formed by the biofilm that restricts the diffusion speed of hazardous substances [144] and acts as a buffer [103]. Wastewater including dye effluent often has elevated temperatures and has been notably treated in MBBR using thermotolerant bacteria [95]. Anaerobic MBBRs are found to be highly resistant against shock loads which are often experienced in dye related industries [62]. Other advantages of using MBBR include high rate of nitrification, uniform oxygen transfer, stable operation and increased surface area for biological activities [142, 173]. Comparing all aspects of dye effluent treatment in activated sludge reactor, MBR and MBBR, it was observed that with same operational cost as MBR, capital expenditure is 68.4% lower in case of MBBR along with comparatively less environmental impacts as observed by Life Cycle Assessment (LCA) analysis. It was also confirmed that the water treated by MBBR can be reused in industrial sector [170].

Albeit all these advantages, a basic drawback for MBBR is lower sludge settle ability in comparison to that of a suspended reactor which can be solved by adding coagulants if necessary [169]. High fluid flow rate is also not desirable in MBBR as it reduces HRT and increases the risk of biomass washout [142]. In view of increasing efficiency of dye removal and decrease the load on MBBR, several other technologies are associated with the process. The present review focusses on various perspectives of dye removal in an MBBR system with or without additional treatment processes, various parameters to be considered during the operations and biological aspects of this treatment process. In this regard, the present chapter discusses the mechanism of dye degradation under various operational conditions in both bacterial and fungal MBBR, followed by the different instances of using MBBR as dye treatment unit coupled with other treatment technologies and the effect of different microorganisms and influencing parameters for dye degradation in MBBR and areas for further investigation. Over the years, several experiments have been conducted but little effort has been made to recapitulate the investigations which is attempted in the present work, thereby pointing out the research gaps in this field.

2 Mechanism of Dye Degradation in MBBR

Considering the different microorganisms that have been attached with carriers, biomass in MBBR can be grouped into bacteria-based carriers and fungi-based carriers. Both of these microbes have different enzymes responsible for decolourization, thereby has different approaches towards it. For example, bacterial enzymes are more substrate specific, whereas that in case of fungi is non-substrate specific which makes it easier to degrade a wide range of dyes. The approach towards degradation by bacteria is usually by breaking the azo bond in azo dyes by reductases whereas in fungi cellulase and peroxidases comes into action. The basic advantage of using MBBR over suspended systems reflects in removal efficiencies. Under identical experimental conditions, it is established that the high concentration of active biomass in carriers contribute to higher COD and colour removal efficiencies in MBBR (61.2% and 51.6%, respectively) as compared to those in activated sludge system (34.1% and 26.8%, respectively) in considerably lower HRT [144].

2.1 *Bacterial Dye Degradation*

Different microorganisms in a consortium responsible for dye degradation often require different environmental conditions for optimum performance and thus, the need for understanding the mechanism of dye degradation under different conditions is extremely necessary. Studies show that anaerobic–aerobic system of biodegradation of dyes is often more effective than aerobic treatment [90]. Azo dye reduction may occur in three probable mechanisms: azo bond cleavage, bioadsorption [113, 174] and hepatic microsomal reduction [51] although dye degradation following the latter is almost negligible [100]. Bioadsorption also contributes to a low as 14%

of decolourization [113]. In general, the azo bond cleavage takes place in anaerobic conditions and is almost impossible to occur in aerobic environment. Azo bonds are characterized by strong electron withdrawing nature which supports easily cleavage in oxygen deprived environment [113]. Azoreductase enzymes are usually responsible for initiating this degradation of azo dyes including Orange II [176].

A number of researches have in conducted confirming the efficiency of using anaerobic/aerobic profile in dye removal in MBBR [112]. The main reductase enzyme needed for the azo bond cleavage is functional in anaerobic environment [70] in the presence of NADH, NADPH and FADH [157]. The transition of anaerobic to aerobic environment is necessary for complete degradation of dyes because the intermediates formed as a result of azo bond cleavage are recalcitrant in absence of oxygen [29]. Thus, the colour in dye wastewater is removed in anaerobic stage, whereas a large proportion of COD is removed in aerobic stage [1, 64, 85, 112, 143]. However, COD removal does take place in anaerobic chambers, although it is often within the range of 1–40% in a single staged reactor. In case of double staged reactor, the removal is quite high, measuring up to 70%. It was experimentally determined that a small proportion of decolourization may take place in the aerobic phase, as low as 20% [13]. In a study conducted by Dong et al. [44] using anaerobic and aerobic MBBRs showed, 20–35% COD being reduced in anaerobic phase. The COD reduced in aerobic phase are contributed by the anaerobically recalcitrant amines formed in absence of oxygen [62]. Complete degradation of azo dyes takes place in two steps: (as shown in Fig. 3).

Step1: Cleavage of azo bond under anaerobic conditions.

After the cleavage of azo bond, the nitro groups react with protons to form aromatic amines [51]. Experimental studies confirm that a large proportion of the dyes are biodegraded rather than being mineralised at this stage [62].

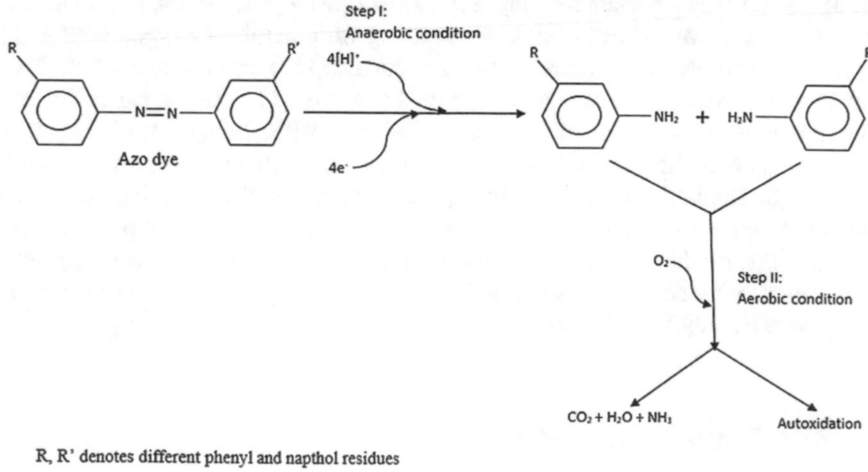

R, R' denotes different phenyl and napthol residues

Fig. 3 Schematic representation showing bacterial azo dye degradation via anaerobic/aerobic pathway [159]

Step2: The intermediate amines are degraded via hydroxylation and ring cleavage in aerobic conditions.

This step is necessary for transformation of amines to CO_2, H_2O and organic acids [8]. Another possibility of aromatic amine transformation, especially when ortho-substituted hydroxyl groups are present, is autoxidation in aerobic environment [88, 158]. This autoxidation is one of the main reasons behind recalcitrant nature of aromatic amines [159]. Even though it is theoretically considered that anaerobically formed amines are easily degraded in aerobic conditions, it has been observed aromatic amines containing sulfonates such as naphthyl amine sulfonate and aminobenzene sulfonate are not mineralized in conventional aerobic suspended growth systems due to hindrance in transport across cell membrane [120]. Considering the stratification of biofilms in MBBR carriers, anoxic and aerobic environment simultaneously exists in a single reactor that will facilitate both the processes at different biofilm depths [26, 87, 113].

2.2 Fungal Dye Degradation

Unlike bacterial species, fungi are capable of degrading a wide range of organic pollutants and are not dye specific [135] due to non-specific nature of enzymes that aid in dye degradation. Enzymes like lignin and manganese peroxidases and laccase catalyse oxidation of even complex azo dyes which are both phenolic and nonphenolic [58, 140]. Interaction and relative contribution of these enzymes vary with fungal species. Similar to bacterial dye degradation pathways, biodegradation, biosorption and bioaccumulation are the noted mechanisms for fungal biodecolourization by both live and dead cells [78] however bioadsorption is limited to a maximum dye removal of 50% [83]. In case of white rot fungi *Phanerochaete chrysosporium*, Lignin Modifying Enzyme (LME) is responsible for reacting with dissolved oxygen producing H_2O_2 [122] and does not require Lignin Peroxidase (LiP) enzyme to degrade dyes including azo, heterocyclic and triphenyl methane dyes [111]. In the case of *Trametes versicolor*, another fungal species studied in MBBR, Manganese Peroxidase (MnP) does not participate in decolourization, which is brought about by oxidation catalysed by ligninase and laccase for majority of the studied dyes including azo and anthraquinone dyes [164, 171]. Laccases that belong to the group of oxidase enzymes catalyse the oxidation of aromatic compounds to their corresponding radicles ultimately reducing to oxygen and water [79]. Adsorption by *T. versicolor* is observed to be only 5–10% [19].

3 MBBR as a Treatment Unit

A comparison between MBBR with other suspended growth processes in the context of dye degradation revealed that the former is far more efficient for removing both

COD and colour from dye wastewater [13, 121, 144]. Comparative studies between moving bed systems and SBR show the difference in COD removal was higher by almost 20% within a HRT of 24 h [13]. MBBR as a sole biological treatment setup for satisfactory degradation of dyes is not frequently studied due to the requirement of multi-staged setup for complete degradation, low BOD: COD ratio along with presence and/or formation of hardly degrading substances. A study involving three MBBR units—two anaerobic followed by an aerobic reactor demonstrated a total removal of 86% COD and 50% colour. Almost all the colour removal took place in the anaerobic reactors, the aerobic unit accounted for mere 1% removal. COD removal efficiency in the two anaerobic reactors was found to be unsatisfactory but was increased to 86% after treatment in aerobic chamber [121]. Anaerobic MBBR operated at different HRTs and temperatures to study the effect of these parameters on removal efficiency showed that under optimum conditions of 48 h HRT and 30 ± 5 °C, rate of decolourization and COD removal was 85% and 78.3%, respectively. Decrease in HRT up to 12 h decreased the COD reduction rate to 45% and temperature decrease also resulted in drastic change for dye decolourization [139]. It is evaluated that up to a certain temperature, there is proportional relation with dye degradation [10]. Dye degradation in MB-SBBR resulted in 100% decolourization when initial dye concentration is increased stepwise. Inducing shock load, decreased the removal efficiency by 40%. Degradation of aromatic intermediates, formed during anaerobic phase, was efficiently completed in the aerobic period along with 10% decolourization [113]. Experimental study confirmed high temperature dye effluent treatment using aerobic MBBR and substantial COD removal was achieved in the single system [95]. Albeit these performances, a number of other treatment technologies coupled with MBBR are also experimented, which are discussed in detail in the following sections (Table 1).

3.1 SBR Along with MBBR Setup

To facilitate the aerobic degradation of dye intermediates formed anaerobically, aerobic MBBR setups can be installed in sequential mode for better performances that aid in complete mineralisation by alternating anaerobic and aerobic phase. Further improvement can be investigated by applying an anaerobic SBR prior to MBSBBR. For an instance, it has been observed that up to an initial acid dye (AR18) concentration of 100 mg/L, 100% removal of dye intermediate 1-Napthylamine 4-sulfonate can be achieved in aerobic moving bed sequential biofilm reactor (MB-SBBR) with corresponding COD removal of at least 71.5%. Increasing dye concentration even up to 1000 mg/L, the removal of 1-N 4-S was never less than 83.9% with an HRT of 2.75 d which conforms that aerobic moving bed reactors can efficiently degrade dye intermediates which are otherwise challenging. Using this setup, 98% colour removal was obtained with initial dye concentration ranging between 100 and 1000 mg/L [85].

Table 1 Studies demonstrating the role of MBBR as dye effluent treatment unit

Reactor configuration	Experimental conditions	Wastewater characteristics	Dye(s) considered	Responsible Microorganisms	Removal obtained in MBBR unit(s)	Other remarks	References
Coagulation with alum + Polyurethane-based fluidized reactor	* Carrier: polyurethane media * CFR[a]: 15% (v/v) * Attached biomass: 2800 mg/L * Suspended biomass: 2000 mg/L	* COD: 654–1092 mg/L * SS: 46–152 mg/L * pH: 12.3 * temp: 32.6 ±2.1 °C	Not specific (dye effluent from the polyester deweighted process)	Bacteria (Mixed culture)	* 92% COD removal	* Coagulation followed by biological treatment performed better than biological pretreatment followed by coagulation * Alum dose required (600 mg/L) was less in the former trial with less sludge production	Park and Lee [123]

(continued)

Table 1 (continued)

Reactor configuration	Experimental conditions	Wastewater characteristics	Dye(s) considered	Responsible Microorganisms	Removal obtained in MBBR unit(s)	Other remarks	References
MBBR + chemical coagulation with $FeCl_3.6H_2O$ + electrochemical oxidation using NaCl	* Carrier: hexahedronal media * CFR[a]:10% (v/v) * HRT: 48 h * DO: 3–4 mg/L * Attached biomass: 570 mg/L	* COD: 870 mg/L * colour: 1340 Pt–Co unit * pH: 13	Not specified (synthetic textile dyeing factory wastewater)	Bacteria (*Aeromonas salmonicida* and *Pseudomonas vesicularis*)	* 68.8% COD removal * 54.5% colour removal	* 3.25×10^{-3} mol/L $FeCl_3.6H_2O$ used as coagulant * 25 mM NaCl was used for oxidation with a current density of 2.1 mA/cm^2 * 95.4% overall COD removal * 98.5% overall colour removal	Kim et al. [80]
3 MBBR reactors (anaerobic, aerobic 1, aerobic 2) + coagulation with $FeCl_2$	* Carrier: polyurethane activated carbon * CFR[a]: 20% (v/v) * HRT: 44 h	* COD: 900 mg/L * colour: 3200 Pt–Co unit * pH: 13 (adjusted to 7) * temp: 42 °C	Not specified (synthetic textile dyeing factory wastewater)	Bacteria (Mixed culture)	* 85% COD removal * 70% colour removal	* 14% $FeCl_2$ used for coagulation * 95% COD removal obtained after total treatment * 97% overall colour removal efficiency	Shin et al. [144]

(continued)

Table 1 (continued)

Reactor configuration	Experimental conditions	Wastewater characteristics	Dye(s) considered	Responsible Microorganisms	Removal obtained in MBBR unit(s)	Other remarks	References
3 MBBR reactors (anaerobic 1, anaerobic 2, aerobic)	* Carrier: polyurethane activated carbon * CFR[a]:20% (v/v) * HRT: 44 h * Attached biomass: 3000 mg/L	* COD: 608 mg/L * colour: 553 Pt–Co unit * T-N: 33 mg/L *T-P: 3.5 mg/L * pH: 12.5 * temp: 40 °C	Not specified (sample collected from dyeing wastewater treatment plant)	Bacteria (Mixed culture)	* 86% COD removal * 50% colour removal	* COD degradation rate in aerobic reactor was higher than that in anaerobic reactors * colour removal in aerobic reactor was 1%	Park et al. [121]
SBR (anaerobic) + MB-SBBR (aerobic)	* Carrier: polyethylene carriers * CFR[a]: 50% (v/v) * DO > 3 mg/L * HRT: 2.75 d * Attached biomass: 1318–1614 mg/L	* dye conc: 100, 5000, 1000 mg/L * COD: 3040–3620 mg/L * Temp: 22 ± 2 °C	Acid Red 18	Bacteria (collected from municipal wastewater treatment plant)	* 65–72% total anaerobically formed dye metabolites were degraded * more than 80% 1-napthylamine 4-sulfonate was degraded	* Glucose and lactose as co-substrate * 98% dye decolorization and above 80% COD removal accoutred in anaerobic SBR	Koupaie et al. [85]

(continued)

Table 1 (continued)

Reactor configuration	Experimental conditions	Wastewater characteristics	Dye(s) considered	Responsible Microorganisms	Removal obtained in MBBR unit(s)	Other remarks	References
2 MBBRs (both aerobic) + chemical coagulation by alum or $FeCl_2$	* Carrier: Polyurethane-dyeing sludge carbonaceous material (PU-DSCM) foam * CFR[a]: 10–30% (v/v) * Attached biomass: 2900 mg/L * HRT: 48 h * DO: 2.3 and 5.2 mg/L	* COD: 539 mg/L * colour: 622 Pt–Co unit * pH: 12.5 * Temp: 40 °C *T-N: 33 mg/L *T-P: 3.5 mg/L	Not specified (effluent collected from synthetic textile dyeing factory)	White rot fungus (*Phanerochaete chrysosporium)*	* 79% COD removal * 54% colour removal	* Both alum and $FeCl_2$ was tested for coagulation * Optimum alum dose: 1.55 mg alum/mg COD * $FeCl_2$ dose: 11 mmol/L * 95.7% COD removal * 73.4% colour removal * COD and colour removal was much higher in the first reactor in compared to the second	Park et al. [122]
2 MBBRs (anaerobic, aerobic) + membrane filtration	* Carrier: polypropylene cylindrical carriers * CFR[a]: 37% (v/v) * HRT (anaerobic): 11 h (aerobic): 5 h	* COD: 500 mg/L * SS: 310 mg/L * dye conc.: 400 Pt–Co * pH: 11	Azo dye: Reactive Brilliant Red X-3B	Bacteria (Mixed culture)	* 90% colour removal * 85% COD removal * 94% SS removal	* Hollow fibre PVDF membrane of pore size 0.02 μm was used for membrane filtration	Dong et al. [44]

(continued)

Table 1 (continued)

Reactor configuration	Experimental conditions	Wastewater characteristics	Dye(s) considered	Responsible Microorganisms	Removal obtained in MBBR unit(s)	Other remarks	References
Photocatalytic oxidation with TiO_2 + MBBR (aerobic)	* Carrier: LECA (special clay granules) * CFR[a]: 50% (v/v) * HRT: 8–20 h	* COD: 1650 mg/L * BOD: 390 mg/L * pH: 11 * colour: 0.81 A_{335} * Temp: 22–26 °C	Not specified (textile plant effluent)	Bacteria (mixed culture)	* 64% COD removal * 72% colour removal	* TiO_2 Degussa P-25 was used as photocatalyst (conc: 0.125 and 0.25 g/L) * maximum removal at 20 h HRT * 79% COD removal * 87% colour removal	Ahmadi et al. [4]
2 MBBR reactors (aerobic)	* Carrier: Anox Kaldnes K1 carriers * HRT: 24 h * SRT: 15 d * DO: 2.0 ± 0.4 mg/L * MLSS: 12.5 g/L	* COD: 650 ± 80 mg/L * NH_3-N: 18 ± 2.2 mg/L * Temp: 30–55 °C	Not specified (effluent from dyeing industry)	Thermotolerant bacteria (genus *Caldilinea, Rubellimicrobium* and *Pseudoxanthomonas)*	* 70.1% COD removal * 39.1% NH_3 removal	* Maximum COD removal was at 40 and 50 °C * Maximum ammonia removal was at 30 and 40 °C	Li et al. [95]

(continued)

Table 1 (continued)

Reactor configuration	Experimental conditions	Wastewater characteristics	Dye(s) considered	Responsible Microorganisms	Removal obtained in MBBR unit(s)	Other remarks	References
Fluidized bed Fenton oxidation + MBBR	* Carrier: corrugated PVC cylinders * CFR[a]: 40–80% (v/v) * contact time: 1–3 d	* COD: 600–800 mg/L * BOD: 180–240 mg/L * dye conc.: 100 mg/L * T-N: 33 mg/L * T-P: 3.5 mg/L *pH: 12.5 * temp: 40 °C	Reactive sulphur dye: Chemistar Turq Blue	Bacteria (*Microbacterium marinilacus*)	* 86% COD removal * 81.5% BOD removal	* Conditions for oxidation: pH: 3, Fe^{+2}: 3 mg/L, H_2O_2: 5 mM * Optimum removal efficiency was obtained at a pH of 7.33, HRT 2.25 d and carrier filling ratio of 67.07%	Francis and Sosamony [53]
2 MBBR (anaerobic, aerobic) + ozonation + MBBR (aerobic)	* Carrier: Polyethylene carriers * CFR[a]: 60% (v/v) * HRT: (for first two MBBRs): 14 h (for last MBBR): 10 h * SRT: 10 days * DO (aerobic MBBRs): 2–4 mg/L	* COD: 824 mg/L * SS: 691 mg/L * NH_3: 40 mg/L * colour: 165 degree * pH: 7.9–8.5	Not specified (sample collected from textile dyeing factory)	Bacteria (Mixed culture)	Anaerobic unit: * COD: 23.1% * colour: 72.6% First aerobic MBBR: * COD: 68.9% * colour: 54%	* 14 min of ozonation of 1 mg/L * 94.3% COD, 97.8% SS, 85.3% ammonia and 96% colour was removed during treatment	Gong [62]

(continued)

Table 1 (continued)

Reactor configuration	Experimental conditions	Wastewater characteristics	Dye(s) considered	Responsible Microorganisms	Removal obtained in MBBR unit(s)	Other remarks	References
Ozonation + MBBR (aerobic)	* Carrier: Anox Kaldnes K1 * CFR[a]: 40% (v/v) * HRT: 6 h	* COD: 400 mg/L * Dye conc.: 25–100 mg/L *NH_3: 30 mg/L	Azo dye: Reactive Orange 16	Bacteria (mixed culture sludge collected from municipal sewage treatment plant)	* 90 ± 1% COD removal * 97 ± 2% ammonium removal	* 5 min ozonation with dose 51.09 ± 0.76 mg/L * More than 97% of dye removal with 93 ± 1% COD and 97 ± 2% ammonium removal * Glucose used as co-substrate	Castro et al. [28]
Ozonation + MBBR (aerobic)	* Carrier: Anox Kaldnes K1 * CFR[a]: 20% (v/v)	* COD: 1000 mg/L * Dye conc.: 100 mg/L	Azo dye: Remazol Black 5	Bacteria (mixed culture isolated from textile wastewater)	* 81.21% colour removal * 83.63% COD removal	* Ozonation: 0.4 gO_3/h, determine time 120 min * With ozonation as pretreatment, colour removal increased up to 86.74% at 1 h batch time	Pratiwi et al. [129]

(continued)

Table 1 (continued)

Reactor configuration	Experimental conditions	Wastewater characteristics	Dye(s) considered	Responsible Microorganisms	Removal obtained in MBBR unit(s)	Other remarks	References
MBBR (anaerobic)	* Carrier: Anox Kaldnes K1 * CFR[a]: 50% (v/v) * HRT: 12, 14, and 48 h	* COD: 2347 mg/L * pH: 7 * Temp: 30 ±5 °C and 21 ±2 °C	Azo dye: Direct Red 75	Bacteria from rice husks	* Optimum performance at 48 h HRT and 30 ±5 °C * 85% colour removal * 78.3% COD removal	* Sensitive to HRT changes * Decrease in HRT resulted in reduction of efficiencies * Bacteria were more effective at temperature around 30 °C	Santos Pereira et al. [138]
Granular activated carbon (GAC) + MBBR (aerobic)	* Carrier: Polypropylene type * CFR[a]: 67.07% (v/v) * HRT: 4–10 h *pH: 7.33	* COD: 1682 mg/L * BOD: 399 mg/L * Colour: 1813 Co–Pt	Not specified (effluent from dye treatment plant)	Bacteria (culture developed from dairy animal faeces and sludge of sewage treatment plant)	* 87.22% COD removal * 80% BOD removal	* Total COD removal 90% * BOD removal 95% * 20 cm GAC bed used	Vaidhegi et al. [155]
Ozonation + MBBR	* Carriers: Anox Kaldnes K1 * CFR[a]: 40% (v/v) * DO: 5 mg/L * HRT: 6 h * Attached biomass: 1300–2100 mg/L	* COD: 375 mg/L * NH_4-N: 40 mg/L * dye conc.: 50 mg/ * Temp: 25 ±2 °C	Reactive Red 239	Bacteria (Mixed culture)	* 88% COD removal * 41% ammonium removal	* Glucose used as co-substrate * 12 min ozonation time of 50 mg/L dose * No colour removal in MBBR	Dias et al. [43]

(continued)

Table 1 (continued)

Reactor configuration	Experimental conditions	Wastewater characteristics	Dye(s) considered	Responsible Microorganisms	Removal obtained in MBBR unit(s)	Other remarks	References
MB-SBBR (anaerobic) + MBR (aerobic)	* Carrier: Polyurethane foam * * CFR[a]: 30% (v/v) * HRT: 48 h	* COD: 900 ± 325 mg/L * colour: 300 ± 90SU * TDS: 9500 ± 1500 mg/L	Black B, Black WNN, Red 3BS	Bacteria (mixed culture from poultry slaughterhouse wastewater)	* 78% dye removal * 70.5 ± 5% COD removal	* Total COD removal 77.1 ± 7.9% * Total colour removal 79.9 ± 1.5% * Two-staged system had better performance than single staged	Azimi et al. [13]
Ozonation + 2 MBBR (both aerobic)	* Carriers: Anox Kaldnes K1 * CFR[a]: 40% (v/v) * DO: 5 mg/L * HRT: 3 h	* COD: 400 mg/L * NH_4-N: 40 mg/L * pH: 7 * Temp: 20 ±2 °C	Reactive Red 239	Bacteria (Mixed culture)	* 94% COD removal in total * 40% ammonium removal	* 12 min and 20 min ozonation time of 20 mg/L dose * Formation of 5 identifiable oxidation resistant intermediates after ozonation * Nitrification was inhibited by 4-amino 6-chloro 1,3,5-triazine 2-ol	Dias et al. [42]

(continued)

Table 1 (continued)

Reactor configuration	Experimental conditions	Wastewater characteristics	Dye(s) considered	Responsible Microorganisms	Removal obtained in MBBR unit(s)	Other remarks	References
MB-MBR (moving bed membrane bioreactor)	* Carriers: hollow cylinders * CFR[a]: 50% (v/v) * DO > 3 mg/L * HRT: 15.25 h, 16.54 h * SRT: 30 d * MLSS: 1120 mg/L	* COD: 857.2 ± 10.5 mg/L * colour: 505.3 ± 3 Pt–Co * pH: 8.07–8.14 * Dye conc: 10 mg/L * Temp: 20 ±1 °C	Reactive Red 390	Bacteria (Mixed culture)	* 89.2% colour removal * 98.5% COD removal	* In terms of COD removal, carrier filling ratio did not have any impact * Decrease in carrier filling ratio to 10% had almost negligible decrease in colour removal	Erkan et al. [48]
MB-SBBR	* Carriers: K1 carriers * CFR[a]: 5 and 10% (v/v) *HRT: 16.8 h and 33.6 h * Biomass wt: 21.12–35.19 g	* Dye conc.: 50–1000 mg/L * COD (dye + sucrose): 649–1398 mg/L * Temp: 24–26 °C	Reactive Orange 16 (RO16)	Bacteria (Mixed culture)	* 100% colour removal and 97% COD removal when dye conc. is increased stepwise	* Shock loading greatly affected both COD and colour removal, dropping the efficiency to 40% * Sucrose was given as co-substrate	Ong et al. [113]
MBBR	* Carriers: Polyurethane foam-polypropylene carriers * CFR[a]: 10–60%	* Dye conc.: 10–100 mg/L * pH: 5–9	Congo Red Dye	Bacteria (*Bacillus sp.*)	* 95.7% dye removal	* Optimum pH, dye concentration and carrier filling ratio was obtained to be 7, 50 mg/L and 45% respectively	Sonwani et al. [148]

(continued)

Table 1 (continued)

Reactor configuration	Experimental conditions	Wastewater characteristics	Dye(s) considered	Responsible Microorganisms	Removal obtained in MBBR unit(s)	Other remarks	References
MBBR	* Carriers: Plastic Biofill C2 carriers * CFR[a]: 30% * Average attached biomass: 3.5 g/L * DO > 2.2 mg/L * HRT: 1 d	* COD: 2000 mg/L * colour: 700 Pt–Co * BOD: 400 mg/L * pH: adjusted to 8.6 * TSS: 940 mg/L * Temp: 25 °C	Not specified (Sample collected from local textile industry)	Bacteria (aerobic sludge collected from textile industry effluent treatment plant)	* 82% COD removal * 61% colour removal * 78.8% TSS removal * 90% T-N and T-P removal	* 100% of MBBR treated effluent can be reused in dyeing process of Yellow, crimson and Navy Procion HEXL *	Yang et al. [170]

CFR[a]: Carrier Filling Ratio

3.2 Coupling Membranes with MBBR

Membrane filtration when used along with a series of anaerobic and aerobic MBBR reactors exhibits a number of advantages including lower HRT and high loading rates. Membrane fouling is often a problem encountered in membrane bioreactor which Erkan et al. [48] attempted to solve by using a novel moving bed membrane bioreactor (MB-MBR). It was suggested that no cleaning—physical or chemical was required for the process. In comparison to colour and COD removal percentages in MBR (87.1% and 93.1%, respectively), MB-MBR showed slightly better results with respect to both parameters (89.2% and 98.5%, respectively) but the major advantage lies in avoiding membrane fouling. Degradation of Reactive Brilliant Red in such a treatment process with hollow fibre PVDF membranes showed 90% colour removal [44]. A hybrid system employing MBBR along with membrane bioreactor (MBR) is a striking option in terms of reactor footprint and energy consumption for treating dye wastewater [169]. Degradation of dye wastewater containing three different azo dyes using anaerobic MBSBBR—aerobic MBR setup, resulted in a degradation efficiency of around 79.9 ± 1.5% and 77.1 ± 7.9% in terms of colour and COD, respectively. The lower colour removal is attributed to autoxidation of dye intermediates formed in the anaerobic phase [13].

3.3 MBBR and Chemical Coagulation

Chemical coagulation is one of the commonly used treatment technologies for dyeing wastewater [20] because the effluent emitted from the related industries often contains a high concentration of solids in suspended as well as colloidal form [22]. The innovation of biological processes made a promising aspect in combining the two technologies to get efficient removal discarding the disadvantages of high sludge production from coagulation process and high organic loading on biological process. Shin et al. [144] studied the efficiency of treating textile wastewater using a three-staged MBBR coupled with coagulation using $FeCl_2$. A maximum removal efficiency of 95% and 68.5% was achieved in terms of COD and colour, respectively, during the combined treatment process, whereas, maximum removal efficiencies for COD and colour was obtained as 55.3% and 56%, respectively, individually in MBBR. The results showed majority of colour removal took place in anaerobic reactor where electrons were provided by the dyes in reducing environment. A combined treatment process was adopted that involved MBBR as biological pretreatment unit, followed by chemical coagulation and electrochemical oxidation. The combined treatment yielded a colour removal of 98.5% along with 95.4% COD removal using 3.25×10^{-3} mol/L $FeCl_3.6H_2O$ along with 25 mM NaCl at a current density of 2.1 mA/cm^2 and a flowrate of 0.7 l/min. The support media resulted in an increase of 19.7% decolourization and 13% COD removal on the overall process. [80].

In an attempt to understand the better reactor configuration with chemical coagulation and biological treatment, studies were performed as biological pretreatment followed by chemical coagulation (FBAS-C) and chemical coagulation followed by biological fluidized bed reactor (C-FBAS). It was observed that biological treatment followed by coagulation with alum performed better with higher COD removal (92%) with lower alum dosage (600 mg/L) and 20% less sludge production. In C-FBAS configuration, a COD removal of 82% was obtained with 1000 mg/L alum dosage [123]. In place of using bacteria as attached microorganisms, Park et al. [122] attempted dye removal in two MBBRs with attached white rot fungus with chemical coagulation as post-treatment in order to improve COD removal. The study investigated the efficiency of alum and $FeCl_2$ as coagulants. COD removal was observed to be higher in case of alum (81%) with an optimum dose of 1.55 mg alum/mg COD as compared to 11 mmol/L $FeCl_2$ (79%). Colour removal was also observed to be higher in alum (42%) than $FeCl_2$ (30%). Almost 96% COD removal was obtained during this treatment with optimized parameters.

3.4 MBBR Coupled with Advanced Oxidation Processes

Advanced oxidation processes can mineralise a wide range of recalcitrant and toxic substances [9] although high cost of this method restricts its solitary application in large industries and is recommended to be applied in case of BOD/COD ratio less than 0.2 [125]. Dye effluents contain a huge proportion of recalcitrant organic compounds that characterizes the low BOD: COD ratio of the wastewater making it hard to biodegrade completely. To increase the biodegradability, the effluents are often subjected to chemical oxidation that improves BOD: COD ratio [55]. Oxidative pretreatment of textile wastewater using Fenton oxidation resulted in the increased BOD to COD ratio from 0.25 to 0.52 [53]. However, dose and contact time of most of these oxidation processes must be optimized as they may form by-products toxic to microorganisms undergoing further biological treatment [94, 130].

3.4.1 Oxidation with Ozone

Dissolved ozone is able to decompose refractory materials into easily biodegradable substances thereby facilitating its biological mineralisation [97, 153]. The negative charge density prevailing around the azo bond attracts ozone to get oxidized [155]. Dye decolourization was studied in a four staged reactor comprising of an aerobic MBBR followed by aerobic MBBR, ozonation and aerobic MBBR where each reactor had its specific function towards the treatment process. The anaerobic chamber facilitated dye degradation with 76.8% colour removal whereas the first aerobic MBBR reduced its intermediates where BOD: COD ratio dropped from 0.33 to 0.14. The residual recalcitrant intermediates were ozonized reviving the BOD: COD ratio of 0.43 which were easily degraded in the last aerobic chamber [62]. A similar study conducted by Pratiwi et al. [129] showed that while degrading Remazol Black (RB5), an azo dye, ozonation time of 120 min with a dosage of 0.4 g O_3/h resulted in an increase of BOD: COD ratio from 0.23 to 0.42. It was observed that with a mere retention time of 1 h, dye removal increased from 68.6 to 86.74% with ozone treatment. Allowing a 24 h batch period, optimum colour and COD removal efficiencies were obtained as 96.9% and 89.13%, respectively.

A study was conducted by Dias et al. [43] which involved the application of ozone followed by biological treatment in aerobic MBBR to treat Reactive Red (RR239). The removal efficiencies of COD and ammonium were obtained as 88% and 41%, respectively. It was observed that nitrification was partially inhibited which resulted in accumulation of nitrite in the reactor. The cause of this inhibition was attributed to triazine and benzofuran. To make the process more efficient, ozonation was further followed by two-staged aerobic MBBRs as bio-treatment which proved to be highly efficient for removing RR239 although high ozone dose produced an intermediate 4-amino-6-chloro-1,3,5-triazine-2-ol which still inhibited nitrification in the reactor, but to a lesser extent. In this study, ozone dosage was reduced from 50 to 20 mg/L. With lower ozonation dosage, triazine in the dye could not be oxidized. Longer ozonation time (20 min in place of 12 min) resulted in formation of intermediates that were recalcitrant and toxic for biomass and thus COD removal efficiency dropped from 94 to 90% in the reactors along with nitrification inhibition [42]. However, a similar study undertaken by Castro et al. [28] involving the removal of Reactive Orange 16 using ozone and aerobic MBBR did not show any inhibition towards removal of ammonium, where a maximum of 94% COD, 97% colour and 99% ammonium was removed where ozone was applied at a dosage of 51.09 ± 0.76 mg/L for 5 min. Thus, from the previous works it can be concluded that increase in ozonation time produces more recalcitrant compounds that might inhibit nitrification in aerobic MBBR.

3.4.2 Fenton Oxidation

Fluidized Fenton oxidation process is supposed to be advantageous over other processes in the fact that iron crystallization and precipitation results in production of less sludge [104]. The quantity of hydroxyl radicals produced during the process is enough to effectively oxidize most of the compounds present in textile effluent but a major disadvantage of this process is the production of ferric hydroxide sludge which needs further handling and disposal. Further, it has been observed that impact of variation in influent wastewater characteristics can be reduced with Fenton oxidation process [82]. Using immobilized *Microbacterium marinilacus* isolated from textile effluents, fluidized bed Fenton process followed by MBBR was capable of removing 87.22% COD at 67.07% filling ratio. Using Box-Behnken statistics, the optimum pH and HRT were suggested as 2.25d and 7.33, respectively. Sludge production combining the two processes resulted in considerable decrease in sludge production thus being more economical [53].

3.4.3 Photocatalytic Oxidation

Photocatalysis with TiO_2 is a non-toxic, easily available chemical with robust optical properties, large surface area and its use in treatment operations is cost effective as compared to several other AOPs [59, 66]. Using polyaniline-TiO_2 nanocomposites immobilized in polystyrene cubes, 89% of degradation of Acid Yellow 17 was obtained which shows the efficacy of photocatalytic action of TiO_2 alone for dye decolourization [106]. Combination of photocatalysis with biological process helps in efficient treatment of colour using TiO_2 and biological removal of COD. The removal efficiency using TiO_2 is comparable with that using Fenton oxidation process and within 48 h of biological treatment nearly 98% of colour removal could be achieved even using suspended growth reactors [38]. With all these advantages and various studies involving photocatalysis to degrade dyes, an experimental study was undertaken to degrade textile wastewater with a BOD: COD ratio of 0.23 using a combination of photocatalytic oxidation using TiO_2 and aerobic MBBR [4]. The researchers compared the removal efficiencies between activated sludge process, MBBR and MBBR coupled with photocatalysis. Oxidation with photocatalyst proved to be the most effective treatment process with considerably better removal efficiencies in terms of both COD and colour than the other two reactors. Modified Stover-Kincannon model was used to determine the COD removal kinetics for the reactors.

3.5 MBBR Along with Adsorption

Bioadsorption is one of the mechanisms of dye effluent decolourization which has very minimal contribution. However, installing an adsorption medium along with MBBR system decreases the organic load, as also improves decolourization efficiency. Often along with biological action, the biomedia used takes part in adsorption of dyes causing a removal of 96% [87]. Moreover, the presence of high concentration of complex recalcitrant compounds vacillates the use of MBBR without a prior treatment that would decrease the loading on biological system. Granular activated carbon has high retaining capacity along with high surface area suitable for adsorption [167]. A reactor setup comprising of a 20 cm granular activated carbon (GAC) bed and an aerobic MBBR was used to study the performance regarding dye wastewater treatment which showed a maximum of 90%, 95%, and 72% COD, BOD removal and colour removal, respectively. Maximum colour removal occurred in the GAC by adsorption whereas MBBR increased the quality of the effluent by decreasing the organic matters. One important advantage of using GAC is it can be reused after corrosive washing thus minimizing the economic input [155].

4 Microorganisms in MBBR for Dye Removal

Microorganisms including bacteria, fungi, yeast have been recorded to degrade different dyes in various environmental conditions [101]. The selection of microorganisms for specific dye removal is quite challenging for high loading rates as it may often inhibit the activities of respective organisms [148]. Mixed culture is observed to perform better than pure culture bacteria and is also more resistant to toxic effects of recalcitrant dyes and is also easy to maintain in industrial treatment plants [73]. A large number of microbial species have been identified to effectively degrade different dyes [52]. Toxicity of the dyes may have adverse effect on the development of biofilms which is why it is often preferred to carry out the process of biofilm formation on carriers in absence of dyes [69]. Bacterial cultures have been reported to show better results than fungal biomass in the treatment of dye wastewater [128]. Thirteen different diazobenzene dyes have been identified that have antifungal properties [116].

4.1 Bacteria

A number of facultative bacterial species are able to degrade dyestuff wastewater in both aerobic and anaerobic conditions [89]. Bacteria isolated from textile effluents are observed to be capable of decolourising dyestuff effluent over a pH range of 6–10, temperature range of 25–40 °C and an initial dye concentration of 50–200 mg/L [97]. Often, filamentous bacteria are noticed to remain in suspended form in MBBR systems during dye degradation [86]. These filamentous bacteria are also the dominant cultures as observed to be attached to the carriers [68]. Thermotolerant bacteria when used as biomass to treat high temperature dyestuff wastewater showed the abundance of genera *Caldilinea, Rubellimicrobium,* and *Pseudoxanthomonas* at different temperature ranges and a number of species has been identified to successfully treat dye effluent [95]. Details about the pathway of bacterial degradation and consequent conditions are discussed in other sections.

4.2 Fungi

This paper has mainly focused on bacterial degradation of azo dyes. However, fungal immobilization is especially favoured due to its morphologic specificity [125]. Fungal degradation of azo dyes in aerobic condition occurs due to their ligninolytic activity using lignin peroxidase [54]. White rot fungi are the most studied species for biodecolourization [89] among which *Phanerochaete chrysosporium* and *Trametes versicolor* has been reported to exhibit satisfactory dye removal when immobilized [118, 122]. *Phanerochaete chrysosporium* has been observed to remove as high as 96%

COD removal and 73.4% colour removal in MBBR along with several other biofilm-based reactors undergoing dye removal. The ligninolytic enzyme system of this particular species can degrade a number of recalcitrant dyes [111]. Besides these, a number of other species of white rot fungi and several other fungal species are recorded to decolourise various dyes in different reactors [54]. Details about the pathway of fungal biodecolourization and consequent experimental conditions are discussed in the previous sections.

5 Factors Affecting Dye Degradation in MBBR

Biological process in reactors is generally influenced by a number of parameters including pH, temperature, substrate available for biological uptake, retention times, biomass concentration. Along with these, support biocarrier surface, hydrodynamics, flow of water, turbulence also directly affect the formation of biofilms on media [11]. For treatment of recalcitrant compounds like wastewater containing dyes, chemical structure as well as inhibitory concentration of those chemicals [21]. Temperature also has direct effect on biomass community and their growth. It was observed for thermotolerant bacteria, optimum EPD yield was obtained at 45 °C majorly produced from humic acid [95]. Interaction of all these factors influence the optimum removal efficiency of dye wastewater treated in MBBR. To develop a robust process design, all these factors must be considered and a thorough understanding of their effects is required from economical as well as an operation point of consideration.

5.1 Effect of Initial Dye Concentration

Increasing dye concentration often limits removal efficiency as the dyes are often inhibitory to enzymatic activities of microbes at elevated levels [117, 151, 168]. However, too low concentration will limit the enzyme binding capability with the dyes, considered as substrate [56]. Thus, for optimum biodecolourization, knowledge about threshold influent dye concentration is crucial [142]. By varying initial concentration of Reactive Orange (RO16) over a range of 50–1000 mg/L in moving bed sequential biofilm reactor (MB-SBBR), it was observed that increasing concentration beyond a certain limit (300 mg/L dye concentration) significantly decreases the degradation in terms of both colour and COD. The removal efficiencies with initial concentration of 50 mg/L were obtained as 100% and 99.19% for colour and COD, whereas that with dye concentration 1000 mg/L was 26.82% and 36.98%, respectively, under other identical conditions. The study also confirms that shock loads with more than 300 mg/L dye concentration caused considerable effect on the reactor performance due to inhibition by toxic RO16 [113]. Often increasing dye concentration to an extent cause similar decolourisation but with lower decolourization rates, which is due to the decrease in growth cultures brought about by inhibition [64].

5.2 Effect of Dye Structure

The fused aromatic structures of anthraquinone-based dyes make them most resistant to biodegradation over a long period of time. High colour intensity of basic dyes makes them difficult to degrade [15]. Lower potential to decolourize dyes can also be attributed to redox potential of the dye [47, 51]. The presence of azo bonds directly influences degradation efficiency, more the azo bonds present, degradation is harder [56, 156] and so, more dyes are resistant to microbial action [24]. The reason for difficulty in degradation of azo dyes is the presence of NH_2-triazine in the meta position of the compound [147]. Dyes having low molecular weights are easily biodegraded than high molecular weight, complex dyes that often are more recalcitrant [137]. Reactive dyes like Reactive Black 5 (RB5) and Reactive Violet 5 (RV5) often produces intermediates that are extremely hard to biodegrade even in aerobic environment which will produce water unsafe for further use or disposal [88, 96]. Reactive Orange 16 (RO16) has sulfonated group in the second position of naphthol ring which makes it harder to degrade [6]. Presence of sulphonic groups creates an electron deficient condition which makes it even harder to be biodegraded [17], for example, vinyl sulfone groups prevents dye adsorption and substrate fixation in biomass [63]. Azimi et al. [13] compared the removal efficiencies of three different azo dyes with variable number of azo bonds under identical experimental conditions. It was observed that Red 3BS, having 1 azo bonds and 1 triazine had a removal efficiency of 91.4 which in case of Black B (2 azo bonds and 2 sulfato ethyl sulfone groups) was 85.7%. Azo dyes having methyl, sulfo, nitro or methoxy groups are difficult to degrade than those having hydroxyl or amino groups in their molecular structure [107]. Wong and Yu [164] observed decolourization by fungus *T. versicolor* was dependent on dye structure as azo and indigo dyes are not substrates of laccase enzyme. Another fungal species *P. chrysosporium* was able to mineralise compounds with hydroxyl, nitro or amino attached with aromatic ring better than with unsubstituted rings [149].

5.3 Effect of Co-Substrate

Azo dyes being hard to biodegrade substances are often degraded by co-metabolism where the dye is treated as a secondary substrate for microorganisms along with an easier biodegradable primary substrate [141, 152]. During anaerobic degradation, the dyes are decolourized for being electron acceptors for electron transport chain. Thus, an easily degradable carbon source is needed [27]. The simple organic carbon provides the energy source to catabolize the dyes in biological cycle during anaerobic phase [159]. It has been observed that *Bacillus subtilis* uses glucose as a co-substrate to reduce azo bond by providing essential reducing factors including NADH and FAD [27]. The composition of wastewater to be treated has to be considered in this aspect because effluents from industries like distillery and pulp and paper contain sufficient concentration of carbon whereas those from dyeing and selected chemical plants is characterized by low organic carbon concentration which has to be supplied externally [89].

It has been observed that acetate is a relatively poor electron donor whereas ethanol, glucose and sucrose are used extensively for better dye uptake [113, 156]. In case of fungal dye degradation, glucose, maltose, cellobiose and starch is observed to show better results in comparison to sucrose or lactose [54]. A combination of glucose and lactose when used as co-substrate was able to degrade about 98% of Acid Red 18 even in high concentration up to 1000 mg/L [85, 86]. Sodium benzoate and sodium acetate also yielded satisfactory removal results for Reactive Black (RB5) removal in biofilm reactor which resulted in 94% COD removal and 99% dye removal [114]. Endogenous metabolism also may serve as a source of organic carbon in case of necessities for dye reduction [81, 115]. A study conducted by Ong et al. [115] showed a sharp decrease of MLSS from 7200 mg/L to 4700 mg/L which supplied the organic carbon required for azo bond cleavage thus making complete mineralization of Acid Orange 7 even without external carbon. Similar conclusion could be drawn in case of fungal biomass where dye degradation improved from 5 mg/L/h to 8 mg/L/h with addition of glucose at a concentration of 1000 mg/L for 50 mg/L dye. In suspended biomass, absence of glucose restricted any decolourization whatsoever [145].

5.4 *Effect of HRT*

Adequate HRTs when provided in the reactor results in effective degradation of a wide range of dyes [16]. Shorter HRTs correspond to the utilization of more dyes as co-substrate along with a primary substrate due to higher biomass activity at high influent dye concentration although total dye removal efficiencies were found to be higher in longer HRTs [141]. Longer retention times in the anaerobic phase is observed to have positive influence on efficiency of dye degradation [84]. Short HRT system requires longer time for biofilm growth in the carriers [68]. In comparison to suspended growth reactors, MBBR can perform similar removal in half of the HRT required for activated sludge process [170]. In the anaerobic MBBR, the recalcitrant compounds formed throughout the process are not completely hydrolyzed at shorter HRTs [61]. However, a much higher HRT in anaerobic reactor decreases acidification efficiency due to the utilization of solubilized volatile fatty acids (VFA) by methanogens [175]. Decrease in HRT from 48 to 12 h resulted in a drop of COD reduction from 78.3 to 54% although the sensitivity of this reduction in HRT is hypothesized to depend on the type of biocarriers [139].

Analysing the effect of HRT on removal, it was observed that increasing flow rates from 25 to 100 mL/h decreased HRT which in turn reduced removal efficiencies from 95.7 to 72.9% [148]. A decrease in HRT from 33 h to 16.8 h resulted in decrease in both COD and colour removal efficiency even though carrier filling ratio was increased from 5 to 10% [113] which is also established in dye degradation studies in other reactors [92]. Optimum HRTs also provide maximum BOD: COD ratio that facilitates optimum aerobic degradation thereby directly effecting removal efficiency [62]. Pratiwi et al. [129] studied the effect of different HRTs varying between 6 and 48 h in the aerobic MBBR and observed that optimum removal efficiencies were obtained at 24 h with maximum removal of both COD and colour. Colour removal efficiency increased linearly from 75.74 to 81.21% from 6 to 24 h HRT and then reduced to 68.6% at 48 h. However, in case of COD removal where efficiency increased from 37.55 to 85.86% linearly with increase in HRT from 6 to 48 h.

5.5 *Effect of Biomass Concentration*

In case of MBBR, a high removal efficiency can be achieved with a comparatively less biomass concentration thus limiting the cost of waste sludge handling. a conventional suspended growth reactor system usually employs an MLSS concentration of around 3000–4000 mg/L which in case of MBBR systems can be reduced to as low as 570 mg/L [80, 144]. For this reason, while it has been claimed that lower biomass concentration has relative lower decolourization rate due to toxicity, MBBR shows satisfactory removal efficiencies [98, 159]. Since, the majority of biomass concentration is attributed to biofilms attached to media, volumetric filling of biocarriers directly affects the removal efficiency of the process. It is observed that there is an optimum biocarrier filling percentage, below which concentration of active biomass in the reactor could be increased to enhance the removal efficiency. However, increasing carrier filling in the reactor beyond optimum ratio results in decrease in mobility of the carriers thus limiting proper mixing and efficient DO and substrate diffusion in biofilms [18, 148]. The adjustable biocarrier filling ratio is one of the most attracting features of MBBR that helps in keeping desired biomass in the system [146]. Considering the effect of biocarriers in between wide range of filling ratio over 40–80%, the optimum carrier filling ratio was determined to be exact 67.07% which removed maximum COD from the wastewater [53]. Using PU-AC carriers for anaerobic and aerobic reactors for dye wastewater treatment, the optimum filling ratios were identified as 30% and 20%, respectively [121]. Using polyurethane foam-polypropylene carriers (PUF-PP), it was observed that beyond a carrier filling ratio of 35%, almost negligible positive effect on dye removal was noticed [148]. However, in the study of Ong et al. [113], increasing biocarrier filling ratio from 5 to 10%, there was a significant increase in decolourization efficiency. Support media is observed to increase biomass concentration even up to three times as that in suspended biomass systems which might be due to both the increase in active biomass concentration as well as limiting inhibitory effect of toxic substances on biomass [80]. Similar to bacterial biomass, fungi attached biocarrier filling have similar effect on dye degradation. Increasing filling ratio of Polyurethane-dyeing sludge carbonaceous material (PU-DSCM) foam from 10 to 20% had sharp increase in dye removal efficiency, however the same in case of increasing the ratio to 30% had no notable impact on removal [122].

5.6 *Effect of Biocarrier*

Addition of biocarriers in the reactor ensures the presence of active biomass thereby increasing removal efficiencies in terms of both dye concentration and COD in comparatively short retention times and high influent dye concentration [65]. Polyurethane foam is often used as support materials in dye degradation due to the inward matrices that serve as sites of anaerobic environment and help in the retention of microorganisms [121, 144, 160] and the pores aid in the formation of stable biofilms [117]. Using simple hexahedral carrier, it was observed that biomedia increased COD and colour removal up to 38.2% and 27.4%, respectively. Addition of biocarriers also resulted in higher SRT, almost 3 times as that activated sludge process effects in dye degradation in MBBR [80]. *Shewanella indica* when immobilized on biocarriers showed high tolerance for Reactive Black 5 (RB5) with stability of 90% on an average [174].

Alginate often proves to be a good immobilization media which was studied by a number of researchers. Using *Orchis mascula* as organic biocarriers for immobilization of mixed cells, almost 100% decolourization was obtained for different Reactive azo dyes including Red (RR2), Blue (RB4), and Yellow (RY15) in anaerobic environment [65]. It took a retention time of 40 h to completely degrade 40 mg/L of RR2 and RB4. Other Reactive dyes like Red (RR195), Orange (RO72), Yellow (RY17), and Blue (RB36) were effectively degraded using immobilized *Pseudomonas putida* and *Bacillus licheniformis* immobilized in polyacrylamide and sodium alginate [150]. Sodium alginate along with PVA was used to entrap mixed activated sludge which completely degraded Reactive Blue with the concentration varying between 10 and 40 mg/L. 87–88% COD removal was observed in each case. Cheng et al. [33] immobilized *Burkholderia vietnamiensis* on alginate-PVA-kaolin gel beads for efficient removal of crystal violet. Both starch and sodium alginate entrapped biomass showed similar results towards Reactive Red (RR2) azo dye degradation where complete decolourization was achieved in anaerobic phase [63]. Using polyethylene glycol media to entrap biological cells, 50% of anaerobically formed recalcitrant intermediates were degraded in aerobic environment at an HRT of 2–8 h [14]. Using polyurethane foam to immobilize *Bacillus subtilis* in a novel multi-staged fluidized bed reactor, about 92% of COD was reduced from real textile wastewater containing Congo Red dye with an initial concentration of 100 mg/L, PUF weight 5 g and pH of 8 [142].

In the light of fungal azo dye degradation, a series of cost-effective biological materials were experimented as biocarriers for growing *Trametes versicolor* and it was observed that jute showed best result with respect to biomass growth, decolourization, and material integrity. Straw however showed better decolourizing property but was easily disintegrated within 2 weeks. Toxicity of amaranth dye towards the fungus was less when immobilized in any carrier which showed the advantage of using attached biomass [145]. *Phanerochaete chrysosporium* or white rot fungus was immobilized on Polyurethane-dyeing sludge carbonaceous material (PU-DSCM) foam carriers which showed impressive removal efficiency [122] (Table 2).

5.7 *Effect of pH*

pH of wastewater is a basic factor that is responsible for the transport of dyes through cell membranes of microorganisms thereby facilitating degradation [32]. Raw dyeing wastewater is usually characterized by high pH ranging around 12–13. A very few bacterial species are capable of growing at such a high alkaline condition. For biological treatments, pH should generally be kept close to neutral, however, azo dye degradation is reported to take place within a pH range of 7–10 [12, 127]. Experimental studies showed that at pH 7, COD removal efficiency was higher than that obtained at pH 12 [121]. Optimum pH for the removal of Congo Red dye was also obtained at pH 7 [148]. Francis and Sosamony [53] conducted a study to observe the effect of pH over a range of 6–9. The results showed that maximum removal efficiency was obtained in between 7 and 7.5 because the concerned microorganism showed maximum growth within that pH range. Since biological azo bond cleavage is enzyme mediated and enzyme ionization will be directly affected by pH change [56]. Setty [142] observed that slight alkaline conditions favoured degradation of azo dyes [93] and increasing pH from 7 to 8 resulted in 5% increment in dye removal efficiency. Moreover, anaerobic cleavage of azo bonds results in basic aromatic amines which increases the pH of the effluent if sufficient buffering action of other chemicals is not present [56].

This, however, is not the case for fungal dye degradation. Fungi often grow in acidic environment and have optimum dye degradation around a pH of 4.5 [75]. Raw dye effluent is often characterized by a high pH at which enzymatic activity of fungi is absent. A comparison between fungal activity towards dye degradation at pH of 7 and 12 showed removal at both pH. It was stated that removal at pH 7 was due to enzymatic degradation whereas that at pH 12 was due to biosorption [122].

6 Research Gaps and Scope of Future Work

A number of the experimental works and kinetics are developed using simulated dye wastewater considering the presence and concentration of that dyestuff only. However, real dye effluents contain salts, sulphur compounds, several nutrients, heavy metals including zinc, chromium, copper, arsenic besides organic nutrients

Table 2 Characteristics of different biocarriers and their effects in dye degradation in MBBR

Carrier type	Dimensional specifications	Physical properties	Carrier filling ratio	COD removal efficiency[a]	Colour removal efficiency[a]	References
Polyurethane media	2 × 2 × 2 cm with 1 cm hole through the cubes	Specific gravity: 1.03 Pores/cm: 30–40	15% (v/v)	92%	–	Park and Lee [123]
Regular hexahedron biomedia	1.3 × 1.3 × 1.3 cm	Density: 0.21 g/cm^3 ESA: 8.51 m^2/g	10% (v/v)	68.8	54.5	Kim et al. [80]
Polyurethane activated carbon (PU-AC)	0.5 × 0.5 × 0.5 cm	Porosity: 0.82 Density: 1.064 g/cm^3 Surface area: 7.4 m^3/kg	20% (v/v)	55.3%	56%	Shin et al. [144]
Polyurethane activated carbon (PU-AC)	1 × 1 × 1 cm	Density: 1.064 g/cm^3 ESA: 59.7 m^2/g	20% (v/v)	86%	50%	Park et al. [121]
Polyethylene carriers	1.5 × 1.5 × 1.5 cm	Density: 0.95 g/cm^3 SSA: 415 m^2/m^3 No. of carriers: 900	50% (v/v)	71.5%	98%	Koupaie et al. [8586]
Polyurethane-dyeing sludge carbonaceous material (PU-DSCM) foam	Height: 18 cm Diameter: 11 cm	Density: 0.85–0.93 g/cm^3 ESA: 55.7 to 58.3 m^2/g	10–30% (v/v)	79%	54%	Park et al. [122]
Polypropylene cylindrical carriers	–	Density: 0.95–0.98 g/cm^3	37% (v/v)	85%	90%	Dong et al. [44]
LECA (special clay granules)	Grain size: 4–10 mm	SSA: 525 m^2/m^3	50% (v/v)	79%	87%	Ahmadi et al. [4]
Anox Kaldnes polypropylene K1 carriers	Height: 7.2 mm Diameter: 9.1 mm	Density: 150 kg/m^3 SSA: 500 m^2/m^3	–	70.1%	–	Li et al. [95]

(continued)

Table 2 (continued)

Carrier type	Dimensional specifications	Physical properties	Carrier filling ratio	COD removal efficiency[a]	Colour removal efficiency[a]	References
Poly-vinyl chloride (PVC) corrugated cylinders	Height: 14 mm Diameter: 15 mm	–	40–80% (v/v)	86%	–	Francis and Sosamony [53]
Cylindrical polyethylene carriers	Height: 10 mm Diameter: 10 mm	Density: 0.98 g/cm^3	60% (v/v)	68.9%	76.8%	Gong [62]
Anox Kaldnes K1 carriers	Height: 7.2 mm Diameter: 9.1 mm	Density: 150 kg/m^3 SSA: 500 m^2/m^3	40% (v/v)	94%	–	Castro et al. [28]
Anox Kaldnes K1 carriers	Height: 7.2 mm Diameter: 9.1 mm	Density: 150 kg/m^3 SSA: 500 m^2/m^3	20% (v/v)	89.13%	96.95%	Pratiwi et al. [129]
Anox Kaldnes K1 carriers	Height: 7.2 mm Diameter: 9.1 mm	Density: 150 kg/m^3 SSA: 500 m^2/m^3	50% (v/v)	78.3%	85%	Santos Pereira et al. [138]
Polypropylene carriers	–	–	67.07%	87.22%	72.14%	Vaidhegi et al. [155]
Anox Kaldnes K1 carriers	Height: 7.2 mm Diameter: 9.1 mm	Density: 150 kg/m^3 SSA: 500 m^2/m^3	40% (v/v)	91%	–	Dias et al. [43]
Polyurethane foam (PU)	1 cm x 1 cm x 1 cm	–	30% (v/v)	70.5%	78%	Azimi et al. [13]
Anox Kaldnes K1 carriers	Height: 7.2 mm Diameter: 9.1 mm	Density: 150 kg/m^3 SSA: 500 m^2/m^3	40% (v/v)	94%	–	Dias et al. [42]
Hollow cylinders	Height: 5 mm Diameter: 16.75 mm	Density: 170 kg/m^3 SSA: 1036 m^2/m^3	20% (v/v)	89.2%	98.5%	Erkan et al. [48]
			10% (v/v)	88.9%	98.5%	
Plastic carriers	–	–	5–10% (v/v)	99.19%	100%	Ong et al. [113]
Polyurethane foam-polypropylene (PUF-PP) carriers	–	Average wt: 1.19 ± 0.03 g	45% (v/v)	–	95.7%	Sonwani et al. [148]

(continued)

Table 2 (continued)

Carrier type	Dimensional specifications	Physical properties	Carrier filling ratio	COD removal efficiency[a]	Colour removal efficiency[a]	References
Plastic Biofill C-2 carriers	Diameter: 25 mm	Free volume: 90% Density: <1 kg/m^3 SSA: 590 m^2/m^3	30% (v/v)	82%	61%	Wang et al. [161]

* Efficiency[a]: maximum removal efficiency in one single reactor
SSA: Specific surface area
ESA: Effective surface area

which might affect biological process of dye degradation [165]. This limitation will result in erroneous development of process kinetics and might not always show similar degradation when applied in industrial cases [56]. In case of photocatalytic oxidation, experiments have been conducted using TiO_2, but recently, ZnO has started to take its place due to lower cost and higher removal efficiencies for azo dye degradation [7, 38, 45]. These have been used with activated sludge processes but not investigated in MBBR. Focussing on the advantages of the later over suspended process, it can be hypothesized that more efficient removal can be obtained with lower HRT and may be for higher input concentrations.

From the discussion it is clear that different dyes have variable structural complexities and thus optimum removal cannot be theoretically predicted. More thorough experiments are needed to be done in that case. Moreover, since bacterial enzymes are substrate specific, the strains used in other biological reactors may be immobilized or grown on carriers for better biodecolourization. Developing successful process design requires optimization and quantification of parameters through the development of kinetics. A very few works have been undertaken so far to estimate the rate of reactions taking place in MBBR for dyestuff removal. One of them investigated the various models that fit the experimental data obtained from dye effluent in MBBR and inferred the use of Grau second order and modified Stover-Kincannon kinetics justified the results of COD removal. However, colour removal kinetics was not considered.

7 Conclusion

MBBR is one of the advanced water treatment technologies that are applied in treating municipal as well as a variety of industrial wastewater. Keeping in view of the different results obtained from the studies undertaken in MBBR, it can be established that MBBR as a sole treatment technology may require more than one unit to efficiently biodecolourize as well as remove COD and nutrients. Employing multi-staged reactor is often more welcome. Since the BOD: COD ratio of these effluents is quite low, pretreatments like ozonation processes improve biodegradability. Coagulation and adsorption highly increases both colour and COD removal. The biofilm attached to carriers improves resistant against recalcitrant compounds thereby decreasing the effect of substrate inhibition. Determination of influencing parameters as well as optimizing them to get efficient as well as economical removal of dyes is essential for process design which requires determination of kinetics as well as models to estimate the quality parameters of effluent to be safe for disposal or reuse.

References

1. Abiri F, Fallah N, Bonakdarpour B (2017) Sequential anaerobic–aerobic biological treatment of colored wastewaters: case study of a textile dyeing factory wastewater. Water Sci Technol 75:1261–1269. https://doi.org/10.2166/wst.2016.531
2. Acuner E, Dilek FB (2004) Treatment of tectilon yellow 2G by Chlorella vulgaris. Process Biochem 39:623–631. https://doi.org/10.1016/S0032-9592(03)00138-9
3. Afroze S, Sen T, Ang M (2015) Agricultural solid wastes in aqueous phase dye adsorption: a review. Agric Wastes Charact Types Manag 169–213
4. Ahmadi M, Amiri P, Amiri N (2015) Combination of TiO_2-photocatalytic process and biological oxidation for the treatment of textile wastewater. Korean J Chem Eng 32:1327–1332. https://doi.org/10.1007/s11814-014-0345-3
5. Al-Alwani MA, Ludin NA, Mohamad AB, Kadhum AAH, Mukhlus A (2018) Application of dyes extracted from Alternanthera dentata leaves and Musa acuminata bracts as natural sensitizers for dye-sensitized solar cells. Spectrochim Acta Part A Mol Biomol Spectrosc 192:487–498. https://doi.org/10.1016/j.saa.2017.11.018
6. Al-Amrani WA, Lim PE, Seng CE, Ngah WSW (2014) Factors affecting bio-decolorization of azo dyes and COD removal in anoxic–aerobic REACT operated sequencing batch reactor. J Taiwan Inst Chem Eng 45:609–616. https://doi.org/10.1016/j.jtice.2013.06.032
7. Almeida Guerra WN, Teixeira Santos JM, Raddi de Araujo LR (2012) Decolorization and mineralization of reactive dyes by a photocatalytic process using ZnO and UV radiation. Water Sci Technol 66:158–164. https://doi.org/10.2166/wst.2012.154
8. Amaral FM, Kato MT, Florêncio L, Gavazza S (2014) Color, organic matter and sulfate removal from textile effluents by anaerobic and aerobic processes. Biores Technol 163:364–369. https://doi.org/10.1016/j.biortech.2014.04.026
9. Andreozzi R, Caprio V, Insola A, Marotta R (1999) Advanced oxidation processes (AOP) for water purification and recovery. Catal Today 53:51–59. https://doi.org/10.1016/S0920-5861(99)00102-9
10. Angelova B, Avramova T, Stefanova L, Mutafov S (2008) Temperature effect on bacterial azo bond reduction kinetics: an Arrhenius plot analysis. Biodegradation 19:387–393. https://doi.org/10.1007/s10532-007-9144-4
11. Ansari FA, Jafri H, Ahmad I, Abulreesh HH (2017) Factors affecting biofilm formation in in vitro and in the rhizosphere. Biofilms Plant Soil Health 15:275–290. https://doi.org/10.1002/9781119246329.ch15
12. Asad S, Amoozegar MA, Pourbabaee A, Sarbolouki MN, Dastgheib SMM (2007) Decolorization of textile azo dyes by newly isolated halophilic and halotolerant bacteria. Bioresour Technol 98:2082–2088. https://doi.org/10.1016/j.biortech.2006.08.020
13. Azimi B, Abdollahzadeh-Sharghi E, Bonakdarpour B (2020) Anaerobic-aerobic processes for the treatment of textile dyeing wastewater containing three commercial reactive azo dyes: Effect of number of stages and bioreactor type. Chin J Chem Eng. https://doi.org/10.1016/j.cjche.2020.10.006. (In Press)
14. Bae W, Han D, Cui F, Kim M (2014) Microbial evaluation for biodegradability of recalcitrant organic in textile wastewater using an immobilized-cell activated sludge process. KSCE J Civ Eng 18:964–970. https://doi.org/10.1007/s12205-014-2193-4
15. Banat IM, Nigam P, Singh D, Marchant R (1996) Microbial decolorization of textile-dye containing effluents: a review. Biores Technol 58:217–227. https://doi.org/10.1016/S0960-8524(96)00113-7
16. Banerjee A, Ghoshal AK (2017) Biodegradation of an actual petroleum wastewater in a packed bed reactor by an immobilized biomass of Bacillus cereus. J Environ Chem Eng 5:1696–1702. https://doi.org/10.1016/j.jece.2017.03.008
17. Barragan BE, Costa C, Marquez MC (2007) Biodegradation of azo dyes by bacteria inoculated on solid media. Dyes Pigments 75:73–81. https://doi.org/10.1016/j.dyepig.2006.05.014

18. Barwal A, Chaudhary R (2015) Impact of carrier filling ratio on oxygen uptake & transfer rate, volumetric oxygen transfer coefficient and energy saving potential in a lab-scale MBBR. J Water Process Eng 8:202–208. https://doi.org/10.1016/j.jwpe.2015.10.008
19. Benito GG, Miranda MP, De Los Santos DR (1997) Decolorization of wastewater from an alcoholic fermentation process with Trametes versicolor. Biores Technol 61:33–37. https://doi.org/10.1016/S0960-8524(97)84695-0
20. Beulker S, Jekel M (1993) Precipitation and coagulation of organic substances in bleachery effluents of pulp mills. Water Sci Technol 27:193–199
21. Bhatia D, Sharma NR, Singh J, Kanwar RS (2017) Biological methods for textile dye removal from wastewater: a review. Crit Rev Environ Sci Technol 47:1836–1876. https://doi.org/10.1080/10643389.2017.1393263
22. Bidhendi GN, Torabian A, Ehsani H, Razmkhah N (2007) Evaluation of industrial dyeing wastewater treatment with coagulants and polyelectrolyte as a coagulant aid. J Environ Health Sci Eng 4:29–36. https://doi.org/10.2166/wst.1993.0277
23. Borkar R, Gulhane M, Kotangale AJ (2013) Moving bed biofilm reactor: a new perspective in wastewater treatment. J Environ Sci Toxicol Food Technol 6:15–21
24. Brás R, Ferra MIA, Pinheiro HM, Gonçalves IC (2001) Batch tests for assessing decolourisation of azo dyes by methanogenic and mixed cultures. J Biotechnol 89:155–162. https://doi.org/10.1016/S0168-1656(01)00312-1
25. Brown MA, De Vito SC (1993) Predicting azo dye toxicity. Crit Rev Environ Sci Technol 23:249–324. https://doi.org/10.1080/10643389309388453
26. Buitrón G, Quezada M, Moreno G (2004) Aerobic degradation of the azo dye acid red 151 in a sequencing batch biofilter. Biores Technol 92:143–149. https://doi.org/10.1016/j.biortech.2003.09.001
27. Carliell CM, Barclay SJ, Buckley CA (1996) Treatment of exhausted reactive dyebath effluent using anaerobic digestion: laboratory and full-scale trials. Water S A 22:225–233. https://doi.org/10.10520/AJA03784738_1419
28. Castro FD, Bassin JP, Dezotti M (2017) Treatment of a simulated textile wastewater containing the reactive orange 16 azo dye by a combination of ozonation and moving-bed biofilm reactor: evaluating the performance, toxicity, and oxidation by-products. Environ Sci Pollut Res 24:6307–6316. https://doi.org/10.1007/s11356-016-7119-x
29. Chakrabarti T, Subrahmanyam PVR, Sundaresan BB (1988) Biodegradation of recalcitrant industrial wastes. Biotreat Syst II:171–234
30. Chang JS, Chou C, Lin YC, Lin PJ, Ho JY, Hu TL (2001) Kinetic characteristics of bacterial azo-dye decolorization by pseudomonas luteola. Water Res 35:2841–2850. https://doi.org/10.1016/S0043-1354(00)00581-9
31. Chen HL, Burns LD (2006) Environmental analysis of textile products. Cloth Text Res J 24:248–261. https://doi.org/10.1177/0887302X06293065
32. Chen Y, Feng L, Li H, Wang Y, Chen G, Zhang Q (2018) Biodegradation and detoxification of direct black G textile dye by a newly isolated thermophilic microflora. Biores Technol 250:650–657. https://doi.org/10.1016/j.biortech.2017.11.092
33. Cheng Y, Lin H, Chen Z, Megharaj M, Naidu R (2012) Biodegradation of crystal violet using Burkholderia vietnamiensis C09V immobilized on PVA–sodium alginate–kaolin gel beads. Ecotoxicol Environ Saf 83:108–114. https://doi.org/10.1016/j.ecoenv.2012.06.017
34. Chequer FMD, Dorta DJ, de Oliveira DP (2011) Azo dyes and their metabolites: does the discharge of the azo dye into water bodies represent human and ecological risks. Adv Treat TextE Effl 28–48. https://doi.org/10.5772/19872
35. Chung KT, Fulk GE, Egan M (1978) Reduction of azo dyes by intestinal anaerobes. Appl Environ Microbiol 35:558–562
36. Chung KT, Stevens SE, Cerniglia CE (1992) The reduction of azo dyes by the intestinal microflora. Crit Rev Microbiol 18:175–190. https://doi.org/10.3109/10408419209114557
37. Chung YC, Chen CY (2009) Degradation of azo dye reactive violet 5 by TiO_2 photocatalysis. Environ Chem Lett 7:347–352. https://doi.org/10.1007/s10311-008-0178-6

38. da Silva LS, Gonçalves MMM, Raddi de Araujo LR (2019) Combined photocatalytic and biological process for textile wastewater treatments. Water Environ Res 91:1490–1497. https://doi.org/10.1002/wer.1143
39. Das P, Banerjee P, Zaman A, Bhattacharya P (2016) Biodegradation of two Azo dyes using Dietzia sp. PD1: process optimization using response surface methodology and artificial neural network. Desalin Water Treat 57:7293–7301. https://doi.org/10.1080/19443994.2015.1013993
40. De Angelis DF, Rodrigues GS (1987) Azo dyes removal from industrial effluents using yeast biomass. Arquivos de Biologia e Tecnologia 30:301–309
41. De Gisi S, Lofrano G, Grassi M, Notarnicola M (2016) Characteristics and adsorption capacities of low-cost sorbents for wastewater treatment: a review. Sustain Mater Technol 9:10–40. https://doi.org/10.1016/j.susmat.2016.06.002
42. Dias NC, Alves TL, Azevedo DA, Bassin JP, Dezotti M (2020) Metabolization of by-products formed by ozonation of the azo dye Reactive Red 239 in moving-bed biofilm reactors in series. Braz J Chem Eng 37:495–504. https://doi.org/10.1007/s43153-020-00046-6
43. Dias NC, Bassin JP, Sant'Anna Jr GL, Dezotti M, (2019) Ozonation of the dye reactive red 239 and biodegradation of ozonation products in a moving-bed biofilm reactor: revealing reaction products and degradation pathways. Int Biodeterior Biodegrad 144:104742–104751. https://doi.org/10.1016/j.ibiod.2019.104742
44. Dong B, Chen H, Yang Y, He Q, Dai X (2014) Treatment of printing and dyeing wastewater using MBBR followed by membrane separation process. Desalin Water Treat 52:4562–4567. https://doi.org/10.1080/19443994.2013.803780
45. Donkadokula NY, Kola AK, Naz I, Devendra S (2020) A review on advanced physico-chemical and biological textile dye wastewater treatment techniques. Rev Environ Sci Biotechnol 19:543–560. https://doi.org/10.1007/s11157-020-09543-z
46. Dos Santos AB, Cervantes FJ, Van Lier JB (2007) Review paper on current technologies for decolourisation of textile wastewaters: perspectives for anaerobic biotechnology. Biores Technol 98:2369–2385. https://doi.org/10.1016/j.biortech.2006.11.013
47. Dubin P, Wright KL (1975) Reduction of azo food dyes in cultures of Proteus vulgaris. Xenobiotica 5:563–571. https://doi.org/10.3109/00498257509056126
48. Erkan HS, Çağlak A, Soysaloglu A, Takatas B, Engin GO (2020) Performance evaluation of conventional membrane bioreactor and moving bed membrane bioreactor for synthetic textile wastewater treatment. J Water Process Eng 38:101631–101638. https://doi.org/10.1016/j.jwpe.2020.101631
49. Farhadian M, Duchez D, Vachelard C, Larroche C (2008) Monoaromatics removal from polluted water through bioreactors—a review. Water Res 42:1325–1341. https://doi.org/10.1016/j.watres.2007.10.021
50. Field JA, Stams AJ, Kato M, Schraa G (1995) Enhanced biodegradation of aromatic pollutants in cocultures of anaerobic and aerobic bacterial consortia. Antonie Van Leeuwenhoek 67:47–77. https://doi.org/10.1007/BF00872195
51. Fitzgerald SW, Bishop PL (1995) Two stage anaerobic/aerobic treatment of sulfonated azo dyes. J Environ Sci Health Part A 30:1251–1276. https://doi.org/10.1080/10934529509376264
52. Forgacs E, Cserhati T, Oros G (2004) Removal of synthetic dyes from wastewaters: a review. Environ Int 30:953–971. https://doi.org/10.1016/j.envint.2004.02.001
53. Francis A, Sosamony KJ (2016) Treatment of pre-treated textile wastewater using moving bed bio-film reactor. Procedia Technol 24:248–255. https://doi.org/10.1016/j.protcy.2016.05.033
54. Fu Y, Viraraghavan T (2001) Fungal decolorization of dye wastewaters: a review. Biores Technol 79:251–262. https://doi.org/10.1016/S0960-8524(01)00028-1
55. García-Montaño J, Torrades F, García-Hortal JA, Domenech X, Peral J (2006) Combining photo-fenton process with aerobic sequencing batch reactor for commercial hetero-bireactive dye removal. Appl Catal B 67:86–92. https://doi.org/10.1016/j.apcatb.2006.04.007
56. Garg SK, Tripathi M (2017) Microbial strategies for discoloration and detoxification of azo dyes from textile effluents. Res J Microbiol 12:1–19. https://doi.org/10.3923/jm.2017.1.19

57. Ghaly AE, Ananthashankar R, Alhattab MVVR, Ramakrishnan VV (2014) Production, characterization and treatment of textile effluents: a critical review. J Chem Eng Process Technol 5:1–18. https://doi.org/10.4172/2157-7048.1000182
58. Glenn JK, Akileswaran L, Gold MH (1986) Mn (II) oxidation is the principal function of the extracellular Mn-peroxidase from Phanerochaete chrysosporium. Arch Biochem Biophys 251:688–696. https://doi.org/10.1016/0003-9861(86)90378-4
59. Goel M, Chovelon JM, Ferronato C, Bayard R, Sreekrishnan TR (2010) The remediation of wastewater containing 4-chlorophenol using integrated photocatalytic and biological treatment. J Photochem Photobiol, B 98:1–6. https://doi.org/10.1016/j.jphotobiol.2009.09.006
60. Golka K, Kopps S, Myslak ZW (2004) Carcinogenicity of azo colorants: influence of solubility and bioavailability. Toxicol Lett 151:203–210. https://doi.org/10.1016/j.toxlet.2003.11.016
61. Gong D, Qin G (2012) Treatment of oilfield wastewater using a microbial fuel cell integrated with an up-flow anaerobic sludge blanket reactor. Desalin Water Treat 49:272–280. https://doi.org/10.1080/19443994.2012.719336
62. Gong XB (2016) Advanced treatment of textile dyeing wastewater through the combination of moving bed biofilm reactors and ozonation. Sep Sci Technol 51:1589–1597. https://doi.org/10.1080/01496395.2016.1165703
63. Hameed BB, Ismail ZZ (2018) Decolorization, biodegradation and detoxification of reactive red azo dye using non-adapted immobilized mixed cells. Biochem Eng J 137:71–77. https://doi.org/10.1016/j.bej.2018.05.018
64. Hameed BB, Ismail ZZ (2019) Decolorization, biodegradation and detoxification of reactive blue azo dye using immobilized mixed cells. Journal of Engineering 25:53–66. https://doi.org/10.31026/j.eng.2019.06.05
65. Hameed BB, Ismail ZZ (2020) New application of Orchis mascula as a biocarrier for immobilization of mixed cells for biodegradation and detoxification of reactive azo dyes. Environ Sci Pollut Res 27:38732–38744. https://doi.org/10.1007/s11356-020-09984-7
66. Harrelkas F, Paulo A, Alves MM, El Khadir L, Zahraa O, Pons MN, van der Zee FP (2008) Photocatalytic and combined anaerobic-photocatalytic treatment of textile dyes. Chemosphere 72:1816–1822. https://doi.org/10.1016/j.chemosphere.2008.05.026
67. Holkar CR, Jadhav AJ, Pinjari DV, Mahamuni NM, Pandit AB (2016) A critical review on textile wastewater treatments: possible approaches. J Environ Manage 182:351–366. https://doi.org/10.1016/j.jenvman.2016.07.090
68. Hosseini SH, Borghei SM (2005) The treatment of phenolic wastewater using a moving bed bio-reactor. Process Biochem 40:1027–1031. https://doi.org/10.1016/j.procbio.2004.05.002
69. Hu Q, Zhou N, Rene ER, Wu D, Sun D, Qiu B (2019) Stimulation of anaerobic biofilm development in the presence of low concentrations of toxic aromatic pollutants. Biores Technol 281:26–30. https://doi.org/10.1016/j.biortech.2019.02.076
70. Huang CR, Lin YK, Shu HY (1994) Wastewater decolorization and TOC-reduction by sequential treatment. Am Dye Report 83:15–17. https://citeseerx.ist.psu.edu/viewdoc/download?doi=10.1.1.597.2227&rep=rep1&type=pdf
71. Huang J, Ling J, Kuang C, Chen J, Xu Y, Li Y (2018) Microbial biodegradation of aniline at low concentrations by Pigmentiphaga daeguensis isolated from textile dyeing sludge. Int Biodeterior Biodegrad 129:117–122. https://doi.org/10.1016/j.ibiod.2018.01.013
72. Jamee R, Siddique R (2019) Biodegradation of synthetic dyes of textile effluent by microorganisms: an environmentally and economically sustainable approach. Eur J Microbiol Immunol 9:114–118. https://doi.org/10.1556/1886.2019.00018
73. Jiang LL, Zhou JJ, Quan CS, Xiu ZL (2017) Advances in industrial microbiome based on microbial consortium for biorefinery. Bioresources and bioprocessing 4:1–10. https://doi.org/10.1186/s40643-017-0141-0
74. Kant R (2012) Textile dyeing industry an environmental hazard. Nat Sci 4:22–26. https://doi.org/10.4236/ns.2012.41004
75. Kapdan I, Kargi F, McMullan G, Marchant R (2000) Comparison of white-rot fungi cultures for decolorization of textile dyestuffs. Bioprocess Eng 22:347–351. https://doi.org/10.1007/s004490050742

76. Kapdan IK, Alparslan S (2005) Application of anaerobic–aerobic sequential treatment system to real textile wastewater for color and COD removal. Enzyme Microb Technol 36:273–279. https://doi.org/10.1016/j.enzmictec.2004.08.040
77. Katheresan V, Kansedo J, Lau SY (2018) Efficiency of various recent wastewater dye removal methods: a review. J Environ Chem Eng 6:4676–4697. https://doi.org/10.1016/j.jece.2018.06.060
78. Kaushik P, Malik A (2009) Fungal dye decolourization: recent advances and future potential. Environ Int 35:127–141. https://doi.org/10.1016/j.envint.2008.05.010
79. Khlifi R, Belbahri L, Woodward S, Ellouz M, Dhouib A, Sayadi S, Mechichi T (2010) Decolourization and detoxification of textile industry wastewater by the laccase-mediator system. J Hazard Mater 175:802–808. https://doi.org/10.1016/j.jhazmat.2009.10.079
80. Kim S, Park C, Kim TH, Lee J, Kim SW (2003) COD reduction and decolorization of textile effluent using a combined process. J Biosci Bioeng 95:102–105. https://doi.org/10.1016/S1389-1723(03)80156-1
81. Kim SY, An JY, Kim BW (2008) The effects of reductant and carbon source on the microbial decolorization of azo dyes in an anaerobic sludge process. Dyes Pigm 76:256–263. https://doi.org/10.1016/j.dyepig.2006.08.042
82. Kim YO, Nam HU, Park YR, Lee JH, Park TJ, Lee TH (2004) Fenton oxidation process control using oxidation-reduction potential measurement for pigment wastewater treatment. Korean J Chem Eng 21:801–805. https://doi.org/10.1007/BF02705523
83. Knapp JS, Newby PS, Reece LP (1995) Decolorization of dyes by wood-rotting basidiomycete fungi. Enzyme Microb Technol 17:664–668. https://doi.org/10.1016/0141-0229(94)00112-5
84. Koçyigit H, Ugurlu A (2015) Biological decolorization of reactive azo dye by anaerobic/aerobic-sequencing batch reactor system. Glob NEST J 17:210–219
85. Koupaie EH, Moghaddam MA, Hashemi SH (2011) Post-treatment of anaerobically degraded azo dye acid red 18 using aerobic moving bed biofilm process: enhanced removal of aromatic amines. J Hazard Mater 195:147–154. https://doi.org/10.1016/j.jhazmat.2011.08.017
86. Koupaie EH, Moghaddam MA, Hashemi SH (2012) Investigation of decolorization kinetics and biodegradation of azo dye acid red 18 using sequential process of anaerobic sequencing batch reactor/moving bed sequencing batch biofilm reactor. Int Biodeterior Biodegradation 71:43–49. https://doi.org/10.1016/j.ibiod.2012.04.002
87. Koupaie EH, Moghaddam MA, Hashemi SH (2013) Evaluation of integrated anaerobic/aerobic fixed-bed sequencing batch biofilm reactor for decolorization and biodegradation of azo dye acid red 18: comparison of using two types of packing media. Biores Technol 127:415–421. https://doi.org/10.1016/j.biortech.2012.10.003
88. Kudlich M, Hetheridge MJ, Knackmuss HJ, Stolz A (1999) Autoxidation reactions of different aromatic o-aminohydroxynaphthalenes that are formed during the anaerobic reduction of sulfonated azo dyes. Environ Sci Technol 33:896–901. https://doi.org/10.1021/es9808346
89. Kuhad RC, Sood N, Tripathi KK, Singh A, Ward OP (2004) Developments in microbial methods for the treatment of dye effluents. Adv Appl Microbiol 56:185–213. https://doi.org/10.1016/S0065-2164(04)56006-9
90. Kumar AN, Reddy CN, Mohan SV (2015) Biomineralization of azo dye bearing wastewater in periodic discontinuous batch reactor: effect of microaerophilic conditions on treatment efficiency. Biores Technol 188:56–64. https://doi.org/10.1016/j.biortech.2015.01.098
91. Kumar K, Devi SS, Krishnamurthi K, Gampawar S, Mishra N, Pandya GH, Chakrabarti T (2006) Decolorisation, biodegradation and detoxification of benzidine based azo dye. Biores Technol 97:407–413. https://doi.org/10.1016/j.biortech.2005.03.031
92. Kumar K, Singh GK, Dastidar MG, Sreekrishnan TR (2014) Effect of mixed liquor volatile suspended solids (MLVSS) and hydraulic retention time (HRT) on the performance of activated sludge process during the biotreatment of real textile wastewater. Water Resour Ind 5:1–8. https://doi.org/10.1016/j.wri.2014.01.001
93. Lade H, Govindwar S, Paul D (2015) Mineralization and detoxification of the carcinogenic azo dye Congo red and real textile effluent by a polyurethane foam immobilized microbial consortium in an upflow column bioreactor. Int J Environ Res Public Health 12:6894–6918. https://doi.org/10.3390/ijerph120606894

94. Ledakowicz S, Solecka M, Zylla R (2001) Biodegradation, decolourisation and detoxification of textile wastewater enhanced by advanced oxidation processes. J Biotechnol 89:175–184. https://doi.org/10.1016/S0168-1656(01)00296-6
95. Li C, Zhang Z, Li Y, Cao J (2015) Study on dyeing wastewater treatment at high temperature by MBBR and the thermotolerant mechanism based on its microbial analysis. Process Biochem 50:1934–1941. https://doi.org/10.1016/j.procbio.2015.08.007
96. Libra JA, Borchert M, Vigelahn L, Storm T (2004) Two stage biological treatment of a diazo reactive textile dye and the fate of the dye metabolites. Chemosphere 56:167–180. https://doi.org/10.1016/j.chemosphere.2004.02.012
97. Lotito AM, De Sanctis M, Di Iaconi C, Bergna G (2014) Textile wastewater treatment: aerobic granular sludge vs activated sludge systems. Water Res 54:337–346. https://doi.org/10.1016/j.watres.2014.01.055
98. Lourenco ND, Novais JM, Pinheiro HM (2001) Effect of some operational parameters on textile dye biodegradation in a sequential batch reactor. J Biotechnol 89:163–174. https://doi.org/10.1016/S0168-1656(01)00313-3
99. Lu K, Zhang XL, Zhao YL, Wu ZL (2010) Removal of color from textile dyeing wastewater by foam separation. J Hazard Mater 182:928–932. https://doi.org/10.1016/j.jhazmat.2010.06.024
100. Martin CN, Kennely JC (1981) Rat liver microsomonal azo reductase activity on four azo dyes derived from benzidine, 3,3'-Dimethylbenzidine or 3,3'-Demethoxybenzidine. Carcinogenisis 2:307–312. https://doi.org/10.1093/carcin/2.4.307
101. McMullan G, Meehan C, Conneely A, Kirby N, Robinson T, Nigam P, Banat I, Marchant R, Smyth WF (2001) Microbial decolourisation and degradation of textile dyes. Appl Microbiol Biotechnol 56:81–87. https://doi.org/10.1007/s002530000587
102. Mishra G, Tripathy M (1993) A critical review of the treatments for decolourization of textile effluent. Colourage 40:35–35
103. Mohan SV, Babu PS, Srikanth S (2013) Azo dye remediation in periodic discontinuous batch mode operation: evaluation of metabolic shifts of the biocatalyst under aerobic, anaerobic and anoxic conditions. Sep Purif Technol 118:196–208. https://doi.org/10.1016/j.seppur.2013.06.037
104. Muangthai I, Ratanatamsakul C, Lu MC (2010) Removal of 2, 4-dichlorophenol by fluidized-bed Fenton process. Sustain Environ Res 20:325–331
105. Muda K, Aris A, Salim MR, Ibrahim Z (2013) Sequential anaerobic-aerobic phase strategy using microbial granular sludge for textile wastewater treatment. Biomass Now Sustain Growth Use 231-264. https://doi.org/10.5772/54458
106. Nair VR, Kodialbail VS (2020) Floating bed reactor for visible light induced photocatalytic degradation of acid yellow 17 using polyaniline-TiO2 nanocomposites immobilized on polystyrene cubes. Environ Sci Pollut Res 27:1–13. https://doi.org/10.1007/s11356-020-07959-2
107. Nigam P, Banat IM, Singh D, Marchant R (1996) Microbial process for the decolorization of textile effluent containing azo, diazo and reactive dyes. Process Biochem 31:435–442. https://doi.org/10.1016/0032-9592(95)00085-2
108. Ødegaard H, Rusten B, Westrum T (1994) A new moving bed biofilm reactor-applications and results. Water Sci Technol 29:157–165. https://doi.org/10.2166/wst.1994.0757
109. Ogawa T, Shibata M, Yatome C, Idaka E (1988) Growth inhibition of Bacillus subtilis by basic dyes. Bull Environ Contam Toxicol 40:545–552. https://doi.org/10.1007/BF01688379
110. Oliveira DP, Carneiro PA, Sakagami MK, Zanoni MVB, Umbuzeiro GA (2007) Chemical characterization of a dye processing plant effluent—identification of the mutagenic components. Mutat Res/Genet Toxicol Environ Mutagen 626:135–142. https://doi.org/10.1016/j.mrgentox.2006.09.008
111. Ollikka P, Alhonmäki K, Leppänen VM, Glumoff T, Raijola T, Suominen I (1993) Decolorization of azo, triphenyl methane, heterocyclic, and polymeric dyes by lignin peroxidase isoenzymes from Phanerochaete chrysosporium. Appl Environ Microbiol 59:4010–4016. https://doi.org/10.1128/aem.59.12.4010-4016.1993

112. O'neill C, Lopez A, Esteves S, Hawkes FR, Hawkes DL, Wilcox S, (2000) Azo-dye degradation in an anaerobic-aerobic treatment system operating on simulated textile effluent. Appl Microbiol Biotechnol 53:249–254. https://doi.org/10.1007/s002530050016
113. Ong C, Lee K, Chang Y (2020) Biodegradation of mono azo dye-reactive orange 16 by acclimatizing biomass systems under an integrated anoxic-aerobic REACT sequencing batch moving bed biofilm reactor. Journal of Water Process Engineering 36:101268–101282. https://doi.org/10.1016/j.jwpe.2020.101268
114. Ong SA, Ho LN, Wong YS, Raman K (2012) Performance and kinetic study on bioremediation of diazo dye (reactive black 5) in wastewater using spent GAC–biofilm sequencing batch reactor. Water Air Soil Pollut 223:1615–1623. https://doi.org/10.1007/s11270-011-0969-4
115. Ong SA, Toorisaka E, Hirata M, Hano T (2008) Granular activated carbon-biofilm configured sequencing batch reactor treatment of CI acid orange 7. Dyes Pigm 76:142–146. https://doi.org/10.1016/j.dyepig.2006.08.024
116. Oros G, Cserhati T, Forgacs E (2001) Strength and selectivity of the fungicidal effect of diazobenzene dyes. Fresenius Environ Bull 10:319–322
117. Padmanaban VC, Geed SR, Achary A, Singh RS (2016) Kinetic studies on degradation of Reactive Red 120 dye in immobilized packed bed reactor by Bacillus cohnii RAPT1. Biores Technol 213:39–43. https://doi.org/10.1016/j.biortech.2016.02.126
118. Pallerla S, Chambers RP (1997) Characterization of a Ca-alginate-immobilized Trametes versicolor bioreactor for decolorization and AOX reduction of paper mill effluents. Biores Technol 60:1–8. https://doi.org/10.1016/S0960-8524(96)00171-X
119. Pan Y, Wang Y, Zhou A, Wang A, Wu Z, Lv L, Li X, Zhang K, Zhu T (2017) Removal of azo dye in an up-flow membrane-less bioelectrochemical system integrated with bio-contact oxidation reactor. Chem Eng J 326:454–461. https://doi.org/10.1016/j.cej.2017.05.146
120. Pandey A, Singh P, Iyengar L (2007) Bacterial decolorization and degradation of azo dyes. Int Biodeterior Biodegrad 59:73–84. https://doi.org/10.1016/j.ibiod.2006.08.006
121. Park HO, Oh S, Bade R, Shin WS (2010) Application of A2O moving-bed biofilm reactors for textile dyeing wastewater treatment. Korean J Chem Eng 27:893–899. https://doi.org/10.1007/s11814-010-0143-5
122. Park HO, Oh S, Bade R, Shin WS (2011) Application of fungal moving-bed biofilm reactors (MBBRs) and chemical coagulation for dyeing wastewater treatment. KSCE J Civ Eng 15:453–461. https://doi.org/10.1007/s12205-011-0997-z
123. Park YK, Lee CH (1996) Dyeing wastewater treatment by activated sludge process with a polyurethane fluidized bed biofilm. Water Sci Technol 34:193–200. https://doi.org/10.1016/0273-1223(96)00646-4
124. Park YK, Lee CH, Yoon TH, Jang IH (1994) Treatment of dyeing industrial complex wastewater by fluidized biofilm media process biological treatment following on coagulation process. J Korean Soc Environ Eng 16:43–50
125. Paździor K, Bilińska L, Ledakowicz S (2019) A review of the existing and emerging technologies in the combination of AOPs and biological processes in industrial textile wastewater treatment. Chem Eng J 376:120597. https://doi.org/10.1016/j.cej.2018.12.057
126. Platzek T, Lang C, Grohmann G, Gi US, Baltes W (1999) Formation of a carcinogenic aromatic amine from an azo dye by human skin bacteria in vitro. Hum Exp Toxicol 18:552–559. https://doi.org/10.1191/096032799678845061
127. Ponraj M, Gokila K, Zambare V (2011) Bacterial decolorization of textile dye-Orange 3R. Int J Adv Biotechnol Res 2:168–177
128. Pourbabaee AA, Malekzadeh F, Sarbolouki MN, Najafi F (2006) Aerobic decolorization and detoxification of a disperse dye in textile effluent by a new isolate of Bacillus sp. Biotechnol Bioeng 93:631–635. https://doi.org/10.1002/bit.20732
129. Pratiwi R, Notodarmojo S, Helmy Q (2018) Decolourization of remazol black-5 textile dyes using moving bed bio-film reactor. In IOP Conf Ser: Earth Environ Sci 106:012089–012095. https://doi.org/10.1088/1755-1315/106/1/012089
130. Punzi M, Nilsson F, Anbalagan A, Svensson BM, Jönsson K, Mattiasson B, Jonstrup M (2015) Combined anaerobic–ozonation process for treatment of textile wastewater: removal

of acute toxicity and mutagenicity. J Hazard Mater 292:52–60. https://doi.org/10.1016/j.jhazmat.2015.03.018
131. Puvaneswari N, Muthukrishnan J, Gunasekaran P (2006) Toxicity assessment and microbial degradation of azo dyes. Indian J Exp Biol 44:618–626
132. Rai HS, Bhattacharyya MS, Singh J, Bansal TK, Vats P, Banerjee UC (2005) Removal of dyes from the effluent of textile and dyestuff manufacturing industry: a review of emerging techniques with reference to biological treatment. Crit Rev Environ Sci Technol 35:219–238. https://doi.org/10.1080/10643380590917932
133. Raj DS, Prabha RJ, Leena R (1970) Analysis of bacterial degradation of azo dye Congo Red using HPLC. J Ind Pollut Control 28:57–62
134. Rauf MA, Ashraf SS (2012) Survey of recent trends in biochemically assisted degradation of dyes. Chem Eng J 209:520–530. https://doi.org/10.1016/j.cej.2012.08.015
135. Reddy CA (1995) The potential for white-rot fungi in the treatment of pollutants. Curr Opin Biotechnol 6:320–328. https://doi.org/10.1016/0958-1669(95)80054-9
136. Robinson T, McMullan G, Marchant R, Nigam P (2001) Remediation of dyes in textile effluent: a critical review on current treatment technologies with a proposed alternative. Biores Technol 77:247–255. https://doi.org/10.1016/S0960-8524(00)00080-8
137. Sani RK, Banerjee UC (1999) Decolorization of triphenylmethane dyes and textile and dyestuff effluent by Kurthia sp. Enzyme Microb Technol 24:433–437. https://doi.org/10.1016/S0141-0229(98)00159-8
138. Santos-Pereira GC, Corso CR, Forss J (2019) Evaluation of two different carriers in the biodegradation process of an azo dye. J Environ Health Sci Eng 17:633–643. https://doi.org/10.1007/s40201-019-00377-8
139. Saratale RG, Saratale GD, Chang JS, Govindwar SP (2011) Bacterial decolorization and degradation of azo dyes: a review. J Taiwan Inst Chem Eng 42:138–157. https://doi.org/10.1016/j.jtice.2010.06.006
140. Schliephake K, Mainwaring DE, Lonergan GT, Jones IK, Baker WL (2000) Transformation and degradation of the disazo dye Chicago Sky Blue by a purified laccase from Pycnoporus cinnabarinus. Enzyme Microb Technol 27:100–107. https://doi.org/10.1016/S0141-0229(00)00181-2
141. Seshadri S, Bishop PL, Agha AM (1994) Anaerobic/aerobic treatment of selected azo dyes in wastewater. Waste Manage 14:127–137. https://doi.org/10.1016/0956-053X(94)90005-1
142. Setty YP (2019) Multistage fluidized bed bioreactor for dye decolorization using immobilized polyurethane foam: a novel approach. Biochem Eng J 152:107368–107376. https://doi.org/10.1016/j.bej.2019.107368
143. Shah MP, Patel KA, Nair SS, Darji AM (2013) Microbial decolorization of Remazol Brilliant Orange 3R, Remazol black B & Remazol brilliant violet dyes in a sequential anaerobic–aerobic system. Int J Environ Bioremediation Biodegrad 1:6–13
144. Shin DH, Shin WS, Kim YH, Ho Han M, Choi SJ (2006) Application of a combined process of moving-bed biofilm reactor (MBBR) and chemical coagulation for dyeing wastewater treatment. Water Sci Technol 54:181–189. https://doi.org/10.2166/wst.2006.863
145. Shin M, Nguyen T, Ramsay J (2002) Evaluation of support materials for the surface immobilization and decoloration of amaranth by Trametes versicolor. Appl Microbiol Biotechnol 60:218–223. https://doi.org/10.1007/s00253-002-1088-3
146. Shore JL, M'Coy WS, Gunsch CK, Deshusses MA (2012) Application of a moving bed biofilm reactor for tertiary ammonia treatment in high temperature industrial wastewater. Biores Technol 112:51–60. https://doi.org/10.1016/j.biortech.2012.02.045
147. Solís M, Solís A, Pérez HI, Manjarrez N, Flores M (2012) Microbial decolouration of azo dyes: a review. Process Biochem 47:1723–1748. https://doi.org/10.1016/j.procbio.2012.08.014
148. Sonwani RK, Swain G, Giri BS, Singh RS, Rai BN (2020) Biodegradation of Congo red dye in a moving bed biofilm reactor: performance evaluation and kinetic modeling. Biores Technol 302:122811–122820. https://doi.org/10.1016/j.biortech.2020.122811
149. Spadaro JT, Gold MH, Renganathan V (1992) Degradation of azo dyes by the lignin-degrading fungus Phanerochaete chrysosporium. Appl Environ Microbiol 58:2397–2401

150. Suganya K, Revathi K (2016) Decolorization of reactive dyes by immobilized bacterial cells from textile effluents. Int J Curr Microbiol Appl Sci 5:528–532
151. Talha MA, Goswami M, Giri BS, Sharma A, Rai BN, Singh RS (2018) Bioremediation of Congo red dye in immobilized batch and continuous packed bed bioreactor by Brevibacillus parabrevis using coconut shell bio-char. Biores Technol 252:37–43. https://doi.org/10.1016/j.biortech.2017.12.081
152. Tan L, Li H, Ning S, Xu B (2014) Aerobic decolorization and degradation of azo dyes by suspended growing cells and immobilized cells of a newly isolated yeast Magnusiomyces ingens LH-F1. Biores Technol 158:321–328. https://doi.org/10.1016/j.biortech.2014.02.063
153. Tchobanoglus G, Burton F, Stensel HD (2003) Wastewater engineering: treatment and reuse. Am Water Work Assoc J 95:201
154. Turhan K, Ozturkcan SA (2013) Decolorization and degradation of reactive dye in aqueous solution by ozonation in a semi-batch bubble column reactor. Water Air Soil Pollut 224:1–13. https://doi.org/10.1007/s11270-012-1353-8
155. Vaidhegi K, Selvam SA, Kumar AM (2018) Treatment of dye waste water using moving bed biofilm reactor & granular activated carbon [MBBR-GAC]. J Adv Res Dyn Control Syst 10:979–985
156. Van Der Zee FP, Bouwman RH, Strik DP, Lettinga G, Field JA (2001) Application of redox mediators to accelerate the transformation of reactive azo dyes in anaerobic bioreactors. Biotechnol Bioeng 75:691–701. https://doi.org/10.1002/bit.10073
157. Van der Zee FP, Cervantes FJ (2009) Impact and application of electron shuttles on the redox (bio) transformation of contaminants: a review. Biotechnol Adv 277:256–277. https://doi.org/10.1016/j.biotechadv.2009.01.004
158. Van der Zee FP, Lettinga G, Field JA (2001) Azo dye decolourisation by anaerobic granular sludge. Chemosphere 44:1169–1176. https://doi.org/10.1016/S0045-6535(00)00270-8
159. Van der Zee FP, Villaverde S (2005) Combined anaerobic–aerobic treatment of azo dyes- a short review of bioreactor studies. Water Res 39:1425–1440. https://doi.org/10.1016/j.watres.2005.03.007
160. Varesche MB, Zaiat M, Vieira LGT, Vazoller RF, Foresti E (1997) Microbial colonization of polyurethane foam matrices in horizontal-flow anaerobic immobilized-sludge reactor. Appl Microbiol Biotechnol 48:534–538. https://doi.org/10.1007/s002530051092
161. Wang Z, Xue M, Huang K, Liu Z (2011) Textile dyeing wastewater treatment. Adv Treat TextE Effl 5:91–116. https://doi.org/10.5772/22670
162. Weber EJ, Lee Wolfe N (1987) Kinetic studies of the reduction of aromatic azo compounds in anaerobic sediment/water systems. Environ Toxicol Chem Int J 6:911–919. https://doi.org/10.1002/etc.5620061202
163. Wijetunga S, Li X-F, Jian C (2010) Effect of organic load on decolourization of textile wastewater containing acid dyes in upflow anaerobic sludge blanket reactor. J Hazard Mater 177:792–798. https://doi.org/10.1016/j.jhazmat.2009.12.103
164. Wong Y, Yu J (1999) Laccase-catalyzed decolorization of synthetic dyes. Water Res 33:3512–3520. https://doi.org/10.1016/S0043-1354(99)00066-4
165. Wu H, Wang S, Kong H, Liu T, Xia M (2007) Performance of combined process of anoxic baffled reactor-biological contact oxidation treating printing and dyeing wastewater. Biores Technol 98:1501–1504. https://doi.org/10.1016/j.biortech.2006.05.037
166. Wuhrmann K, Mechsner KL, Kappeler TH (1980) Investigation on rate—determining factors in the microbial reduction of azo dyes. Eur J Appl Microbiol Biotechnol 9:325–338. https://doi.org/10.1007/BF00508109
167. Xu D, Cheng F, Zhang Y, Song Z (2014) Degradation of methyl orange in aqueous solution by microwave irradiation in the presence of granular-activated carbon. Water Air Soil Pollut 225:1–7. https://doi.org/10.1007/s11270-014-1983-0
168. Yadav M, Srivastva N, Singh RS, Upadhyay SN, Dubey SK (2014) Biodegradation of chlorpyrifos by Pseudomonas sp. in a continuous packed bed bioreactor. Biores Technol 165:265–269. https://doi.org/10.1016/j.biortech.2014.01.098

169. Yang X, Crespi Rosell M, López Grimau V (2018) A review on the present situation of wastewater treatment in textile industry with membrane bioreactor and moving bed biofilm reactor. Desalin Water Treat 103:315–322. https://doi.org/10.5004/dwt.2018.21962
170. Yang X, López-Grimau V, Vilaseca M, Crespi M (2020) Treatment of textile wastewater by CAS, MBR, and MBBR: a comparative study from technical, economic, and environmental perspectives. Water 12:1306–1322. https://doi.org/10.3390/w12051306
171. Young L, Yu J (1997) Ligninase-catalysed decolorization of synthetic dyes. Water Res 31:1187–1193. https://doi.org/10.1016/S0043-1354(96)00380-6
172. Zaoyan Y, Ke S, Guangliang S, Fan Y, Jinshan D, Huanian M (1992) Anaerobic–aerobic treatment of a dye wastewater by combination of RBC with activated sludge. Water Sci Technol 26:2093–2096. https://doi.org/10.2166/wst.1992.0669
173. Zhang W, Tang B, Bin L (2017) Research progress in biofilm-membrane bioreactor: a critical review. Ind Eng Chem Res 56:6900–6909. https://doi.org/10.1021/acs.iecr.7b00794
174. Zhang X, Song H, Chen Y, Zhuang M, Liu W (2021) Microbially mediated cleavage of reactive black 5 in attached growth bioreactors with immobilized Shewanella indica strain on different carriers. Int Biodeterior Biodegrad 157:105142–105152. https://doi.org/10.1016/j.ibiod.2020.105142
175. Zhang Y, Gao B, Lu L, Yue Q, Wang Q, Jia Y (2010) Treatment of produced water from polymer flooding in oil production by the combined method of hydrolysis acidification-dynamic membrane bioreactor–coagulation process. J Petrol Sci Eng 74:14–19. https://doi.org/10.1016/j.petrol.2010.08.001
176. Zimmermann T, Kulla HG, Leisinger T (1982) Properties of purified Orange II azoreductase, the enzyme initiating azo dye degradation by Pseudomonas KF46. Eur J Biochem 129:197–203. https://doi.org/10.1111/j.1432-1033.1982.tb07040.x
177. Zissi U, Lyberatos G (1996) Azo-dye biodegradation under anoxic conditions. Water Sci Technol 34:495–500. https://doi.org/10.1016/0273-1223(96)00684-1

Up-Flow Anaerobic Sludge Blanket Reactors in Dye Removal: Mechanisms, Influence Factors, and Performance

Ronei de Almeida and Claudinei de Souza Guimarães

Abstract Society faces eco-environmental challenges when it comes to managing industrial wastewaters. In particular, textile effluents are one of the main threats to living being due to toxic dyes. Technologies used to remove dye compounds include physicochemical and membrane filtration processes. However, since current technologies have several limitations, including high investment and energy demand, researchers have investigated cheaper and eco-friendly alternatives. Among them, anaerobic processes have been an effective method to decolourise dye-containing effluents. In that direction, the up-flow anaerobic sludge blanket (UASB) technology stands out in terms of high cost-effectiveness. The authors reviewed the published literature on UASB reactors in dye compounds removal. Mechanisms, merits, demerits, and technical aspects of UASB reactors are introduced. Challenges and opportunities are discussed. The major points are (1) mechanisms of dye removal in UASB reactors comprise mainly dye adsorption onto sludge granules and azo bond cleavage (biodegradation), (2) dye structure and concentration, external organic carbon source, redox mediators, and bioreactor operating conditions play a key role in the treatment performance, and (3) UASB technology exhibits high decolourisation rates. Removal efficiencies of chemical oxygen demand and colour lie within the range of 60–85% and 75–96%, respectively. However, anaerobic treatments may not be able to mineralise by-products of anaerobic metabolisms. Consequently, post-treatment of the anaerobically treated effluent is required, and 4) the energy production during the decolourisation process in UASB reactors is estimated at 22 kWh per m^3 of treated wastewater. Bio-energy recovery can promote wastewater valorisation and decrease the economic burdens of dye-containing effluents treatment. Future

R. de Almeida (✉)
School of Chemistry, Federal University of Rio de Janeiro, Cidade Universitária, Av. Athos da Silveira Ramos, 149, Bl. I, Laboratory I-124, Ilha do Fundão, Rio de Janeiro 21941-909, Brazil
e-mail: ronei@eq.ufrj.br

C. de Souza Guimarães
School of Chemistry, Biochemical Engineering Department, Federal University of Rio de Janeiro, Cidade Universitária, Av. Athos da Silveira Ramos, 149, Bl. E, room E-203, Ilha do Fundão, Rio de Janeiro 21941-909, Brazil
e-mail: claudinei@eq.ufrj.br

A. Khadir et al. (eds.), *Biological Approaches in Dye-Containing Wastewater*, Sustainable Textiles: Production, Processing, Manufacturing & Chemistry,
https://doi.org/10.1007/978-981-19-0545-2_7

studies should focus on optimising influence parameters of full-scale UASB reactors and biogas recovery from dye-containing wastewater treatment.

Keywords Anaerobic process · Biodegradation · Bio-energy · Decolourisation · Dye · Granular sludge · Industrial wastewater · Methanogens · Resource recovery · UASB

1 Introduction

The ecological and social impacts caused by dye compounds from dyeing, pharmaceutical, pesticides, cosmetics, and food industries have been among the most significant environmental sanitation challenges. In particular, textile industries used about 50% of the total dyes produced and consume a considerable quantity of water; hence, it is considered one of the largest activities responsible for aquatic pollution [86]. Due to the presence of chemicals and non-biodegradable substances, textile effluents have genotoxicity, mutagenicity, and carcinogenicity potentials, which brings attention to public health security and safety of terrestrial environments [37].

Several methods for decolourisation of textile wastewaters have been reported in the literature, such as coagulation-flocculation [95], chemical oxidation [1], adsorption [58], and membrane-based technologies [34]. However, physical–chemical processes are associated with high installation and operational costs, high chemical demands, and the generation of polluted sludge and membrane concentrates [78]. Furthermore, it should be noted that the change in toxicity during the treatment by chemical oxidation methods and the possible generation of by-products also represents a significant drawback [47], which has pushed dye industries to investigate cheaper and eco-friendly options for full-scale applications.

Biological techniques stand out in terms of simplicity and high cost-effectiveness. Based on oxygen requirement, biological methods are classified into aerobic and anaerobic. In aerobic processes, microorganisms use oxygen as an oxidising agent to mineralise pollutants, while anaerobic biotransformation consists of removing contaminants in the absence of oxygen. In an anaerobic environment, sulphate, nitrate, and carbon dioxide act as oxidising agents [2]. The anaerobic process is a reliable and cost-effective method for textile effluent treatment due to its several advantages, such as low energy and chemical demand and low polluted sludge generation [32, 46, 70]. Besides, energy demand increase has motivated studies in the field of anaerobic technology. During anaerobic metabolism, biogas—a high calorific energy source—is produced and can be converted to thermal and/or electrical power [101].

Among the various anaerobic bioreactors, up-flow anaerobic sludge blanket (UASB) technology, developed by Lettinga et al. [59] in the late 1970s, has been applied to treat a broad of industrial effluents and has achieved maturity in the treatment of domestic wastewater [19]. UASB reactors can be (1) applied for high-strength organic wastewater [91], (2) employed on small and large scales [67], and (3) used for recovery resource proposes [75]. Concerning treatment efficiency, they have high pollutants removal rates and the capacity to withstand organic shock loads [45].

In light of these facts, the present chapter provides an overview of the application of UASB reactors in dye wastewaters treatment. Mechanisms of dye removal and technical aspects of UASB reactors are discussed. Besides, the authors investigate the factors that determine dye removal in UASB reactors. In the end, challenges and opportunities are summarised.

2 UASB Reactors

2.1 *Bioreactor Concept*

The UASB reactors have been applied in wastewater treatment systems due to their reliability, simplicity, and high cost-effectiveness. UASB reactors have a significant position in sewage treatment plants in emerging economies such as Brazil and Mexico and have successfully been used in distilleries, dairy industries, slaughterhouses, and chemical companies for industrial effluents [67]. Data analysis of different industrial anaerobic treatment plants showed that UASB reactors are the most predominant system. Of 1,215 surveyed facilities installed in 65 nations, UASB reactors were used in 682 of them (56%) [38].

The UASB reactor comprises a rectangular or cylindrical unmixed tank and a three-phase (gas–liquid-solid) separator located on top of it. The typical height-diameter ratio of UASB reactors ranges from 0.2 to 0.5 [62]. Inside the bioreactor, wastewater flows upward, crossing a blanket of active biomass, with good organics biotransformation and settling ability [63]. Biogas produced at the bottom of the reactor and the influent flow cause natural turbulence, keeping efficient contact between active biomass (granular sludge) and wastewater (influent). Biomass concentration in the bioreactor can reach 80 g L^{-1} [91]. The three-phase separator allows the physical removal of suspended solids and guarantees high sludge retention time (SRT), operating the system without the necessity of support media to prevent biomass washout [89]. Figure 1 shows a schematic diagram of the UASB reactor.

The development of a dense granular sludge bed in the UASB reactors was decisive to the success of this technology. In fact, their high treatment performance is attributed to the formation of a dense sludge bed in the bottom of the bioreactor [49]. The granular biomass is an immobilised microbial aggregate with a highly compact structure and huge specific surface area, positive to adsorb and biotransform the pollutants. In contrast, it usually takes 2–8 months to develop anaerobic granular sludge, which requires a long bioreactor start-up period, one of UASB technology's main bottlenecks [61]. Despite several studies having been performed, the formation of anaerobic granules is not comprehensively known. Hulshoff Pol et al. [48] reviewed theories on sludge granulation in UASB reactors. In sum, they found

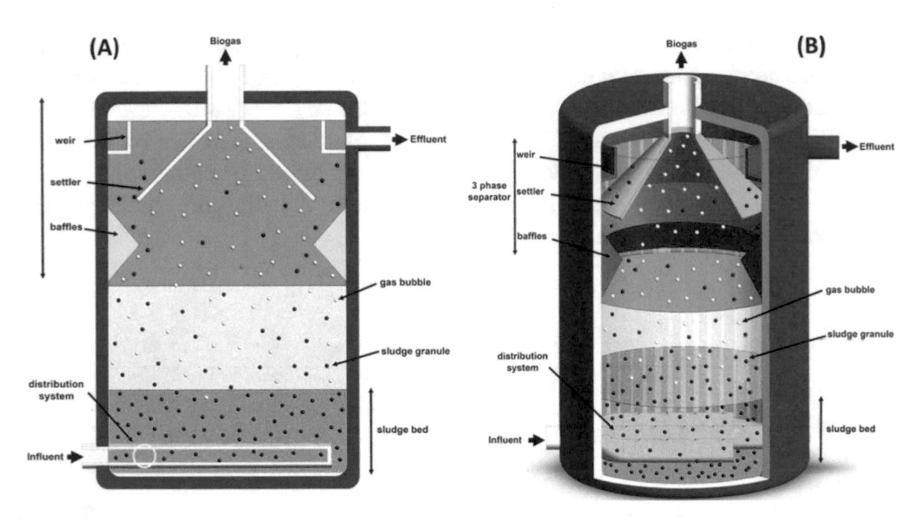

Fig. 1 UASB reactor schematic diagram in two-dimensional (**a**) and three-dimensional (**b**) shapes. Drawn by the authors

that inert support particles jointly with operational conditions play a pivotal role in forming granular sludge.

Design and operational parameters strongly influence biodegradation performance and sludge settling ability of UASB reactors. Several factors, including effluent characteristics, temperature regime, up-flow velocity, organic loading rate (OLR), and hydraulic retention time (HRT), play a vital role in the UASB biodecolourisation. For instance, HRTs ranging from 3 to 10 h appeared optimal, resulting in chemical oxygen demand (COD) removal efficiency in the range of 60 to 85% at temperatures higher than 20 °C [45]. OLR is also a key parameter. Industrial wastewaters are commonly operated with OLR ranging from 4 to 15 kg COD m^3 d^{-1} [91]. Additionally, the up-flow velocity of the liquid is responsible for maintaining mixing and guarantees efficient contact between sludge and influent. In full-scale bioreactors treating high-strength wastewaters, the up-flow velocity is around 2.0 m h^{-1}, while settling velocities range from 20 to 80 $^{m\,h-1}$, ensuring high SRT in the treatment system [18].

Souza [88] introduced the main design features of UASB reactors, while [25] described the effects of process parameters on bioreactors' operational performance. Readers are guided to these contributions for background information. UASB reactors have several advantages, including (1) low investment and operating costs, (2) low footprint, (3) high organic matter removal efficiency, (4) the ability to withstand organic shock loads, and (5) biogas production, which can be recovered as energy input. On the other hand, some issues need to be addressed, such as long bioreactor start-up, acidification by accumulating organic acids, inhibition risks due to toxic substances, insufficient pathogens/nutrients removals, which require effluent post-treatment, and sludge management. The pros and cons of UASB's technological applications are summarised in Table 1.

Table 1 Pros and cons of UASB's technological applications. Based on Refs. [18, 25]

Aspect	Pros	Cons
Operation	Low footprint Easy unit operation High organic load rates Low sludge production Short HRT and high SRT Low nutrients demand	Corrosion and gas leakage problems Odour generation Long start-up Granulation process control Undergoes fast acidification Sludge management
Efficiency	High COD removal (>70%) Ability to withstand organic shock loads	Low pathogens, N, and P removals Effluent post-treatment is required Sensibility to toxic substances and heavy metals
Economic	Low capital and operational costs Simple bioreactor construction Biogas can be recovered as energy input to reduce operational costs	Needs maintenance due to the corrosion and gas leakage Temperature adjustment/control in cold regions

Note COD = Chemical Oxygen Demand, HRT = Hydraulic Retention Time. STR = Sludge Retention Time

2.2 *Mechanisms of Dye Compounds Removal in UASB Reactors*

Dye removal under anaerobic conditions is a reduction process in which literature primarily covers the biochemistry of azo dyes. Azo dyes account for more than half of dyes produced worldwide [31]. The main mechanism of their degradation in anaerobic conditions comprises azo bond (–N = N–) cleavage via extracellular azoreductase enzyme, which involves a transfer of four electrons (reducing equivalents). The decolourisation process occurs through two stages at the –N = N– linkage. Azo dyes act as electron acceptors. In the first stage, intermediates hydrazo are formed. Afterwards, hydrazo undergoes reductive cleavage leading to the formation of aromatic amines—uncoloured by-products, as shown in Eq. 1 [86].

$$R_1{-}{-}N = N{-}{-}R_2 \overset{2e^- + 2H^+}{\longrightarrow} R_1{-}{-}NH{-}{-}NH{-}{-}R_2 \overset{2e^- + 2H^+}{\longrightarrow} R_1NH_2 + R_2NH_2 \tag{1}$$

where R_1 and R_2 are aryls or heteroaryl groups.

However, produced aromatic amines are, in general, anaerobically recalcitrant and have higher toxicity than dye precursors [39]. Consequently, anaerobically treated effluent needs further treatment. Many scholars propose to use hybrid anaerobic–aerobic systems to complete the dye removal effectively [5, 15, 51 56, 69]. Under low oxygen concentration, some facultative aerobes can consume oxygen and introduce hydroxyl groups into polyaromatic compounds, which facilitates subsequent biodegradation pathways [40]. Therefore, the aerobic process acts as a polishing step completing the mineralisation of intermediates of the anaerobic biotransformation [74].

In addition, the adsorption of dyes in the sludge granules can be significant for the decolourisation process in UASB reactors. Haider et al. [46] operated a UASB reactor in the intermittent regime (OLR of 2 kg COD m^{-3} d^{-1}, HRT 24 h). They observed that the non-feeding period of run contributed more for total COD removal than continuous runs, concluding that physical dye entrapment onto biomass granules was preponderant. Indeed, kinetic studies show that the dye removal mechanism in UASB reactors is first abiotic (adsorption) and then biotic (biodegradation) [43]. Therefore, the adsorption mechanism by sludge granules makes an important contribution during decolourisation processes in UASB reactors.

The anaerobic process is divided into four steps: hydrolysis, acidogenesis, acetogenesis, and methanogenesis. In the three former steps named acid fermentation, organic macromolecules are hydrolysed and metabolised by fermentative bacterias and converted to carbon dioxide, hydrogen, and acetic acid. In the last step, acetic acid, carbon dioxide, and hydrogen are converted to carbon dioxide and methane by methanogenic archaeans [8]. The sequential stages of the anaerobic process are shown in Fig. 2.

As stated above, anaerobic decolourisation is a reductive process of azo bond cleavage via extracellular azoreductase enzyme, which involves a transfer of four electrons (reducing equivalents). The reducing equivalents (i.e., electron donors) are formed during the conversion of the organic matter through different stages of anaerobic metabolism. H_2, CO_2, ethanol, and formate are effective electron donors. The syntrophic relationship among microorganisms plays a pivotal role in the anaerobic process, and poor electron transfer inter-species can hamper the treatment performance. It is important to note that biodecolourisation under anaerobic conditions requires additional organic carbon sources since dye-reducing microbial consortia cannot use dye as the growth substrate [27, 30, 81]. Fermentative bacteria and hydrogenotrophic methanogens are the main ones responsible for dye reduction. *Methanosarcina* archaea, *Clostridium*, *Enterococcus*, *Pseudomonas*, *Bacillus*, *Aeromonas*, *Enterococcus*, *Desulfovibrio*, and *Desulfomicrobium* bacteria are reported to be effective in the anaerobic biodecolourisation [82, 85, 105].

2.3 Influence Factors of UASB Reactors in Dye Removal

Dye structure and concentration, electron donors and redox mediators, pH, temperature regime, hydraulic retention time (HRT), and organic loading rate (OLR) are the main influence parameters governing dyes removal in UASB reactors [81, 102] (Fig. 3). It is consensus that monitoring the anaerobic process to ensure the balance among these influencing parameters is pivotal for a stable reactor operation. Thus, the present section describes the main influence parameters during the decolourisation process in UASB reactors.

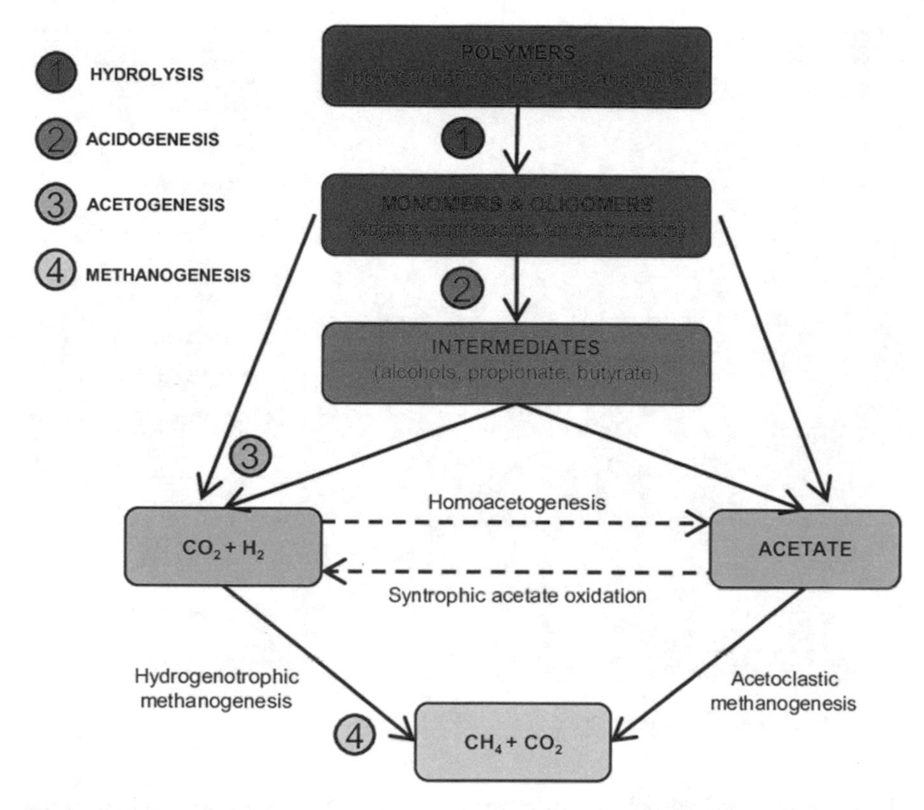

Fig. 2 Different stages of the anaerobic process (based on O'Flaherty et al. [68], Rozzi and Remigi [76])

2.3.1 Dye Structure and Concentration

Dye compounds are heterogeneous chemicals of high molecular weight, complex structures, low biodegradability, and high toxicity. Literature is available stating that high dye concentration leads to poor biodecolourisation efficiencies due to the inhibition of the dye-reducing anaerobes by dye toxicity or blocking azoreductase enzymes [83, 106]. Dai et al. [24] showed that a high azo dye concentration (>450 mg L^{-1}) could decrease the granular sludge porosity and strength, reduce its settling ability, and inhibit methanogenic activity.

Decolourisation of textile effluents was studied in a UASB reactor at 25, 50, 100, 150, and 300 mg dye L^{-1} [87]. Colour removal decreased for the increase of dye concentration. Decolourisation was 94% at 150 mg dye L^{-1} and 89% at 300 ^{mg} dye L^{-1}. Similar findings have been reported by Murali et al. [66].

Furthermore, high dye dosage is usually associated with high salinity, reducing microbial activity, especially methanogens [100]. Excess salts adversely affect granulation and UASB stability. Wang et al. [97] demonstrated that anaerobic granules could tolerate salts concentration up to 10 g L^{-1}. High salinity conditions decreased

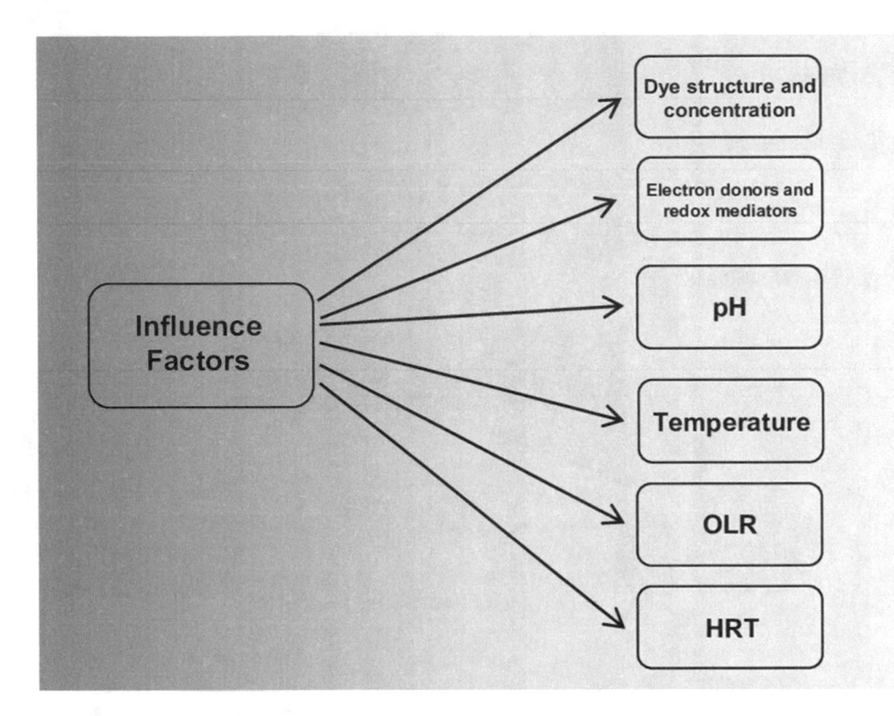

Fig. 3 Main influence factors for dye removal in UASB reactors

biomasses size and hydrophobicity, which hinders biodegradation and sludge settling ability. Sulfuric acid is commonly added to adjust the pH to overcome the salinity of textile effluents. On the other hand, Amaral et al. [5] reported that sulphate dosage higher than 300 mg L^{-1} could also inhibit anaerobic metabolism. In anaerobic conditions, sulphates and dye molecules compete to become the final electron acceptor of the reducing equivalents. As a result, sulphates obstruct the electron transfer to dye compounds, reducing the biodecolourisation efficiency [30]. Despite that, this mechanism is not fully understood, and hence, further studies are required.

Additionally, dye structure variability could also be a significant obstacle to the overall mineralisation of the molecules by microorganisms. Due to the many functional groups of dye compounds, steric hindrance can hamper enzymatic activity. Even minor structural differences can affect biodecolourisation [65]. Chinwetkitvanich [20] studied the anaerobic removal of four different dyes. Decolourisation efficiencies of anthraquinone monochlorotriazinyl, anthraquinone vinylsulphonyl, and bis azo vinylsulphonyl were 66, 64, and 63%, respectively. Moreover, the author stated that different chemical structures imply different removal mechanisms. For example, anthraquinone dye was mainly removed through adsorption into sludge flocs, while biodegradation via azo bond cleavage was prominent in azo dye decolourisation.

2.3.2 Electron Donors and Redox Mediators

The rate of anaerobically dye removal depends on the dye structure and concentration, electron donors (i.e., organic carbon sources), and redox mediators. Redox mediators are essential in the anaerobic decolourisation, accelerating the electron transfer from organic matter to dye compounds; thus, they accelerate the biotransformation kinetic [35]. Riboflavin and sulfonated compounds such as anthraquinone sulfonate and disulfonated anthraquinone are usually employed as redox mediators [16, 29]. Martins et al. [64] investigated the effect of riboflavin and different carbon sources on removing azo dye named Remazol Golden Yellow. Decolourisation without riboflavin was about 30.7% at 25°C during 24 h. The addition of soluble riboflavin (0.0175 mg L^{-1}) led to increased biodecolourisation of more than 50% during 48 h. Similar results were obtained using yeast extract (500 mg L^{-1}) as a carbon source and anthraquinone-2,6-disulfonate as a redox mediator during the anaerobic treatment in UASB reactors [9, 28].

2.3.3 pH

The pH is related to establishing a micro-environment that affects the rate of microbial growth and enzymatic activity; therefore, it strongly influences dye removal efficiency [99]. In anaerobic processes, a very acidic or alkaline environment inhibits the activity of methanogens and the growth of acid-producing bacteria by increasing non-ionic organic acid, which can reduce biodecolourisation efficiency. Literature shows that anaerobes can decolourise at a wide range of pH from 5.0 to 10.0 [17, 44]. On the other hand, methanogens grow efficiently in the pH range of 6.0–8.0 and are very sensitive to pH fluctuation [42]. Chen et al. [17] have observed 97% decolourisation of azo dye Direct Black G at pH 8.0,79% decolourisation at pH 11.0; and 81% decolourisation at pH 4.0 after 48 h of incubation.

2.3.4 Temperature

Temperature significantly affects the dye-reducing microbial consortia, especially methanogens, therefore for stable bioreactor operation, maintenance of specific temperature is pivotal. Generally, an efficient anaerobic process occurs in mesophilic (35–40 °C) and thermophilic (50–65 °C) regimes. Many studies have reported that temperature is an important control parameter in anaerobic treatment and strongly influences the microbial community structure and the effluent treatability [29, 72, 79]. Typically, dye removal is fast at the thermophilic range. However, sky-high temperatures can reduce microbiota diversity and, as a consequence, decrease the biodecolourisation efficiency [12]. A study demonstrated that the optimum temperature range for biodecolourisation ranges from 30 to 55 °C, and exceeding this scope could harm the syntrophic relationship among anaerobic microorganisms [28].

2.3.5 Organic Loading Rate

Organic loading rate (OLR) can be defined as the concentration of dye or chemical oxygen demand (COD) entering the bioreactor in continuous mode per day. Many researchers have investigated the optimal OLR and predicted its maximum tolerable value. For example, COD removal efficiency in dye wastewater treatment was 61% at OLR of 2.40 kg COD m^{-3} d^{-1}, which decreased to 37% when the OLR was increased to 22.5 ^{kg} COD m^{-3} d^{-1} in an anaerobic reactor [54]. In another biodecolourisation study, OLR ranging from 1.03 to 6.65 kg COD m^{-3} d^{-1} was selected as the optimum value. This range provided colour removal efficiencies from 92 to 95% (Işık and Sponza, 2004a). In contrast, Amaral et al. [5] reported that at OLRs of 1.84, 2.42, and 2.70 kg COD m^{-3} d^{-1}, decolourizing rates were only 30%, 37%, and 52%, respectively.

It is important to note that at the start-up stage of the UASB reactor, adding a massive amount of dye can temporarily inhibit microbial activities because the anaerobes are not fully acclimatised. Furthermore, high OLR can affect methanogens metabolism and inhibit methane production due to the acidification of the medium [60]. For example, [54] obtained methane production efficiencies of 75% at OLR of 2.4 kg COD m^{-3} d^{-1} and 38% when OLR was increased to 22.5 ^{kg} COD m^{-3} d^{-1}. It is helpful to mention that thermophilic conditions and effluent recirculation are the potential parameters that minimise the adverse effect of high OLR in anaerobic reactors [77].

2.3.6 Hydraulic Retention Time

Hydraulic retention time (HRT) represents the time that the affluent spend in the bioreactor and is closely associated with the bioreactor's robustness and treatment effectiveness. HRT is linked to microbial growth and dependent on temperature regime, organic loads, and dye structure. Decreasing the HRT can lead to misdeveloping granular sludge and/or acidification, whereas a longer than optimal HRT results in low utilisation of reactor components and biomass washout [25].

In a recent study, the colour removals decrease from 98 to 94% when the HRT decreased from 12 to 3 h [41]. In contrast, Amaral et al. [4] found that increasing the HRT parameter did not improve azo dye removal under anaerobic conditions. Decolourisation efficiency was 67% at an HRT of 16 h and 55% at 96 h. For the treatment of dye-containing wastewaters in UASB reactor, researchers have reported successful operation at 5–20 h of HRT [50, 53, 73]. However, it should be mentioned that differences can be observed depending on several factors, such as operating procedures and wastewater composition.

3 Decolourisation Performance of UASB Reactors

3.1 Dye Removal Efficiency

Coloured effluents contain persistent pollutants, which can exhibit mutagenic, carcinogenic, and toxic effects. Therefore, dye wastewaters need to be adequately treated before being discharged in watercourses [57]. UASB technology has been proven to be a promising method to remove dyes from effluents. Considering the chapter topic, the performance of UASB technology for azo dye removal is the most general information available. As previously discussed, the removal process is achieved through adsorption into granular sludges and anaerobic biodegradation via azo bond cleavage. Accordingly, the authors focused on azo dye biodegradation studies. Table 2 displays COD/colour removals and operational conditions of UASB reactors treating coloured wastewaters.

As shown in the table above, several studies have been conducted to evaluate dye removal in UASB reactors. The treatment efficiency is primarily assessed in terms of organic matter reduction and decolourisation. For instance, Cui et al. [23] used a UASB reactor for azo dye Alizarin Yellow R removal. The UASB reactor was operated under a batch condition, 25 ± 2°C, OLR of 100 g dye m^{-3} d^{-1}, and HRT ranging from 8 to 12 h. Colour and COD removal efficiencies of 96% and 54% were recorded at a dye concentration of 50 mg L^{-1} and HRT of 12 h per batch. The authors also investigated the OLR regime during the UASB treatment. They ranged the OLR from 100 to 800 g dye m^{-3} d^{-1}. As discussed earlier, high dye loading rates can lead to poor decolourisation performance due to the inhibition of the anaerobic microbiota by dye toxicity. The results showed that the colour removal efficiency decreased to 63% at OLR of 800 g dye m^{-3} d^{-1}. However, the inhibition was reversible, and decolourisation efficiency was recovered to 92% when OLR was decreased to 600 g dye m^{-3} d^{-1}.

In a recent study, the effects of the intermittent operation on the biodegradation of dye compounds were tested using a laboratory UASB reactor [46]. The feeding period of 12 h and non-feeding period of 12 h provided decolourisation of 71% and COD removal efficiency of 58% (at HRT of 24 h and OLR of 2 kg COD m^{-3} d^{-1}). During the non-feeding period, anaerobes resisted dye toxicity and handled operational changes in temperature, HRT, and OLR. The authors concluded that the discontinuous operation could be used as a strategy to improve the stability of decolourisation systems [46].

Firmino et al. [36] evaluated the efficacy of the UASB reactor for Direct Red 28 dye removal. The best system performance was at an HRT of 24 h, where 57.1% of colour and 60.3% of COD removals efficiency (on average) were registered. Işik and Sponza [53] studied Direct Red 28 azo dye mineralisation in a UASB reactor operated at continuous mode. Colour disappeared within a few minutes after entering into the UASB reactor due to the adsorption by anaerobic granules. Decolourisation efficiency remained at 99% during 103 d of operation.

Table 2 Treatment performance of UASB reactors in azo dye removal

Scheme	Scale	UASB reactor conditions	Dye compounds			UASB treatment results		References
			Type	Name	Concentration/Amount	COD	Colour	
UASB reactor	Lab	Semi-continuous, 20–30 °C, HRT 8–20 h, OLR 2–2.4 g COD m^{-3} d^{-1}	Azo dyes	Acid Yellow 17, Basic Blue 3, and Basic Red 2	–	> 60%	20–78%	An et al. [6]
UASB reactor	Lab	Continuous, 37ºC, HRT 20 h, OLR 3.86 kg COD m^{-3} d^{-1}	Azo dyes	Reactive Black 5, Direct Red 28, Direct Black 38, Direct Brown 2, and Direct Yellow 12 mixture	about 25 mg L^{-1}	80.5%	100%	Işik [50]
Batch Anaerobic process + **UASB reactor +** Continuous stirred tank reactor	Lab	Continuous, 37ºC, HRT 3–30 h, OLR 2–15 kg COD m^{-3} d^{-1}	Azo dyes	Reactive Black 5 and Direct Red 5	about 100 mg L^{-1} of each dye	26.6–40.8%	87–97%	Işık and Sponza [52]
UASB reactor	Lab	Batch, 27ºC, HRT 2 – 24 h, OLR[a]	Azo dye	Direct Red 28	0.3–1.2 mM	60.3–65.5%	44.8–57.1%	Santos et al. [29]
UASB reactor + Sequencing batch reactor	Lab	Batch, 30ºC, HRT 24 h, OLR 0.3 g L^{-1} d^{-1},	Azo dye	Orange II	–	45%	> 95%	Ong et al. [71]
UASB reactor	Lab	Batch, 37 ºC, HRT 18.3 h, OLR 0.286 kg COD m^{-3} d^{-1}	Azo dye	Direct Red 28	about 100 mg L^{-1}	58%	>99%	Işik and Sponza [53]

(continued)

Table 2 (continued)

Scheme	Scale	UASB reactor conditions	Dye compounds			UASB treatment results		References
			Type	Name	Concentration/Amount	COD	Colour	
UASB reactor	Lab	Batch, 37 ± 2 °C, HRT 24 h, OLR 1.8 ± 0.2 g L^{-1} d^{-1}	Azo dyes	Acid Orange 7 and Direct Red 254	–	>85%	>88%	Brás et al. [13]
UASB reactor	Lab	Continuous, 35 ± 3 °C, HRT 24 h, OLR[a]	Azo dye	Acid Red 131, Acid Blue 204, and Acid Yellow 79 mixture	10–300 mg L^{-1}	85–93%	73–94%	Somasiri et al. [87]
UASB reactor + Continuous stirred tank reactor	Lab	Continuous, 35°C, HRT of 30.12 h, OLR 2.40 kg COD m^{-3} d^{-1}	Azo dye	Reactive Black 5	150 mg L^{-1}	61%	99.8%	Karatas et al. [54]
UASB reactor + Submerged aerated biofilter	Pilot	Continuous, 27 °C, HRT 24 h, OLR 1.3 kg m^{-3} d^{-1}	–	Real textile wastewater	about 305 kg of dyes consumed in 210 d	59%	64%	Ferraz et al. [33]
UASB reactor	Lab	Semi-batch, 25 °C, HRT 24 h, OLR[a]	Azo dye	Methyl Orange	50, 100, and 150 mg L^{-1}	49–69%	90–96%	Murali et al. [66]
Hybrid UASB bioreactor	Lab	Continuous, 33°C, HRT 50 h, OLR 12 kg COD m^{-3} d^{-1}	–	Real textile wastewater	–	94.8%	84.4%	Katal et al. [56]
UASB reactor + Down-flow hanging sponge	Lab	Batch, 22 ± 3°C, HRT 6 h, OLR 3.3 g COD L^{-1} d^{-1}	–	Real textile wastewater	–	>30%	>60%	Tawfik et al. [90]

(continued)

Table 2 (continued)

Scheme	Scale	UASB reactor conditions	Dye compounds			UASB treatment results		References
			Type	Name	Concentration/Amount	COD	Colour	
UASB reactor + Submerged aerated biofilter	Pilot	Continuous, 30 °C, HRT 8–12 h, OLR 1.84–2.70 kg COD m^{-3} d^{-1}	Azo dye	Direct Black 22	779 kg consumed in 335 d	34–43%	30–52%	Amaral et al. [5]
UASB reactor	Lab	Batch, 25 ± 2°C, HRT 8–12 h, OLR 100–800 g dye m^{-3} d^{-1}	Azo dye	Alizarin Yellow R	50–200 mg L^{-1}	>50%	>95%	Cui et al. [23]
Coagulation + **UASB reactor**	Lab	Batch, 22–27°C, HRT 20 h, OLR 2.4 kg COD m^{-3} d^{-1}	Azo dyes	Reactive Black 5, Direct Red 28, and Disperse blue 3	about 200 mg L^{-1}	92%	92%	Verma et al. [94]
UASB reactor	Lab	Batch, 37 ± 2°C, HRT 5 h, OLR[a]	Azo dyes	Remazol Blue, Remazol Yellow, and Remazol Brilliant Yellow	–	85%	98%	Pereira et al. [73]
Ozonation + **UASB reactor**	Lab	Batch, Temperature[a], HRT 40 h, OLR[a]	Azo dye	Reactive Black 5	1500 mg L^{-1}	90%	94%	Venkatesh et al. [93]
UASB reactor + Fenton process	Lab	Batch, HRT 12 h, OLR 1.5 kg COD m^{-3} d^{-1}	Azo dye	Malachite Green	100 mg L^{-1}	82%	96.1%	Santos et al. [80]

(continued)

Table 2 (continued)

Scheme	Scale	UASB reactor conditions	Dye compounds			UASB treatment results		References
			Type	Name	Concentration/Amount	COD	Colour	
UASB reactor + Submerged aerated biofilter	Pilot	Continuous, 23–27 °C, HRT 16–96 h, OLR 1.41–1.77 kg COD m^{-3} d^{-1}	–	Real textile wastewater	6,024 kg of dyes consumed in 622 d	44–62%	55–67%	Amaral et al. [4]
UASB reactor	Lab	Continuous, 27°C, HRT 24 h, OLR[a]	Azo dye	Reactive Red 2	50 mg L^{-1}	89%	51%	de Barros et al. [26]
UASB reactor + Activated sludge process	Lab	Continuous, 16–29 °C, HRT 24 h, OLR[a]	Azo dye	Yellow Gold Remazol	50 mg L^{-1}	67–88%	85%	Bahia et al. [11]
UASB reactor + shallow polishing pond	Lab	Continuous, 16–29 °C, HRT 24 h, OLR[a]	Azo dye	Yellow Gold Remazol	50 mg L^{-1}	67–88%	85%	Bahia et al. [11]
UASB reactor	Lab	Continuous, temperature[a], TRH 24 ^{h}, OLR[a]	Azo dye	Red Bronze	40–325 mg L^{-1}	60–91%	75–94%	Fazal et al. [32]
UASB reactor + Aerated bioreactor	Lab	Continuous, 37 ± 1 °C, HRT 6 h, OLR 12.97 kg COD m^{-3} d^{-1}	Azo dye	2-Naphthol Red	0.1 g L^{-1}	85.6%	96%	Gadow and Li [41]
UASB reactor + microaerated UASB reactor	Lab	Continuous, 25.0 ± 1.4°C, HRT[a], OLR 1.27–1.50 $^{kg\ m-3}$ d^{-1}	Azo dye	Direct Black 22	0.6 mM	67– 72%	70–78%	Carvalho et al. [15]

(continued)

Table 2 (continued)

Scheme	Scale	UASB reactor conditions	Dye compounds			UASB treatment results		References
			Type	Name	Concentration/Amount	COD	Colour	
UASB reactor + shallow polishing pond	Lab	Continuous, 16–29 °C, HRT 24 h, OLR[a]	–	Real textile wastewater	–	80%	50%	Bahia et al. [10]

Note [a]Data not available. COD = Chemical oxygen demand, UASB = Up-flow anaerobic sludge blanket, HRT = Hidraulic retention time, OLR = Organic loading rate

UASB reactors are efficient for dye removal and exhibit a very high decolourisation efficiency (up to 99%). However, anaerobic treatment may not mineralise by-products of anaerobic metabolisms, such as polyaromatic amines and recalcitrant substances. Consequently, post-treatment of the anaerobically treated effluent is needed to achieve the required regulatory disposal standards. In this sense, anaerobic–aerobic combined systems have been proposed to remove aromatic by-products and recalcitrant COD efficiently. Gadow and Li [41] combined continuous UASB and aerobic processes to remove azo dye 2-Naphthol Red from industrial textile wastewater. The system achieved 98.9% and 98.4% of COD and colour removal at optimum conditions (OLR of 12.97 g COD m^{-3} d^{-1}; HRT of 6 h).

In previous research, activated sludge and shallow polishing ponds were used as polishing steps in combined anaerobic–aerobic systems to remove azo dye Yellow Gold Remazol (50 mg L^{-1}) [11]. Despite the low colour and COD removal efficiencies (around 20%), these aerobic processes produced effluents free of toxicity to bioluminescent *Vibrio fisheri* bacteria. In other studies, Ferraz et al. [33], Amaral et al. [4, 5] investigated submerged aerated biofilters as a polishing step for UASB effluents in order to remove anaerobic metabolites. The researchers confirmed the effectiveness of the aerobic process for the oxidation of aromatic amines, obtaining a COD removal efficiency of more the 50%. Furthermore, Ferraz et al. [33] obtained treated effluents with non-toxicity to bioorganism *Daphnia magna*.

In another interesting work, Carvalho et al. [15] proposed a microaerated UASB reactor to remove Direct Black 22 azo dye. The UASB reactor was aerated in the upper part (0.18 ± 0.05 mg O_2 L^{-1}) to mineralise amines generated in the anaerobic process. COD and colour removal ranged from 59 to 78%. Treated effluent of microaerated reactor was 16-fold less toxic when compared to conventional UASB, confirming the effectiveness of microaeration method for removal of anaerobic intermediates. Similar results were reported under aeration condition of 1.0 mL air min^{-1} treating azo dye Reactive Red 2 (50 mg L^{-1}) [26].

Based on the reviewed literature, removal efficiencies of COD and colour in UASB reactors lie within the range of 60–85% and 75–96%, respectively. Typically, high efficiencies are obtained at operational conditions of 30–40 °C, TRH of 20–30 h, and OLR of 2–15 kg COD m^{-3} d^{-1}. Nevertheless, it must be emphasised that despite the excellent UASB descolourising efficiency reported in published studies, its treatment performance is case-specific and depending on wastewater composition and operational conditions. Furthermore, the available literature is mainly based on laboratory investigations and, therefore, more research is needed to scale-up and evaluate UASB techno-economic feasibility in field applications.

3.2 *Bio-Energy Production*

Anaerobic technology has the potential to degrade dye pollutants while at the same time providing a huge potential source of clean energy. Dye-containing wastewaters are loaded with organic chemicals, and in UASB reactors, the organic load is biotransformed and converted to biogas. Depending on the wastewater characteristic, biogas composition typically lies within the ranges $CH_4 = 50–70\%$, $CO_2 = 30–50\%$, and $H_2 = 1–5\%$, along with traces of water vapour, N_2, H_2S, ammonia, and siloxanes [7]. Hence, it has a high calorific value and can produce thermal and/or electrical energy. Biogas also can be processed to produce biomethane—a direct substitute for natural gas [103].

Katal et al. [56] operated a lab-scale UASB reactor to treat textile effluent and determined the biogas production yield. At HRT of 50 h, maximum biogas productivity of 36 L d^{-1} with a biomethane content of 79% was obtained. In other bench studies, biomethane rates from 0.36 to 2.7 L d^{-1} were archived [50–52]. As previously stated, removal efficiencies of COD and colour in UASB reactors treating dye effluents lie within the range of 60–85% and 75–96%, respectively. On the other hand, based on literature references, biomethane production can reach up to 0.30 m^3 CH_4 per kg of COD removed [41, 105]. Ranges of COD/colour removal and CH_4 yield are shown in Fig. 4, which is constructed based on the data compiled in Table 2 and Ref. [41, 105].

From the characterisation of dye-containing effluents, a COD average of 7.3 kg m^{-3} was found by Santos et al. [30]. This value multiplied by biomethane yield (0.30 m^3 CH_4 kg $COD_{removed}$) means CH_4 productivity of 2.19 m^3 per m^3 of wastewater. Assuming a conversion factor of 10 kWh m^{-3} of CH_4 [96], UASB reactors could reach up to 21.90 kWh of bio-energy recovery per m^3 of treated effluent.

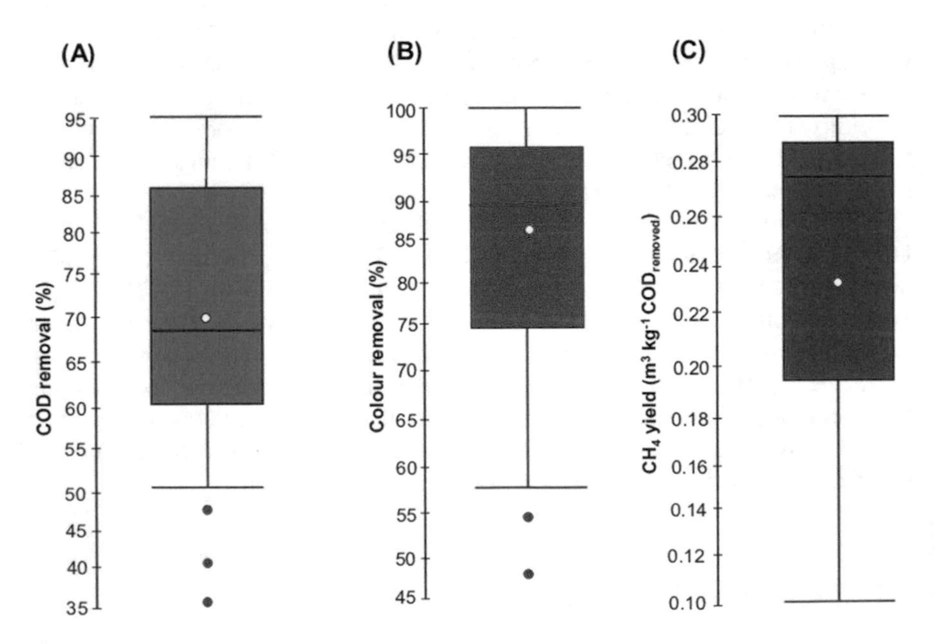

Fig. 4 Ranges of COD removal (**a**) and colour removal (**b**) from literature references cited in the text, and CH_4 yield (**c**) based on Refs. [41, 105]. The circle and the line within the box indicate the average and median values, respectively. Red circles show outliers

Considering that large scale dyers discharge from 70 to 400 m^3 d^{-1} of wastewater [84], UASB reactors could lead to an annual energy saving of 1,878,472.50 kWh at the treatment facility as a result of the electric energy production surplus, corresponding to a saving of US$ 184,090.30 $year^{-1}$. Thus, biogas recovery from UASB reactors can play a strategic role in promoting wastewater valorisation and decreasing textile wastewater treatment's economic burdens.

4 Challenges and Opportunities

Despite the merits of UASB reactors, four issues can be identified as limitations for their implementation in the treatment of colourised effluents. First, the development of granular sludge is a time-consuming process, which requires a long bioreactor start-up period. Furthermore, great efforts are still needed to elucidate granulation mechanisms and ensure process robustness. Second, as discussed in Sect. 2.3, various parameters influence the treatment performance, and hence, an imbalance among them can result in poor decolourising efficiency. Besides, anaerobic microorganisms are susceptible to inhibitory effects by high salinity, sulphate, and dye dosage, for example. These issues, if not adequately monitored and controlled, lead to process deterioration. Third, anaerobic processes may not be able to mineralise by-products of anaerobic metabolisms such as toxic aromatic amines. Therefore, post-treatment of the anaerobically treated effluent is required. As previously discussed, aerobic systems have been coupled to UASB reactors for efficient dye removal. However, studies assessing the treatment performance in pilot and full-scale are scarce, as well as biotoxicity assays to evaluate the quality of treated water. Last, the management of sludge containing dye compounds is a big challenge that must be addressed. Excess sludge from UASB reactors requires treatment in the form of dewatering, drying, stabilisation, disinfection, and disposal [22]. The polluted sludge contains toxic chemicals, and therefore its proper management must be guaranteed. Within a circular economy context, efforts have been made to recover add-value products from sludges (e.g., dyes, energy, salts, metals, and nutrients) [14]. For example, Yildirir and Ballice [104] treated textile biological sludges via hydrothermal gasification to produce fuel gas with a high calorific value.

It is beyond the scope of the present chapter to consider recovered add-value products from dye industry wastes in any detail. Currently, this topic has been investigated even more, and a comprehensive review of resource recovery of coloured effluents was recently published by Varjani et al. [92]. Indeed, the development of sustainable and cost-effective methods for resource recovery and their assessment based on a life-cycle perspective is a promising field of research [55].

Wastewater treatment plants (WWTPs) generally have a high energy demand [21]. UASB technology offers opportunities for renewable energy production and reduction of fossil fuel consumption. In fact, the hierarchy structure for wastewater management (Fig. 5) should be implemented to ensure efficient treatment, wastewater valorisation, and the transition of WWTPs to sustainable facilities. In this respect,

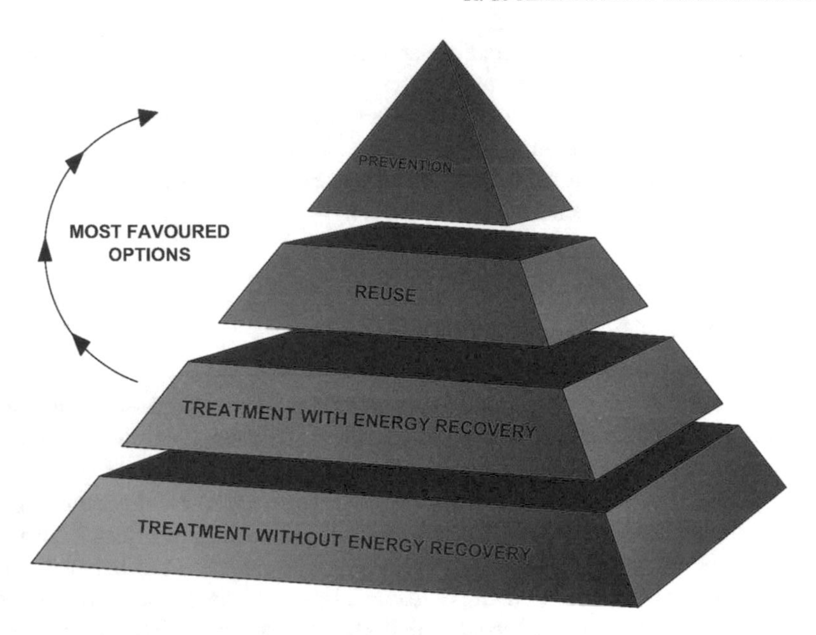

Fig. 5 The hierarchy structure of wastewater management

bio-energy recovery can play a strategic role. It should also be underlined that anaerobic treatments are less energy-intensive and produce lower excess sludge when compared with aerobic processes [3]. Furthermore, among the positive impacts of biogas recovery via UASB technology stand out the mitigation of greenhouse gas emissions such as methane and carbon dioxide, which can foster the carbon neutrality of WWTPs in the middle and long term [98].

5 Conclusions

The present chapter reviewed the published literature about UASB reactors in dye removal. UASB technology exhibits high decolourisation rates with COD and colour removal efficiencies within the range of 60–85% and 75–96%, respectively. However, the available literature is mainly based on laboratory investigations and, therefore, more research is needed to scale up and evaluate UASB techno-economic feasibility in field applications. Four major challenges have been identified for UASB reactors implementation in dye wastewater treatment: (1) long start-up period, (2) inhibitory effects by high dye dosage, salinity, and sulphate, (3) treated effluent needs post-treatment due to the ineffectiveness of UASB reactors in mineralise by-products of anaerobic metabolism, and (4) sludge management. It should be emphasised that this is not an exhaustive overview of UASB reactors in the decolourisation process since unravelling all the detailed mechanisms of dye removal, UASB optimisation strategies, and research opportunities are hard to achieve in a single chapter. The following aspects should be addressed in further studies: (1) strategies to reduce the reactor start-up, (2) mechanisms of sludge granulation, influence factors, and granulation control, (3) optimising of influence parameters of UASB reactors treating real textile wastewater, (4) developing combined systems to boost the treatment performance, (5) biotoxicity assays of treated effluent, and (6) techno-economic assessment of biogas recovery during the treatment of real dye-containing effluents.

6 Conflicts of Interest

The authors declare no conflict of interest.

Acknowledgements Ronei de Almeida acknowledges the financial support received from Conselho Nacional de Desenvolvimento Científico e Tecnológico (CNPq) (Grant Number 165018/2018-6).

References

1. Al-Kdasi A, Idris A, Saed K, Guan CT (2004) Treatment of textile wastewater by advanced oxidation processes–a review. Global NEST J 6(1):222–230
2. Ali H (2010) Biodegradation of synthetic dyes—a review. Water Air Soil Pollut 213(1–4):251–273. https://doi.org/10.1007/s11270-010-0382-4
3. AlSayed A, Soliman M, Eldyasti A (2020) Anaerobic-based water resources recovery facilities: a review. Energies 13(14):3662. https://doi.org/10.3390/en13143662
4. Amaral FM, Florêncio L, Kato MT, Santa-Cruz PA, Gavazza S (2017) Hydraulic retention time influence on azo dye and sulfate removal during the sequential anaerobic–aerobic treatment of real textile wastewater. Water Sci Technol 76(12):3319–3327. https://doi.org/10.2166/wst.2017.378
5. Amaral FM, Kato MT, Florêncio L, Gavazza S (2014) Color, organic matter and sulfate removal from textile effluents by anaerobic and aerobic processes. Biores Technol 163:364–369. https://doi.org/10.1016/j.biortech.2014.04.026
6. An H, Qian Y, Gu X, Tang WZ (1996) Biological treatment of dye wastewaters using an anaerobic-oxic system. Chemosphere 33(12):2533–2542. https://doi.org/10.1016/S0045-6535(96)00349-9
7. Angelidaki I, Treu L, Tsapekos P, Luo G, Campanaro S, Wenzel H, Kougias PG (2018) Biogas upgrading and utilization: current status and perspectives. Biotechnol Adv 36(2):452–466. https://doi.org/10.1016/j.biotechadv.2018.01.011
8. Anukam A, Mohammadi A, Naqvi M, Granström K (2019) A review of the chemistry of anaerobic digestion: methods of accelerating and optimizing process efficiency. Processes 7(8):504. https://doi.org/10.3390/pr7080504
9. Baêta BEL, Aquino SF, Silva SQ, Rabelo CA (2012) Anaerobic degradation of azo dye Drimaren blue HFRL in UASB reactor in the presence of yeast extract a source of carbon and redox mediator. Biodegradation 23(2):199–208. https://doi.org/10.1007/s10532-011-9499-4
10. Bahia M, Borges TA, Passos F, de Aquino SF, de Silva Q (2020) Evaluation of a combined system based on an Upflow Anaerobic Sludge Blanket Reactor (UASB) and Shallow Polishing

Pond (SPP) for textile effluent treatment Braz Arch Biol Technol 63:1–8. https://doi.org/10.1590/1678-4324-2020180130
11. Bahia M, Passos F, Adarme OFH, Aquino SF, Silva SQ (2018) Anaerobic-aerobic combined system for the biological treatment of Azo Dye solution using Residual Yeast. Water Environ Res 90(8):729–737. https://doi.org/10.2175/106143017X15131012153167
12. Boonyakamol A, Imai T, Chairattanamanokorn P, Higuchi T, Sekine M, Ukita M (2009) Reactive Blue 4 Decolorization under Mesophilic and Thermophilic Anaerobic Treatments. Appl Biochem Biotechnol 152(3):405–417. https://doi.org/10.1007/s12010-008-8237-9
13. Brás R, Gomes A, Ferra MIA, Pinheiro HM, Gonçalves IC (2005) Monoazo and diazo dye decolourisation studies in a methanogenic UASB reactor. J Biotechnol 115(1):57–66. https://doi.org/10.1016/j.jbiotec.2004.08.001
14. Bratina B, Šorgo A, Kramberger J, Ajdnik U, Zemljič LF, Ekart J, Šafarič R (2016) From municipal/industrial wastewater sludge and FOG to fertilizer: a proposal for economic sustainable sludge management. J Environ Manag 183:1009–1025. https://doi.org/10.1016/j.jenvman.2016.09.063
15. Carvalho JRS, Amaral FM, Florencio L, Kato MT, Delforno TP, Gavazza S (2020) Microaerated UASB reactor treating textile wastewater: the core microbiome and removal of azo dye Direct Black 22. Chemosphere 242:125157. https://doi.org/10.1016/j.chemosphere.2019.125157
16. Cervantes FJ, Garcia-Espinosa A, Moreno-Reynosa MA, Rangel-Mendez JR (2010) Immobilized redox mediators on anion exchange resins and their role on the reductive decolorization of azo dyes. Environ Sci Technol 44(5):1747–1753. https://doi.org/10.1021/es9027919
17. Chen Y, Feng L, Li H, Wang Y, Chen G, Zhang Q (2018) Biodegradation and detoxification of Direct Black G textile dye by a newly isolated thermophilic microflora. Biores Technol 250:650–657. https://doi.org/10.1016/j.biortech.2017.11.092
18. de Chernicharo CAL (2007) Biological wastewater treatment series: anaerobic reactors, vol. 4. IWA Publishing, London, UK, p. 184. http://iwaponline.com/ebooks/book-pdf/1100/wio9781780402116.pdf
19. Chernicharo CAL, Almeida PGS, Lobato LCS, Couto TC, Borges JM, Lacerda YS (2009) Experience with the design and start up of two full-scale UASB plants in Brazil: enhancements and drawbacks. Water Sci Technol J Int Assoc Water Pollut Res 60(2):507–515. https://doi.org/10.2166/wst.2009.383
20. Chinwetkitvanich S (2000) Anaerobic decolorization of reactive dyebath effluents by a two-stage UASB system with tapioca as a co-substrate. Water Res 34(8):2223–2232. https://doi.org/10.1016/S0043-1354(99)00403-0
21. Chrispim MC, Scholz M, Nolasco MA (2021) Biogas recovery for sustainable cities: a critical review of enhancement techniques and key local conditions for implementation. Sustain Cities Soc 72:103033. https://doi.org/10.1016/j.scs.2021.103033
22. Cieślik BM, Namieśnik J, Konieczka P (2015) Review of sewage sludge management: standards, regulations and analytical methods. J Clean Prod 90:1–15. https://doi.org/10.1016/j.jclepro.2014.11.031
23. Cui M-H, Cui D, Liang B, Sangeetha T, Wang A-J, Cheng H-Y (2016) Decolorization enhancement by optimizing azo dye loading rate in an anaerobic reactor. RSC Adv 6(55):49995–50001. https://doi.org/10.1039/C6RA04665G
24. Dai R, Chen X, Luo Y, Ma P, Ni S, Xiang X, Li G (2016) Inhibitory effect and mechanism of azo dyes on anaerobic methanogenic wastewater treatment: can redox mediator remediate the inhibition? Water Res 104:408–417. https://doi.org/10.1016/j.watres.2016.08.046
25. Daud MK, Rizvi H, Akram MF, Ali S, Rizwan M, Nafees M, Jin ZS (2018) Review of upflow anaerobic sludge blanket reactor technology: effect of different parameters and developments for domestic wastewater treatment. J Chem 2018:1–13. https://doi.org/10.1155/2018/1596319
26. de Barros AN, da Silva MER, Firmino PIM, de Vasconcelos EAF, dos Santos AB (2018) Impact of microaeration and the redox mediator anthraquinone-2, 6-disulfonate on azo dye reduction and by-products degradation. Clean: Soil, Air, Water 46(8):1700518. https://doi.org/10.1002/clen.201700518

27. Demirel B, Scherer P (2008) The roles of acetotrophic and hydrogenotrophic methanogens during anaerobic conversion of biomass to methane: a review. Rev Environ Sci Bio/Technol 7(2):173–190. https://doi.org/10.1007/s11157-008-9131-1
28. dos Santos AB, Traverse J, Cervantes FJ, van Lier JB (2005) Enhancing the electron transfer capacity and subsequent color removal in bioreactors by applying thermophilic anaerobic treatment and redox mediators. Biotechnol Bioeng 89(1):42–52. https://doi.org/10.1002/bit.20308
29. dos Santos AB, Bisschops IAE, Cervantes FJ, van Lier JB (2004) Effect of different redox mediators during thermophilic azo dye reduction by anaerobic granular sludge and comparative study between mesophilic (30ºC) and thermophilic (55ºC) treatments for decolourisation of textile wastewaters. Chemosphere 55(9):1149–1157. https://doi.org/10.1016/j.chemosphere.2004.01.031
30. dos Santos AB, Cervantes FJ, van Lier JB (2007) Review paper on current technologies for decolourisation of textile wastewaters: perspectives for anaerobic biotechnology. Biores Technol 98(12):2369–2385. https://doi.org/10.1016/j.biortech.2006.11.013
31. Farooqi IH, Basheer F, Tiwari P (2017) Biodegradation of methylene blue dye by sequential treatment using anaerobic hybrid reactor and submerged aerobic fixed film bioreactor. J Inst Eng (India) Ser A 98(4):397–403. https://doi.org/10.1007/s40030-017-0251-x
32. Fazal S, Huang S, Zhang Y, Ullah Z, Ali A, Xu H (2019) Biological treatment of red bronze dye through anaerobic process. Arab J Geosci 12(13):415. https://doi.org/10.1007/s12517-019-4572-0
33. Ferraz ADN, Kato MT, Florencio L, Gavazza S (2011) Textile effluent treatment in a UASB reactor followed by submerged aerated biofiltration. Water Sci Technol 64(8):1581–1589. https://doi.org/10.2166/wst.2011.674
34. Fersi C, Gzara L, Dhahbi M (2005) Treatment of textile effluents by membrane technologies. Desalination 185(1–3):399–409. https://doi.org/10.1016/j.desal.2005.03.087
35. Field JA, Cervantes FJ, van der Zee FP, Lettinga G (2000) Role of quinones in the biodegradation of priority pollutants: a review. Water Sci Technol 42(5–6):215–222. https://doi.org/10.2166/wst.2000.0516
36. Firmino PIM, da Silva MER, Cervantes FJ, dos Santos AB (2010) Colour removal of dyes from synthetic and real textile wastewaters in one- and two-stage anaerobic systems. Biores Technol 101(20):7773–7779. https://doi.org/10.1016/j.biortech.2010.05.050
37. Florêncio TM, de Godoi LA, Rocha VC, Oliveira JMS, Motteran F, Gavazza S, Vicentine KFD, Damianovic MHRZ (2021) Anaerobic structured-bed reactor for azo dye decolorization in the presence of sulfate ions. J Chem Technol Biotechnol 96(6):1700–1708. https://doi.org/10.1002/jctb.6695
38. Frankin RJ (2001) Full-scale experiences with anaerobic treatment of industrial wastewater. Water Sci Technol J Int Assoc Water Pollut Res 44(8):1–6. https://doi.org/10.2166/wst.2001.0451
39. Frijters CTMJ, Vos RH, Scheffer G, Mulder R (2006) Decolorizing and detoxifying textile wastewater, containing both soluble and insoluble dyes, in a full scale combined anaerobic/aerobic system. Water Res 40(6):1249–1257. https://doi.org/10.1016/j.watres.2006.01.013
40. Fuchs G (2008) Anaerobic metabolism of aromatic compounds. Ann N Y Acad Sci 1125(1):82–99. https://doi.org/10.1196/annals.1419.010
41. Gadow SI, Li Y-Y (2020) Development of an integrated anaerobic/aerobic bioreactor for biodegradation of recalcitrant azo dye and bioenergy recovery: HRT effects and functional resilience. Bioresour Technol Rep 9:100388. https://doi.org/10.1016/j.biteb.2020.100388
42. Garcia J-L, Patel BKC, Ollivier B (2000) Taxonomic, phylogenetic, and ecological diversity of methanogenic archaea. Anaerobe 6(4):205–226. https://doi.org/10.1006/anae.2000.0345
43. Gonzalez-Gutierrez LV, Escamilla-Silva EM (2009) Reactive red azo dye degradation in a UASB bioreactor: mechanism and kinetics. Eng Life Sci 9(4):311–316. https://doi.org/10.1002/elsc.200900036

44. Guan Z-B, Zhang N, Song C-M, Zhou W, Zhou L-X, Zhao H, Xu C-W, Cai Y-J, Liao X-R (2014) Molecular cloning, characterization, and dye-decolorizing ability of a temperature- and pH-stable laccase from Bacillus subtilis X1. Appl Biochem Biotechnol 172(3):1147–1157. https://doi.org/10.1007/s12010-013-0614-3
45. van Haadel A, van der Lubbe J (2019) Chapter 4: UASB reactor design guidelines. In: van Haandel A, van der Lubbe J (eds) Anaerobic sewage treatment: optimization of process and physical design of anaerobic and complementary processes. IWA Publishing, pp. 133–192. https://doi.org/10.2166/9781780409627
46. Haider A, Khan SJ, Nawaz MS, Saleem MU (2018) Effect of intermittent operation of lab-scale upflow anaerobic sludge blanket (UASB) reactor on textile wastewater treatment. Desalin Water Treat 136:120–130. https://doi.org/10.5004/dwt.2018.23231
47. Holkar CR, Jadhav AJ, Pinjari DV, Mahamuni NM, Pandit AB (2016) A critical review on textile wastewater treatments: possible approaches. J Environ Manage 182:351–366. https://doi.org/10.1016/j.jenvman.2016.07.090
48. Hulshoff Pol LW, de Castro Lopes SI, Lettinga G, Lens PNL (2004) Anaerobic sludge granulation. Water Res 38(6):1376–1389. https://doi.org/10.1016/j.watres.2003.12.002
49. Hulshoff Pol LW, de Zeeuw WJ, Velzeboer CTM, Lettinga G (1983) Granulation in UASB-reactors. Water Sci Technol 15(8–9):291–304. https://doi.org/10.2166/wst.1983.0172
50. Işik M (2004) Efficiency of simulated textile wastewater decolorization process based on the methanogenic activity of upflow anaerobic sludge blanket reactor in salt inhibition condition. Enzyme Microb Technol 35(5):399–404. https://doi.org/10.1016/j.enzmictec.2004.04.018
51. Işık M, Sponza DT (2004) Decolorization of azo dyes under batch anaerobic and sequential anaerobic/aerobic conditions. J Environ Sci Health, Part A 39(4):1107–1127. https://doi.org/10.1081/ESE-120028417
52. Işik M, Sponza DT (2004) Anaerobic/aerobic sequential treatment of a cotton textile mill wastewater. J Chem Technol Biotechnol 79(11):1268–1274. https://doi.org/10.1002/jctb.1122
53. Işik M, Sponza DT (2005) Effects of alkalinity and co-substrate on the performance of an upflow anaerobic sludge blanket (UASB) reactor through decolorization of Congo Red azo dye. Biores Technol 96(5):633–643. https://doi.org/10.1016/j.biortech.2004.06.004
54. Karatas M, Dursun S, Argun ME (2010) The decolorization of azo dye Reactive Black 5 in a sequential anaerobic-aerobic system. Ekoloji 19(74):15–23
55. Karimi Estahbanati MR, Kumar S, Khajvand M, Drogui P, Tyagi RD (2021) Environmental impacts of recovery of resources from industrial wastewater. In: Biomass, biofuels, biochemicals, pp 121–162. https://doi.org/10.1016/B978-0-12-821878-5.00003-9
56. Katal R, Zare H, Rastegar SO, Mavaddat P, Darzi GN (2014) Removal of dye and chemical oxygen demand (COD) reduction from textile industrial wastewater using hybrid bioreactors. Environ Eng Manag J 13(1):43–50. https://doi.org/10.30638/eemj.2014.007
57. Kishor R, Purchase D, Saratale GD, Saratale RG, Ferreira LFR, Bilal M, Chandra R, Bharagava RN (2021) Ecotoxicological and health concerns of persistent coloring pollutants of textile industry wastewater and treatment approaches for environmental safety. J Environ Chem Eng 9(2):105012. https://doi.org/10.1016/j.jece.2020.105012
58. Koumanova B, Allen SJ (2005) Decolourisation of water/wastewater using adsorption (review). J Univ Chem Technol Metall 40:175–192
59. Lettinga G, van Velsen AFM, Hobma SW, de Zeeuw W, Klapwijk A (1980) Use of the upflow sludge blanket (USB) reactor concept for biological wastewater treatment, especially for anaerobic treatment. Biotechnol Bioeng 22(4):699–734. https://doi.org/10.1002/bit.260220402
60. Liu T, Schnürer A, Björkmalm J, Willquist K, Kreuger E (2020) Diversity and abundance of microbial communities in UASB reactors during methane production from hydrolyzed wheat straw and lucerne. Microorganisms 8(9):1394. https://doi.org/10.3390/microorganisms8091394
61. Liu Y, Xu H-L, Yang S-F, Tay J-H (2003) Mechanisms and models for anaerobic granulation in upflow anaerobic sludge blanket reactor. Water Res 37(3):661–673. https://doi.org/10.1016/s0043-1354(02)00351-2

62. Mainardis M, Buttazzoni M, Goi D (2020) Up-flow anaerobic sludge blanket (Uasb) technology for energy recovery: a review on state-of-the-art and recent technological advances. Bioengineering 7(2). https://doi.org/10.3390/bioengineering7020043
63. Mao C, Feng Y, Wang X, Ren G (2015) Review on research achievements of biogas from anaerobic digestion. Renew Sustain Energy Rev 45:540–555. https://doi.org/10.1016/j.rser.2015.02.032
64. Martins LR, Baêta BEL, Gurgel LVA, de Aquino SF, Gil LF (2015) Application of cellulose-immobilized riboflavin as a redox mediator for anaerobic degradation of a model azo dye Remazol Golden Yellow RNL. Ind Crops Prod 65(1):454–462. https://doi.org/10.1016/j.indcrop.2014.10.059
65. Methneni N, Morales-González JA, Jaziri A, Mansour HB, Fernandez-Serrano M (2021) Persistent organic and inorganic pollutants in the effluents from the textile dyeing industries: ecotoxicology appraisal via a battery of biotests. Environ Res 196:110956. https://doi.org/10.1016/j.envres.2021.110956
66. Murali V, Ong S-A, Ho L-N, Wong Y-S (2013) Decolorization of methyl orange using upflow anaerobic sludge blanket (UASB) reactor—an investigation of co-substrate and dye degradation kinetics. Desalin Water Treat 51(40–42):7621–7630. https://doi.org/10.1080/19443994.2013.782255
67. Noyola A, Padilla-Rivera A, Morgan-Sagastume JM, Güereca LP, Hernández-Padilla F (2012) Typology of municipal wastewater treatment technologies in Latin America. Clean: Soil, Air, Water 40(9):926–932. https://doi.org/10.1002/clen.201100707
68. O'Flaherty V, Collins G, Mahony T (2006) The microbiology and biochemistry of anaerobic bioreactors with relevance to domestic sewage treatment. Rev Environ Sci Bio/Technol 5(1):39–55. https://doi.org/10.1007/s11157-005-5478-8
69. Oh Y-K, Kim Y-J, Ahn Y, Song S-K, Park S (2004) Color removal of real textile wastewater by sequential anaerobic and aerobic reactors. Biotechnol Bioprocess Eng 9(5):419–422. https://doi.org/10.1007/BF02933068
70. Oktem YA, Palabiyik BB, Selcuk H (2019) Removal of colour and organic matter from textile wastewaters using two anaerobic processes. Desalin Water Treat 172:229–234. https://doi.org/10.5004/dwt.2019.24756
71. Ong S-A, Toorisaka E, Hirata M, Hano T (2005) Decolorization of azo dye (Orange II) in a sequential UASB–SBR system. Sep Purif Technol 42(3):297–302. https://doi.org/10.1016/j.seppur.2004.09.004
72. Panswad T, Luangdilok W (2000) Decolorization of reactive dyes with different molecular structures under different environmental conditions. Water Res 34(17):4177–4184. https://doi.org/10.1016/S0043-1354(00)00200-1
73. Pereira RA, Salvador AF, Dias P, Pereira MFR, Alves MM, Pereira L (2016) Perspectives on carbon materials as powerful catalysts in continuous anaerobic bioreactors. Water Res 101:441–447. https://doi.org/10.1016/j.watres.2016.06.004
74. Popli S, Patel UD (2015) Destruction of azo dyes by anaerobic–aerobic sequential biological treatment: a review. Int J Environ Sci Technol 12(1):405–420. https://doi.org/10.1007/s13762-014-0499-x
75. Rosa AP, Chernicharo CAL, Lobato LCS, Silva RV, Padilha RF, Borges JM (2018) Assessing the potential of renewable energy sources (biogas and sludge) in a full-scale UASB-based treatment plant. Renew Energy 124:21–26. https://doi.org/10.1016/j.renene.2017.09.025
76. Rozzi A, Remigi E (2004) Methods of assessing microbial activity and inhibition under anaerobic conditions: a literature review. Rev Environ Sci Bio/Technol 3(2):93–115. https://doi.org/10.1007/s11157-004-5762-z
77. Ryue J, Lin L, Kakar FL, Elbeshbishy E, Al-Mamun A, Dhar BR (2020) A critical review of conventional and emerging methods for improving process stability in thermophilic anaerobic digestion. Energy Sustain Dev 54:72–84. https://doi.org/10.1016/j.esd.2019.11.001
78. Samsami S, Mohamadi M, Sarrafzadeh M-H, Rene ER, Firoozbahr M (2020) Recent advances in the treatment of dye-containing wastewater from textile industries: Overview and perspectives. Process Saf Environ Prot 143:138–163. https://doi.org/10.1016/j.psep.2020.05.034

79. Samuchiwal S, Gola D, Malik A (2021) Decolourization of textile effluent using native microbial consortium enriched from textile industry effluent. J Hazard Mater 402:123835. https://doi.org/10.1016/j.jhazmat.2020.123835
80. Santos EMA, Nascimento ATP, do Paulino TRS, BarrosoBCS, Aguiar CR.(2016) Reator anaeróbio tipo UASB conjugado com processo Fenton para remoção de cor e demanda química de oxigênio de água residuária sintética de indústria têxtil. Engenharia Sanitaria e Ambiental 22(2):285–292. https://doi.org/10.1590/s1413-41522016148154
81. Saratale RG, Saratale GD, Chang JS, Govindwar SP (2011) Bacterial decolorization and degradation of azo dyes: a review. J Taiwan Inst Chem Eng 42(1):138–157. https://doi.org/10.1016/j.jtice.2010.06.006
82. Sarayu K, Sandhya S (2012) Current technologies for biological treatment of textile wastewater–a review. Appl Biochem Biotechnol 167(3):645–661. https://doi.org/10.1007/s12010-012-9716-6
83. Seyedi ZS, Zahraei Z, Jookar Kashi F (2020) Decolorization of reactive black 5 and reactive red 152 Azo dyes by new Haloalkaliphilic bacteria isolated from the textile wastewater. Curr Microbiol 77(9):2084–2092. https://doi.org/10.1007/s00284-020-02039-7
84. Shuchista D, Ashraful I (2015) A review on textile wastewater characterization in Bangladesh. Resour Environ 5(1):15–44. https://doi.org/10.5923/j.re.20150501.03
85. Silva SQ, Silva DC, Lanna MCS, Baeta BEL, Aquino SF (2014) Microbial dynamics during azo dye degradation in a UASB reactor supplied with yeast extract. Braz J Microbiol 45(4):1153–1160. https://doi.org/10.1590/S1517-83822014000400005
86. Singh K, Arora S (2011) Removal of synthetic textile dyes from wastewaters: a critical review on present treatment technologies. Crit Rev Environ Sci Technol 41(9):807–878. https://doi.org/10.1080/10643380903218376
87. Somasiri W, Ruan W, Xiufen L, Jian C (2006) Decolourization of textile wastewater containing acid dyes in UASB reactor system under mixed anaerobic granular sludge. Elec J Env Agricult Food Chem Title 5(1):1224–1234
88. Souza ME (1986) Criteria for the utilization, design and operation of UASB reactors. Water Sci Technol 18(12):55–69. https://doi.org/10.2166/wst.1986.0163
89. Speece RE (1983) Anaerobic biotechnology for industrial wastewater treatment. Environ Sci Technol 17(9):416A-427A. https://doi.org/10.1021/es00115a001
90. Tawfik A, Zaki D, Zahran M (2014) Use of sequential UASB/DHS processes for the decolorization of reactive dyes wastewater. Sustain Environ Res 24(2):129–138
91. van Lier JB, van der Zee FP, Frijters CTMJ, Ersahin ME (2015) Celebrating 40 years anaerobic sludge bed reactors for industrial wastewater treatment. Rev Environ Sci Bio/Technol 14(4):681–702. https://doi.org/10.1007/s11157-015-9375-5
92. Varjani S, Rakholiya P, Shindhal T, Shah AV, Ngo HH (2021) Trends in dye industry effluent treatment and recovery of value added products. J Water Process Eng 39:101734. https://doi.org/10.1016/j.jwpe.2020.101734
93. Venkatesh S, Venkatesh K, Quaff AR (2017) Dye decomposition by combined ozonation and anaerobic treatment: Cost effective technology. J Appl Res Technol 15(4):340–345. https://doi.org/10.1016/j.jart.2017.02.006
94. Verma AK, Bhunia P, Dash RR (2016) Performance of UASB reactor treating synthetic textile wastewater: effect of physicochemical pretreatment. Desalin Water Treat 57(18):8050–8060. https://doi.org/10.1080/19443994.2015.1017739
95. Verma AK, Dash RR, Bhunia P (2012) A review on chemical coagulation/flocculation technologies for removal of colour from textile wastewaters. J Environ Manage 93(1):154–168. https://doi.org/10.1016/j.jenvman.2011.09.012
96. Volschan Junior I, de Almeida R, Cammarota MC (2021) A review of sludge pretreatment methods and co-digestion to boost biogas production and energy self-sufficiency in wastewater treatment plants. J Water Process Eng 40:101857. https://doi.org/10.1016/j.jwpe.2020.101857
97. Wang W, Wu B, Pan S, Yang K, Hu Z, Yuan S (2017) Performance robustness of the UASB reactors treating saline phenolic wastewater and analysis of microbial community structure. J Hazard Mater 331:21–27. https://doi.org/10.1016/j.jhazmat.2017.02.025

98. Wei J, Hao X, van Loosdrecht MCM, Li J (2018) Feasibility analysis of anaerobic digestion of excess sludge enhanced by iron: a review. Renew Sustain Energy Rev 89:16–26. https://doi.org/10.1016/j.rser.2018.02.042
99. Wuhrmann K, Mechsner K, Kappeler T (1980) Investigation on rate-determining factors in the microbial reduction of azo dyes. Eur J Appl Microbiol Biotechnol 9(4):325–338. https://doi.org/10.1007/BF00508109
100. Xu H, Yang B, Liu Y, Li F, Shen C, Ma C, Tian Q, Song X, Sand W (2018) Recent advances in anaerobic biological processes for textile printing and dyeing wastewater treatment: a mini-review. World J Microbiol Biotechnol 34(11):165. https://doi.org/10.1007/s11274-018-2548-y
101. Yadvika S, Sreekrishnan TR, Kohli S, Rana V (2004) Enhancement of biogas production from solid substrates using different techniques-a review. Biores Technol 95(1):1–10. https://doi.org/10.1016/j.biortech.2004.02.010
102. Yasar A, Tabinda AB (2010) Anaerobic treatment of industrial wastewater by UASB reactor integrated with chemical oxidation processes; an overview. Pol J Environ Stud 19(5):1051–1061
103. Yentekakis IV, Goula G (2017) Biogas management: advanced utilization for production of renewable energy and added-value chemicals. Front Environ Sci 5:7. https://doi.org/10.3389/fenvs.2017.00007
104. Yildirir E, Ballice L (2019) Supercritical water gasification of wet sludge from biological treatment of textile and leather industrial wastewater. J Supercrit Fluids 146:100–106. https://doi.org/10.1016/j.supflu.2019.01.012
105. Zhang W, Liu F, Wang D, Jin Y (2018) Impact of reactor configuration on treatment performance and microbial diversity in treating high-strength dyeing wastewater: anaerobic flat-sheet ceramic membrane bioreactor versus upflow anaerobic sludge blanket reactor. Biores Technol 269:269–275. https://doi.org/10.1016/j.biortech.2018.08.126
106. Zin KM, Effendi Halmi MI, Abd Gani SS, Zaidan UH, Samsuri AW, Abd Shukor MY (2020) Microbial decolorization of triazo dye, direct blue 71: an optimization approach using Response Surface Methodology (RSM) and Artificial Neural Network (ANN). Biomed Res Int 2020:1–16. https://doi.org/10.1155/2020/2734135

Current Biological Approaches in Dye Wastewater Treatment Review on Sequential Aerobic/Anaerobic Batch Reactors for Dye Removal

Diwakar Kumar and Sunil Kumar Gupta

Abstract Due to the abrupt increase in the use of a variety of dyes and their production, the wastewater generated from this industry became major concerns sources of wastewater pollution. In recent years, the biodegradation of dyes was extensively explored. Due to the operational flexibility and the possibilities of excellent process control sequencing aerobic/anaerobic batch reactors (SBR) are extensively used in dye wastewater treatment. This chapter reviews the anaerobic/aerobic condition for the biodegradation of dye wastewater. Furthermore, we summed up the potential dye removal mechanisms in order to lay the foundation for the biological removal of dye wastewater. This chapter summarises the review on (1) the various classifications of dyes, (2) various treatment methods for dye removal, (3) detailed study of an aerobic/anaerobic sequential batch reactor, (4) anaerobic/aerobic system with high-rate bioreactor and their types used in the textile industry, and (5) advantages and disadvantages of aerobic/anaerobic SBR system.

1 Introduction

The dye wastewater production is increasing abruptly in recent years and will be increasing in upcoming years. Every year, over 100,000 tonnes of dyes are produced, approximately 10% of estimated dyes are released to the environment, potentially causing severe harmful effects on human health and aquatic life [1–3]. According to applications of various synthetic dyes, every year approximately 300,000 tonnes of several dyes are used for dyeing operations; thus synthetic dyes play an important role in polluting the water [4]. During the dyeing process, anthraquinone,

D. Kumar
Department of Environmental Science and Engineering, Indian Institute of Technology (Indian School of Mines), Dhanbad, Jharkhand 826004, India

S. K. Gupta (✉)
Professor, Department of Environmental Science and Engineering, Indian Institute of Technology (Indian School of Mines), Dhanbad, Jharkhand 826004, India
e-mail: sunil@iitism.ac.in

A. Khadir et al. (eds.), *Biological Approaches in Dye-Containing Wastewater*, Sustainable Textiles: Production, Processing, Manufacturing & Chemistry,
https://doi.org/10.1007/978-981-19-0545-2_8

naphthalene, benzene, and other constituents are used as raw materials and typically chelate compounds with salts or metals to produce dye wastewater containing acids, alkalis, hydrocarbons, salts, amine, nitro, halogen, the related intermediates, and other compounds. Moreover, few dyes are degraded to generate carcinogenic aromatic amines or dyes containing phenol, pyridine, cyanide, and heavy metals such as cadmium, mercury, and chromium. The constituents of wastewater are toxic and complex [5–7]. Various industries use a large volume of synthetic dyes, including textile dyeing (60%), paper (10%), and plastic matter (10%) [8, 9]. Textile effluents are typically highly coloured, with high BOD, COD, TDS, TSS, and heavy metals traces [10–14]. Because of their carcinogenic nature, even small amounts of dye in water can have negative environmental consequences. To avoid the dye's negative effects, the development of cost-effective and eco-friendly dye degradation technology for wastewater is inevitable. As a result, thorough research has started to develop innovative techniques for the mitigation of wastewater containing dye. The biological methods have relatively low operating costs and produce stable results, whereas the physicochemical methods have few limitations, such as increased operational costs, interference causing compounds to the wastewater, and the production of substantial quantities of sludge. As a result, biological methods for treating dye-containing wastewater have become increasingly popular [15–18]. The biological degradation of dyes can result in high solubility in water, high molecular weight, and fused aromatic ring structures that constrain permeability through biological cellular membranes. Textile dyes are classified as basic, cationic, anthraquinone-based, azo, diazo, or metal complex dyes based on their chemical configuration. Dye molecular structures, microorganisms, enzymatic activities, and other factors that influence dye biodegradation pathways. As a result, the dye biodegradation requires anaerobic, aerobic, combined anaerobic/aerobic, and microbial treatment. Conventional sequencing batch reactor (SBR) treatment systems have various steps of fill, react, settle, decant, and idle with full aeration during the react period to oxidise the organic matter and degrade dye. For efficacious colour and COD removal, the aerobic/anaerobic SBR technology is used in the treatment of dye-containing wastewater [19–21]. The roadmap of the chapter is described in Fig. 1.

2 Dye Classification

Dyes are classified according to their origin, structure, colour, and application [22–24]. Dyes can be organic or synthetic in origin, and they can be categorised chemically into acridine, anthraquinone, chromophores, nitroso, and azin dyes. Dispersive dyes, vat dyes, and azo dyes are examples of dye classification based on application. Dyes can also be categorised as per the particle charge when dissolved into the water, i.e., anionic, cationic, and non-ionic dyes [25]. Due to the complexities of colour nomenclature, it is often advantageous to classify dyes based on application. Anionic and non-ionic dyes typically contain azo groups or anthraquinone dyes as chromophoric groups. Azo dyes containing the –N=N– group are the most frequently

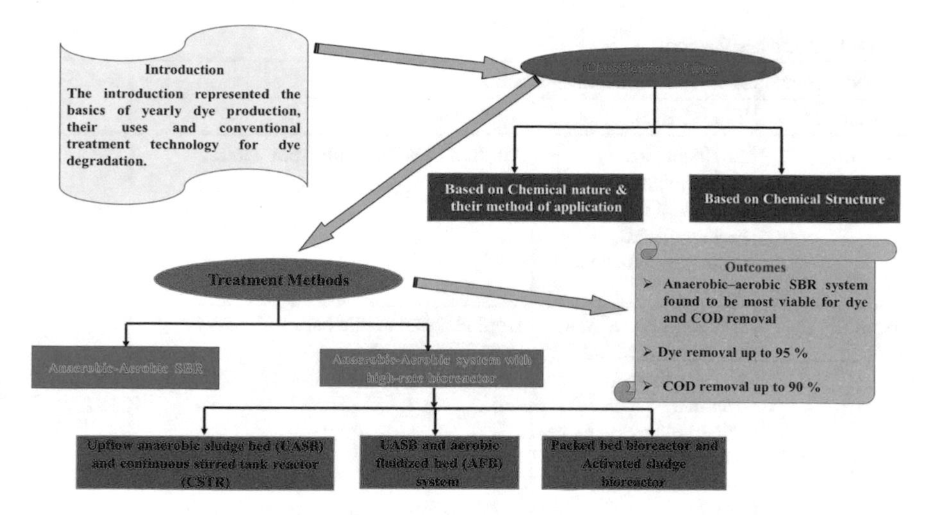

Fig. 1 Roadmap of the book chapter

used dyes in the textile industry, accounting for more than half of the annual global dye production [26]. Anthraquinone is the fundamental unit of this group of dyes. Disperse dyes are a type of dye that is used to colour cellulose acetate, nylon, and hydrophobic fibres. Sulphoricinoleic acid is the most commonly used dispersing agent in these dyes. Dyes are classified on the basis of their chemical nature and application, and the chemical structure is shown in Table 1 [25, 27].

3 Literature Review

Literature is reviewed and some selected literature of the past 15 years on the degradation of dyes are shown and listed in Table 2, which shows the colour removal percentage along with the percentage COD removal of the different dyes by the SBR system or in combination of SBR with the other system. The SBR system is efficient in the removal of colour as well as COD removal for different dye concentration.

4 Treatment Techniques

Biological and physicochemical are two significant methodologies for the treatment of dyes. Oxidation, irradiation/ozonation, flocculation, adsorption, coagulation, membrane filtration, precipitation bleaching, and ion-exchange are the traditional methods for dealing with textile effluents or dye-containing wastewater. These dye remediation methods have the disadvantages of being inefficient, expensive,

Table 1 Dye classifications

Based on chemical composition and application			Reference
Categories	Type of dyes	Application	
Acidic	Xanthene, azo, anthraquinone (including nitroso premetallised), triphenylmethane and nitro	Applied to acidic bath from neutral	[25]
Basic	Anthraquinone, Azo, xanthene, acridine, triarylmethane, cyanine, azinediphenylmethane, diazahemicyanine, hemicyanine, and oxazine	Applied from acidic bath	
Direct	Oxazine, stilbene, phthalocyanine, and azo	Neutral or alkaline bath comprising a supplementary electrolyte	
Disperse	Anthraquinone, azo, styryl, benzodifuranone, and nitro	Applied via pressure/temperature transporter methods	[27]
Reactive	Anthraquinone, formazan, phthalocyanine, azo, oxazine and basic	The reaction takes place between the reactive site of dye and the functional group of fibre	
Sulphur	Structures that are indefinite	Aromatic substrate vetted with Na_2S and deoxidised on fibre to non-soluble sulphur products	
Vat	Anthraquinone and indigoids	The dye can be made water-soluble by dipping it in $NaHSO_3$ and then applied to the fibre	

Chemical structure based		
Dyes	Chemical structure	Reference
Azo dyes	—N=N—	[25]
Nitro dyes	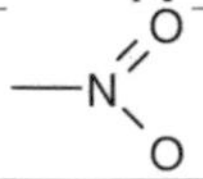 O —N O	[27]
Indigoid dyes	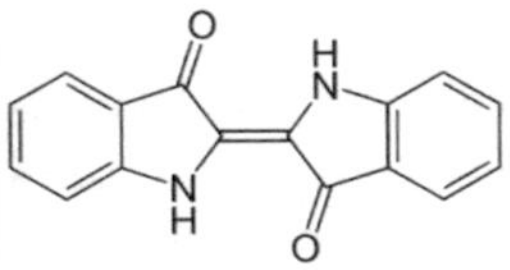 O H N N H O	[28]

(continued)

Table 1 (continued)

Based on chemical composition and application			Reference
Categories	Type of dyes	Application	
Nitroso dyes	—N=O		[27]
Anthraquinone dyes	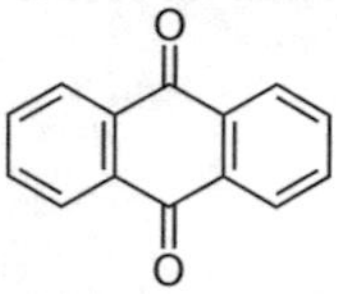		[28]
Triarylmethane dyes	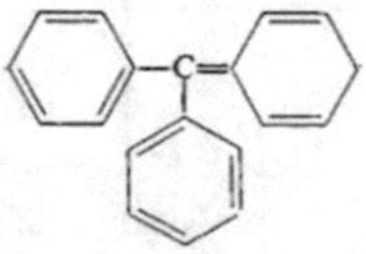		[25]

Table 2 Literature on dyes degradation by SBR

Mechanism	Dye	Dye concentration	% Removal	Reference
SBR	Methylene blue	4–10 mg/l	Color = 56% COD = 93%	[29]
Sequencing batch moving bed biofilm reactor (SBMBBR)	Mono azo dye-reactive orange 16	50–1000 mg/l	Color = 100% COD = 98.54% (Dye conc. 50 mg/L)	[30]
SBR	Azo dyes	50–100 mg/l	Color = 99.0% COD = 92.84% BOD = 90.62%	[31]
Anaerobic sequencing batch reactor (AnSBR)	Reactive red 159	500 mg/l	Color = 97.68%	[32]
SBR	Remazol brilliant violet 5R dye	50–200 mg/l	Color = 92% COD = 75%	[33]
Moving bed sequencing batch biofilm reactor/anaerobic sequencing batch reactor (MBSBBR /AnSBR)	Azo dye acid red 18	100 mg/l and 500 and 1000 mg/l (another reactor)	Color = 97% COD = 83.07%	[34]
SBR	Remazol yellow RR, Remazol blue RR and Remazol red RR	100 and 1000 mg/l	Color = 95% COD = 70%	[35]

requiring a lot of energy, and producing a lot of waste. Aside from these traditional methods, bioremediation has recently received significant attention with comparatively inexpensive and reasonably effective potential treatment for textile effluents. The sub-categorisation of the treatment methods is depicted schematically in Fig. 2. Fig. 3 depicts a classification based on the aerobic/anaerobic system.

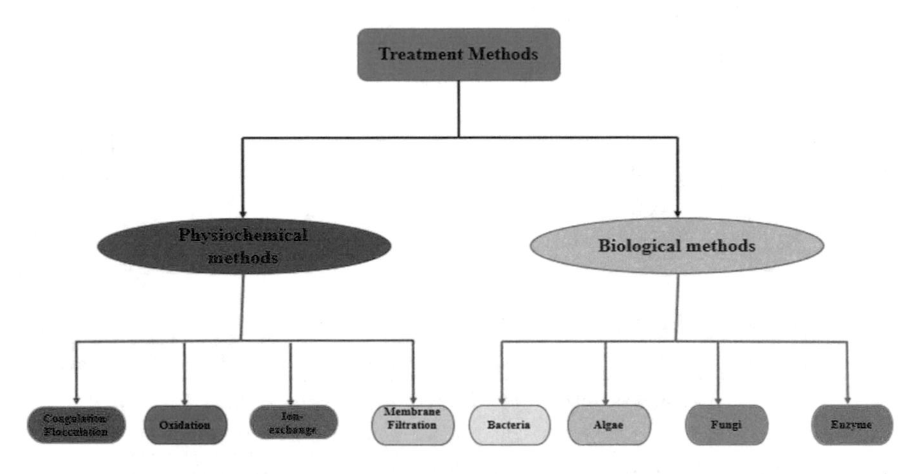

Fig. 2 Schematic diagram of dye treatment techniques. (*Source* Ihsanullah et al. [27] Journal of Water Process Engineering)

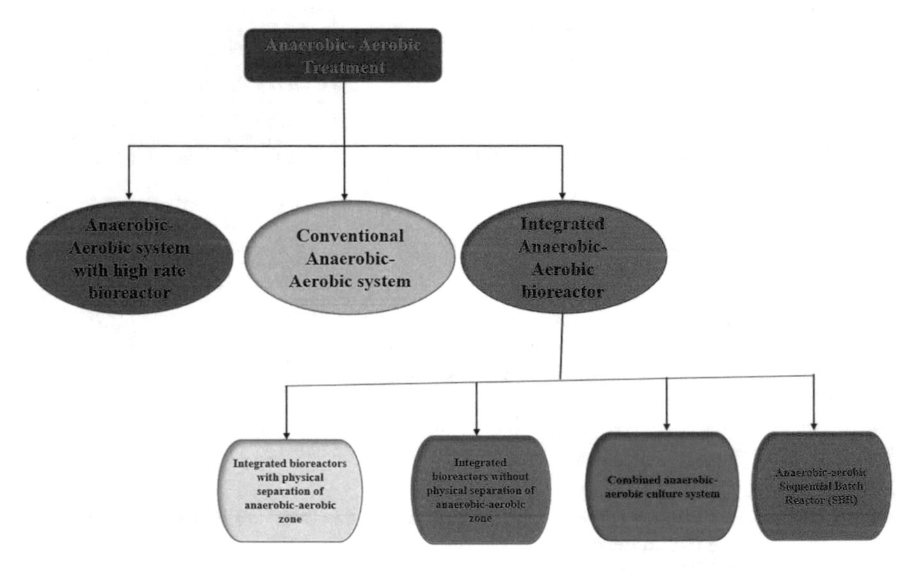

Fig. 3 Schematic presentation of the integrated anaerobic/aerobic process adopted and modified. (*Source* Chan et al. [22] Chemical Engineering Journal)

Many different physicochemical treatment methods include adsorption [36–39], coagulation and flocculation [40, 41], ion-exchange [42], oxidation [43], photodegradation [44–46], and membrane separation [47].

Alongside the textile wastewater treatment, conventional methods are used. Bioremediation is an economical alternative with the advantages of less production of harmless products. However, the treatment process is complicated because of the presence of microorganisms and the complex nature of dyes themselves. The following sections discuss numerous bioremediation techniques for dye effluents remediation. Bioremediation is a biological method that converts organic wastes into non-toxic substances or minimises the concentration to a reasonable standard [21]. In current years, there has been a strong emphasis on compacted high-rate bioreactors for wastewater treatment for biomass production and to meet standards. As a result, incorporated bioreactors that integrate anaerobic and aerobic operations in a single reactor come into consideration. The integration of anaerobic and aerobic decomposition in a single reactor significantly improves degradation efficiency [20]. Such a type of aerobic/anaerobic system is classified as shown in Fig. 3. In these systems one of the treatment systems is anaerobic/aerobic SBR. The extensive study about anaerobic/aerobic sequencing batch reactor will be discussed in segment 4.

4.1 Anaerobic/aerobic SBR

The conventional SBR treatment process uses the five stages, i.e., fill, react, settle, decant, and idle cyclically with full aeration in the react stage to oxidise organic compounds as shown in Fig. 5. SBR systems were lately being modified by modifying the react stages to provide the anaerobic (Fig. 4) and aerobic processes in a specific number and sequence for nutrient (C, N, and P) elimination. Anaerobic/aerobic SBR technologies are now aggressively marketed for the treatment of textile industry wastewater for effective COD and colour removal [19, 48]. The concentration of microbes can be effortlessly increased by interchanging anaerobic/aerobic processes via aeration in the anaerobic/aerobic SBR system. The concentration of oxygen, mixing, and time duration can all be adjusted to meet the needs of the specific treatment plant. In an anaerobic/aerobic SBR, pure nitrogen gas is injected when the anaerobic process is active, and the air is supplied when the aerobic process is active, as shown in Fig. 5. COD removal was achieved at a higher rate of 90% at a cycle of 17.5/2.5 h with anaerobic/aerobic compared to 87% at a cycle of 14/6 h by anaerobic/aerobic [49] in textile wastewater treatment. It revealed that the time duration in the anaerobic phase must be extended sufficient to achieve better COD and colour removal [50]. Mohan and Babu [51] discovered that the degradation rate of dyes in anaerobic conditions was greater than that in aerobic and anaerobic conditions, leading them to believe that anaerobic conditions provided the advantages of both aerobic and anaerobic dye degradation. The SBR diagram is depicted in Fig. 6. Melgoza and Cruz [52] investigated the dye degradation using anaerobic/aerobic process and discovered that disperse blue 79 (DB79) dye was transformed into

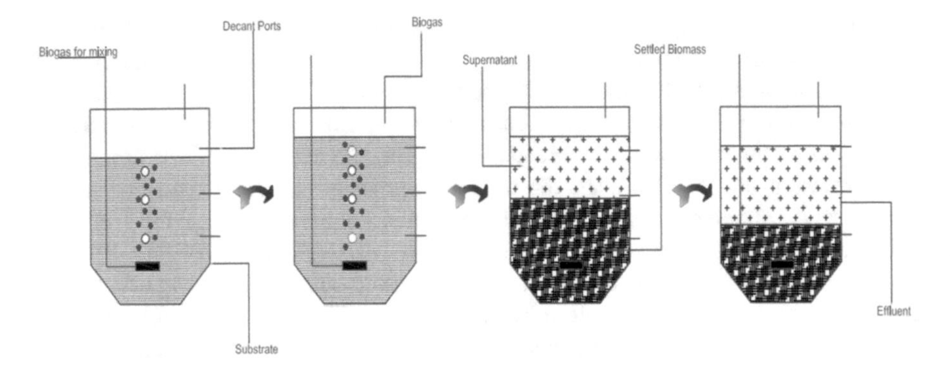

Fig. 4 Process involved in the anaerobic sequential batch reactor adopted and modified. (*Source* S.K. Khanal, Anaerobic Biotechnology for Bioenergy Production, Wiley-Blackwell, Ames, IA, USA, 2008)

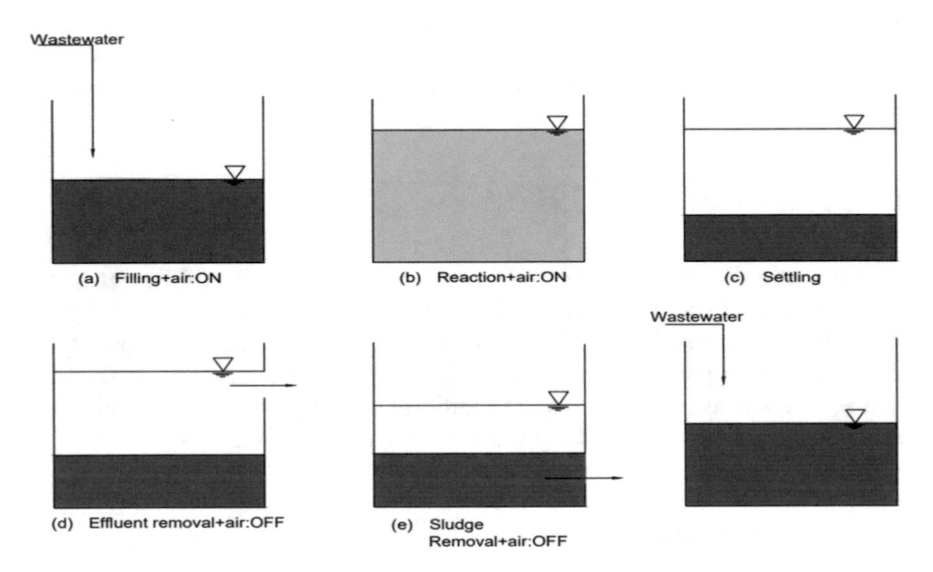

Fig. 5 Process involved in the sequential batch reactor (SBR). (*Source* S. Copelli et al. (2015) Reference Module in Chemistry, Molecular Sciences and Chemical Engineering)

aromatic amines in the anaerobic process, allowing microbes to decolourise dye. The azo dye degradation was aided by a sequential anaerobic/aerobic system [51].

4.2 Anaerobic/aerobic System with High-Rate Bioreactor

The sequence and combination of treatment options are crucial to effective industrial and municipal wastewater management. Reactors for the bioconversion of substrates

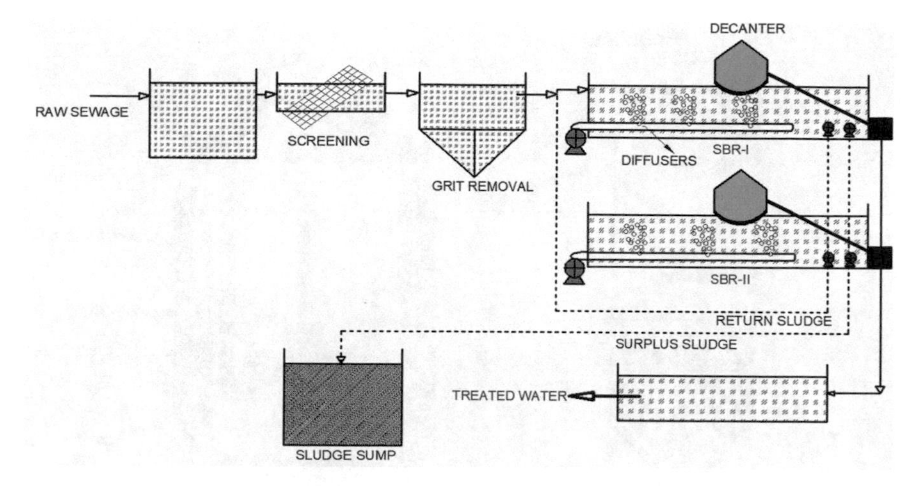

Fig. 6 Flow diagram of the sequential batch reactor (SBR) adopted and modified. (*Source* Shaw et al. 2002) Water Research)

into desirable outcomes can be designed based upon the organism's growth necessities. Textile industry wastewater, starch industry wastewater, slaughterhouse wastewater, wastewater, and primary municipal wastewater have all been treated using combined aerobic and anaerobic bioreactors. High COD removal can be achieved in anaerobic/aerobic process with high-rate bioreactors while using less HRT (from hours to days). As a result, anaerobic/aerobic treatment is a viable method for treating industrial wastewater.

4.2.1 Upflow Anaerobic Sludge Bed (UASB) and Continuous Stirred Tank Reactor (CSTR) System

Biofilm bioreactors are bioreactors that use microorganisms primarily attached on the surface and adhered inside the reactor and might be used for wastewater treatment. Biofilm organisms absorb and degrade toxic substances of the wastewater. Examples include UASB, fluidised bed, airlift, membrane, and packed bed reactors. Suspended growth bioreactors use the metabolism of microbes to use substrates and decompose them into residues under anaerobic, aerobic, or sequential anaerobic/aerobic conditions. Batch reactors, plug flow reactors, CSTRs etc. are a few examples.

UASB technology was developed to treat wastewater with the help of anaerobic digesters. The UASB bioreactor is equipped with a three-phase separator system that assists the reactor in separating water, gas, and sludge mixtures in the presence of high turbulence. Figure 7a depicts a schematic diagram of the UASB reactor, which comprises an upflow of the influent against a closely packed sludge bed with high bacterial metabolism. The most frequently used process in aerobic treatment is the

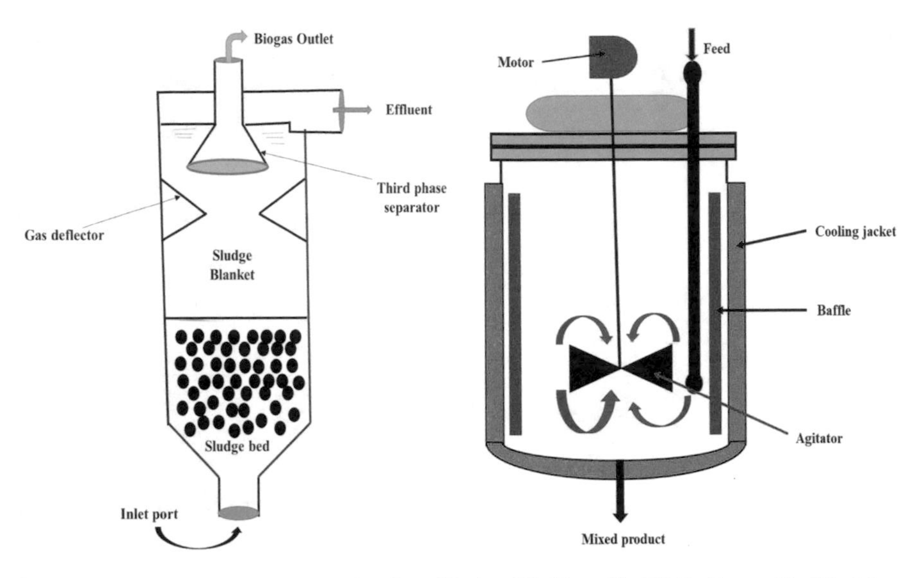

Fig. 7 A schematic diagram adopted and modified. **a** UASB and **b** CSTR. *Source* Modified from van Haandel & Lettinga (1994) and Wang (1994)

activated sludge process, which is aerated and stirred flocculated suspension of a diverse population of bacteria that make contact with wastewater [53]. The activated sludge process can be CSTR, as shown in Fig. 7b, or SBR. The UASB reactor wastes the total sludge produced in the anaerobic/aerobic system. Many researchers have ascertained the effectiveness of the aerobic CSTR/UASB reactor for dealing with an extensive range of industrial effluents with BOD/COD ratios ranging from 0.17 to 0.74 [54]. The aerobic CSTR/UASB systems can remove 83–98% of COD with influent COD varying from 500 to 20,000 mg/l as HRT varied from 11.54 h to 6 days [55–58].

4.2.2 Aerobic Fluidised Bed (AFB) and UASB System

The biomass in the AFB system is fastened to small suspended particles such as sand, high-density plastic beads, anthracite, etc. due to upward flow. Biomass advancement and nutrient uptake enhance because of the greater surface area of granules and a greater degree of mixing. The AFB system eliminates the limitations of the substrate diffusion that are typically implicit in the fixed bed system (Fig. 8). The AFB reactor has several advantages, including high OLR, high concentration of biomass, clogging of bed restricted, less HRT, large surface area for mass transfer, and low mass transfer resistance [59–61]. It was discovered that by operating the three-phase AFB even at a low average HRT of 30 min, 82% of COD removal was achieved having a COD load of 180 mg/l of synthetic wastewater [62]. When the UASB and AFB reactors are combined, 75% of COD removal can be achieved at 14 h of HRT for treating the

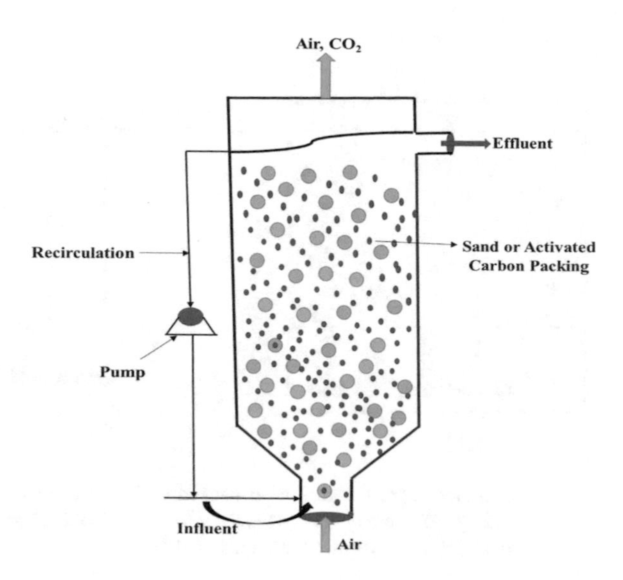

Fig. 8 Schematic representation of AFB system adopted and modified. (*Source* Biological wastewater treatment in warm climate regions, p. 717, ISBN 9,781,843,390,022)

synthetic textile wastewater having COD of 2700 mg/l [63]. According to findings, when the UASB-AFB is combined, they produced 45% less sludge volume than any other aerobic system.

4.2.3 Packed Bed Bioreactor and Activated Sludge Bioreactor

The PBR is frequently used in biochemical processing and comprises a cylindrical vessel packed with catalyst materials or adsorbents such as granular activated carbon, zeolite pellet, etc. The packed bed is acclimated to develop and improve the interface between the two phases involved in this process. To achieve a high surface area, the catalyst must be highly porous. Because of the high surface area of the catalyst, reactants easily diffuse into the pore spaces and are adsorbed at the active site. At the end of the process, the desorption leads to the products that get reverted back into the bulk. To minimise channelling, the bed height-to-diameter ratio should be around 0.5 and the column-to-particle diameter ratio should be 10–20. Figure 9a depicts a schematic diagram of a packed bed bioreactor. The PBR reactor is being used to decompose the mixtures of dye [64]. The PBR is generally preferred due to the less maintenance and cost, but the limitation of this reactor is that the deactivation of catalyst took place when the active sites get clogged.

The proportion of microbes, amount of oxygen, and substrate in an activated sludge bioreactor are all the same because this type of reactor has a homogeneous tank in which the feed is distributed across the reactor. The activated sludge process involves (i) a microbial suspension is added to the wastewater that is fed into the aeration tank, (ii) separating solids and liquids, (iii) processing of waste discharge, and (iv) the remaining biomass are recirculated in the aeration tank. Figure 9b depicts

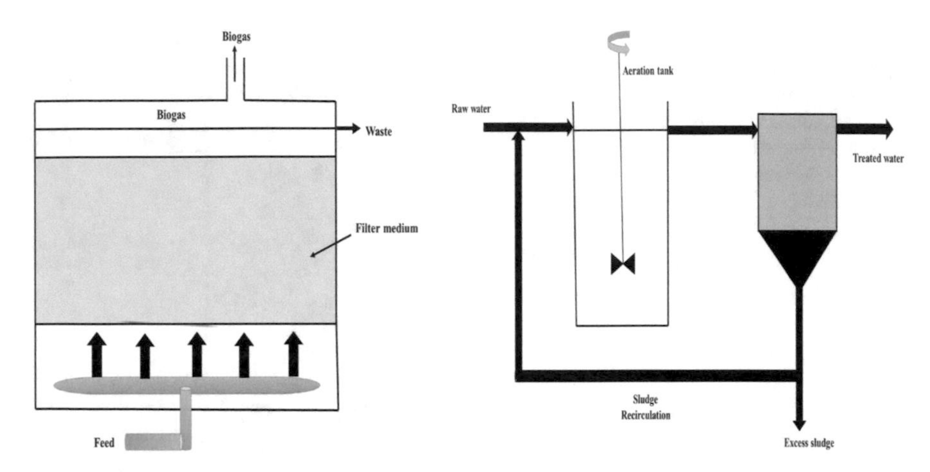

Fig. 9 Schematic representation of **a** packed bed bioreactor and **b** activated sludge bioreactor. Source *Working principle of typical bioreactors* (Sustainable Design and Industrial Applications in Mitigation of GHG Emissions 2020, pp. 145–173)

a schematic representation of an activated sludge bioreactor. The performance of anaerobic/aerobic systems with high-rate bioreactors for textile industry effluent is summarised in Table 3.

Table 3 Anaerobic/aerobic systems with high-rate bioreactors

No	Systems	Wastewater	HRT (hr/day)	COD (Influent) (mg/l)	COD removal (%)	References
1	UASB + AFB	Textile wastewater (synthetic)	20 h	2700	80	[63]
2	UASB + CSTR	Wool acid dying wastewater	3.3 days	499–2000	83–97	[65]
3	UASB + AFB	textile wastewater (synthetic)	2.7–32.7 h	2000–3000		[66]
4	UASB + CSTR	Simulated textile wastewater	19.17 h–1.22 day	4214	91–97	[67]
5	UASB + CSTR	Textile wastewater (Cotton mill)	5.75 days	604–1038	40–85	[55]
6	Packed bed Reactor + AS	Textile industry wastewater	22–82 h	800–1200	50–85	[68]

Table 4 Advantages/disadvantages (anaerobic/aerobic SBR)

Anaerobic/aerobic SBR	Advantages	Disadvantages	Reference
	There is no need for a clarifier though it has a single tank configuration	There is complex anaerobic/aerobic microbial consortium control	[22]
	There are low capital and energy requirements	Special care must be taken to govern anaerobic and aerobic residence time	
	Flexible operation	A quality standard of superiority is necessary	

5 Advantages/Disadvantages (Anaerobic/Aerobic SBR)

Every system has its own set of advantages and disadvantages. Table 4 lists some of the advantages and disadvantages of anaerobic/aerobic SBR system. The combined anaerobic/aerobic and anaerobic/aerobic SBR bioreactors need less amount of energy than either type of combined bioreactors, as there was a need for less aeration in the treatment process.

6 Conclusions

The presented chapter reviewed various types of dyes used and their classifications. Anaerobic/aerobic SBR system demonstrated as the most viable technology for the removal of organic matter from dye owing to its highest efficiency, i.e., colour and COD removal up to 95 and 90%, respectively. Flexibility in operation and low capital cost make the process the most techno-economical than others. However, more research is needed on the regulation of anaerobic/aerobic microbial community, biomass production, and the ability to recover from organic shock loads.

References

1. Castro E, Avellaneda A, Marco P (2014) Combination of advanced oxidation processes and biological treatment for the removal of benzidine-derived dyes. Environ Prog Sustainable Energy 33(3):873–885
2. Shafqat, M., et al (2017) Evaluation of bacteria isolated from textile wastewater and rhizosphere to simultaneously degrade azo dyes and promote plant growth. J Chem Technol Biotechnol 92(10):2760–2768
3. Almeida, F.P., et al. (2017) Mesoporous aluminosilicate glasses: potential materials for dye removal from wastewater effluents. J Solid State Chem 253:406–413
4. Bechtold T, Burtscher E, Turcanu A (2001) Cathodic decolourisation of textile waste water containing reactive dyes using a multi-cathode electrolyser. J Chem Technol Biotechnol 76(3):303–311

5. Pan F, et al (2017) Application of magnetic OMS-2 in sequencing batch reactor for treating dye wastewater as a modulator of microbial community. J Hazard Mater 340:36–46
6. Zeng, Q-F, et al Photooxidation degradation of reactive brilliant red K-2BP in aqueous solution by ultraviolet radiation/sodium hypochlorite. CLEAN–Soil Air Water 37(7):574–580
7. Lumbaque EC, et al (2017) Degradation and ecotoxicity of dye reactive black 5 after reductive-oxidative process. Environ Sci Pollut Res 24(7):6126–6134
8. Hassaan MA, El Nemr A, Madkour FF (2017) Testing the advanced oxidation processes on the degradation of Direct Blue 86 dye in wastewater. Egypt J Aquat Res 43(1):11–19
9. El Nemr A, et al (2009) Removal of direct blue-86 from aqueous solution by new activated carbon developed from orange peel. J Hazard Mater 161(1):102–110
10. Yogalakshmi KN, et al (2020) Nano-bioremediation: a new age technology for the treatment of dyes in textile effluents. In: Bioremediation of Industrial Waste for Environmental Safety. Springer, Singapore, pp 313–347
11. Fazal T, et al (2018) Bioremediation of textile wastewater and successive biodiesel production using microalgae. Renew Sustain Energy Rev 82:3107–3126
12. Khandegar V, Saroha AK (2013) Electrocoagulation for the treatment of textile industry effluent–a review. J Environ Manage 128:949–963
13. Bisschops I, Spanjers H (2003) Literature review on textile wastewater characterisation. Environ Technol 24(11):1399–1411
14. Yaseen DA, Scholz M (2019) Textile dye wastewater characteristics and constituents of synthetic effluents: a critical review. Int J Environ Sci Technol 16(2):1193–1226
15. Wijetunga S, Li X-F, Jian C (2010) Effect of organic load on decolourization of textile wastewater containing acid dyes in upflow anaerobic sludge blanket reactor. J Hazard Mater 177(1–3):792–798
16. Nouren S, et al By-product identification and phytotoxicity of biodegraded direct yellow 4 dye. Chemosphere 169:474–484
17. Bilal M, Asgher M (2015) Sandal reactive dyes decolorization and cytotoxicity reduction using manganese peroxidase immobilized onto polyvinyl alcohol-alginate beads. Chem Cent J 9(1):1–14
18. Holkar CR, et al (2016) A critical review on textile wastewater treatments: possible approaches. J Environ Manag 182:351–366
19. Kapdan IK, Oztekin R (2006) The effect of hydraulic residence time and initial COD concentration on color and COD removal performance of the anaerobic–aerobic SBR system. J Hazard Mater 136(3):896–901
20. Penha S, Matos M, Franco F (2005) Evaluation of an integrated anaerobic/aerobic SBR system for the treatment of wool dyeing effluents. Biodegradation 16(1):81–89
21. Supaka N, et al (2004) Microbial decolorization of reactive azo dyes in a sequential anaerobic–aerobic system. Chem Eng J 99(2):169–176
22. Chan YJ, et al A review on anaerobic–aerobic treatment of industrial and municipal wastewater. Chem Eng J 155(1–2):1–18
23. Pasukphun N, Vinitnantharat S (2003) Degradation of organic substances and reactive dye in an immobilized-cell sequencing batch reactor operation on simulated textile wastewater. J Environ Sci Health Part A 38(10):2019–2028
24. Vikrant K, et al (2018) Recent advancements in bioremediation of dye: current status and challenges. Bioresource Technol 253:355–367
25. Yagub MT, et al (2014) Dye and its removal from aqueous solution by adsorption: a review. Adv Coll Interface Sci 209:172–184
26. Giovanella P, et al (2020) Metal and organic pollutants bioremediation by extremophile microorganisms. J Hazard Mater 382:121024
27. Ihsanullah I, et al (2020) Bioremediation of dyes: current status and prospects. J Water Process Eng 38:101680
28. Ali H (2010) Biodegradation of synthetic dyes—a review. Water Air Soil Pollut 213(1):251–273
29. Ma D-Y, et al (2011) Aerobic granulation for methylene blue biodegradation in a sequencing batch reactor. Desalination 276(1–3):233–238

30. Ong C, Lee K, Chang Y Biodegradation of mono azo dye-reactive orange 16 by acclimatizing biomass systems under an integrated anoxic-aerobic REACT sequencing batch moving bed biofilm reactor. J Water Process Eng 36:101268
31. Sandhya S et al (2005) Microaerophilic–aerobic sequential batch reactor for treatment of azo dyes containing simulated wastewater. Process Biochem 40(2):885–890
32. Srisuwun A, et al (2018) Decolorization of reactive red 159 by a consortium of photosynthetic bacteria using an anaerobic sequencing batch reactor (AnSBR). Preparat Biochem Biotechnol 48(4):303–311
33. Çınar Ö, et al Effect of cycle time on biodegradation of azo dye in sequencing batch reactor. Process Saf Environ Protect 86(6):455–460
34. Koupaie EH, Moghaddam MA, Hashemi SH (2011) Post-treatment of anaerobically degraded azo dye acid red 18 using aerobic moving bed biofilm process: enhanced removal of aromatic amines. J Hazard Mater 195:147–154
35. Jonstrup M, et al (2011) Sequential anaerobic–aerobic treatment of azo dyes: decolourisation and amine degradability. Desalination 280(1–3):339–346
36. Zubair M, et al (2018) Starch-NiFe-layered double hydroxide composites: efficient removal of methyl orange from aqueous phase. J Mol Liq 249:254–264
37. Ghaedi M et al (2015) Modeling of competitive ultrasonic assisted removal of the dyes–Methylene blue and Safranin-O using Fe3O4 nanoparticles. Chem Eng J 268:28–37
38. Mohammadi N, et al (2011) Adsorption process of methyl orange dye onto mesoporous carbon material–kinetic and thermodynamic studies. J Coll Interface Sci 362(2):457–462
39. Regti A, et al (2017) Potential use of activated carbon derived from Persea species under alkaline conditions for removing cationic dye from wastewaters. J Assoc Arab Univ Basic Appl Sci 24:10–18
40. Shi B, et al (2007) Removal of direct dyes by coagulation: the performance of preformed polymeric aluminum species. J Hazard Mater 143(1–2):567–574
41. Moghaddam SS, Moghaddam MRA, Arami M Coagulation/flocculation process for dye removal using sludge from water treatment plant: optimization through response surface methodology. J Hazard Mater 175(1–3):651–657
42. Karcher S, Kornmüller A, Jekel M (2002) Anion exchange resins for removal of reactive dyes from textile wastewaters. Water Res 36(19):4717–4724
43. Advanced chemical oxidation of reactive dyes in simulated dyehouse effluents by ferrioxalate-Fenton/UV-A and TiO2/UV-A processes
44. Gupta VK, et al (2011) Removal of the hazardous dye—tartrazine by photodegradation on titanium dioxide surface. Mater Sci Eng C 31(5):1062–1067
45. Saravanan R et al (2016) Conducting PANI stimulated ZnO system for visible light photocatalytic degradation of coloured dyes. J Mol Liq 221:1029–1033
46. Rajendran S, et al (2016) Ce 3+-ion-induced visible-light photocatalytic degradation and electrochemical activity of ZnO/CeO 2 nanocomposite. Sci Rep 6(1):1–11
47. Yang C, et al (2020) Fabrication and characterization of a high performance polyimide ultrafiltration membrane for dye removal. J Coll Interface Sci 562:589–597
48. Lynch JM, Moffat AJ (2005) Bioremediation–prospects for the future application of innovative applied biological research. Ann Appl Biol 146(2):217–221
49. Tartakovsky B, Manuel M-F, Guiot SR (2005) Degradation of trichloroethylene in a coupled anaerobic–aerobic bioreactor: modeling and experiment. Biochem Eng J 26(1):72–81
50. Mohan SV, Babu PS, Srikanth S (2013) Azo dye remediation in periodic discontinuous batch mode operation: evaluation of metabolic shifts of the biocatalyst under aerobic, anaerobic and anoxic conditions. Separat Purif Technol 118:196–208
51. Sponza DT, Işik M (2002) Decolorization and azo dye degradation by anaerobic/aerobic sequential process. Enzyme Microb Technol 31(1–2):102–110
52. Melgoza RM, Cruz A, Buitrón G (2004) Anaerobic/aerobic treatment of colorants present in textile effluents. Water Sci Technol 50(2):149–155
53. Sheng G-P, Han-Qing Y, Cui H (2008) Model-evaluation of the erosion behavior of activated sludge under shear conditions using a chemical-equilibrium-based model. Chem Eng J 140(1–3):241–246

54. Anaerobic/aerobic treatment of municipal landfill leachate in sequential two-stage up-flow anaerobic sludge blanket reactor (UASB)/completely stirred tank reactor (CSTR) systems
55. Işik M, Sponza DT (2004) Anaerobic/aerobic sequential treatment of a cotton textile mill wastewater. J Chem Technol Biotechnol Int Res Process Environ Clean Technol 79(11):1268–1274
56. Tezel U, et al (2001) Sequential (anaerobic/aerobic) biological treatment of Dalaman SEKA pulp and paper industry effluent. Waste Manag 21(8):717–724
57. Gizgis N, Georgiou M, Diamadopoulos E (2006) Sequential anaerobic/aerobic biological treatment of olive mill wastewater and municipal wastewater. J Chem Technol Biotechnol Int Res Process Environ Clean Technol 81(9):1563–1569
58. Sklyar V, et al (2003) Combined biologic (anaerobic-aerobic) and chemical treatment of starch industry wastewater. Appl Biochem Biotechnol 109(1):253–262
59. Shieh WK, Keenan JD (1986) Fluidized bed biofilm reactor for wastewater treatment. In: Bioproducts. Springer, Berlin, Heidelberg, pp 131–169
60. de Souza AAU, et al (2008) Application of a fluidized bed bioreactor for cod reduction in textile industry effluents. Resources Conserv Recycl 52(3):511–521
61. Heijnen JJ et al (1989) Review on the application of anaerobic fluidized bed reactors in wastewater treatment. Chem Eng J 41(3):B37–B50
62. Tavares CRG, Sant'anna GL Jr, Capdeville B (1995) The effect of air superficial velocity on biofilm accumulation in a three-phase fluidized-bed reactor. Water Res 29(10):2293–2298
63. Yu J et al (2000) Distribution and change of microbial activity in combined UASB and AFB reactors for wastewater treatment. Bioprocess Eng 22(4):315–322
64. Jaibiba P, Naga Vignesh S, Hariharan S (2020) Working principle of typical bioreactors. In: Bioreactors. Elsevier, pp 145–173
65. Işık M, Sponza DT (2006) Biological treatment of acid dyeing wastewater using a sequential anaerobic/aerobic reactor system. Enzyme Microb Technol 38(7):887–892
66. Hong C (2004) ATP content and biomass activity in sequential anaerobic/aerobic reactors. J Zhejiang Univ-Sci A 5(6):727–732
67. Işık M, Sponza DT (2008) Anaerobic/aerobic treatment of a simulated textile wastewater. Separat Purif Technol 60(1):64–72
68. Kapdan IK, Alparslan S (2005) Application of anaerobic–aerobic sequential treatment system to real textile wastewater for color and COD removal. Enzyme Microb Technol 36(2–3):273–279

Aerobic/Anaerobic Membrane Bioreactor in Textile Wastewater

Jiayuan Ji, Yemei Li, and Jialing Ni

Abstract Textile wastewater is typical industrial wastewater that can lead to serious environmental problems if discharged without treatment. Thus, it is meaningful for developing the treatment processes for textile wastewater. At present, many common biological methods are available for the textile wastewater treatment, including activated sludge, granular sludge, membrane bioreactors (MBR), the up-flow anaerobic sludge blanket, sequential batch reactors, and so on. This chapter focuses on the use of MBRs for treating textile wastewater. We reviewed key parameters that had effects on the treatment performance, the enhanced strategies for MBRs treatment and the application of MBRs in pilot-scale and full-scale. With the optimal key parameters, pollutant removal performance could obtain high level with the removal efficiencies of more than 90% for organic matters, dye, and total nitrogen, and 100% for suspended solid. Moreover, some typical membrane cleaning methods were introduced to address the membrane fouling. For achieving a sustainable flux operation, anti-fouling strategies for operational MBRs were explained. Besides, studies related to the microbial community in the biological treatment of textile wastewater were described, and the status of the analysis of microorganisms in the treatment of textile wastewater with MBRs was discussed.

Keywords Membrane bioreactor · Anaerobic membrane bioreactor · Textile Wastewater · Biodegradation · Dye treatment · Membrane filtration · Membrane fouling · Anti-fouling strategy · Microbial community

J. Ji (✉)
Institute of Fluid Science, Tohoku University, 2-1-1 Katahira, Aoba-ku, Sendai 980-8577, Japan
e-mail: kikaen@tohoku.ac.jp

Y. Li
Center for Material Cycles and Waste Management Research, National Institute for Environmental Studies, 16-2 Onogawa, Tsukuba 305-8506, Japan
e-mail: lylym0821@126.com

J. Ni
Graduate School of Engineering, Department of Chemical Engineering, Tohoku University, 6-6-07 Aoba, Aoba-ku, Aramaki 980-8579, Japan
e-mail: jialing.ni.b8@tohoku.ac.jp

A. Khadir et al. (eds.), *Biological Approaches in Dye-Containing Wastewater*, Sustainable Textiles: Production, Processing, Manufacturing & Chemistry,
https://doi.org/10.1007/978-981-19-0545-2_9

Abbreviations

A2O	Anaerobic anoxic oxic
ABR	Anoxic baffled reactor
AeMBR	Aerobic membrane bioreactor
AnMBR	Anaerobic membrane bioreactor
ASR	Assimilatory sulfate reduction
BOD	Biochemical oxygen demand
CFV	Cross-flow velocity
COD	Chemical oxygen demand
DSR	Dissimilatory sulfate reduction
EPS	Extracellular polymeric substance
GO	Graphene oxides
GPC	Gel permeation chromatography
HRT	Hydraulic retention time
HUs	Hazen units
MBR	Membrane bioreactor
MF	Microfiltration
NF	Nanofiltration
PACl	Poly-aluminum chloride
PES	Polyethersulfone resin
PP	Polypropylene
PSU	Polysulfone
PTFE	Polytetrafluoroethylene
PVDF	Polyvinylidene difluoride
RO	Reverse osmosis
SEM	Scanning electron microscopy
SMP	Soluble microbial product
SRT	Sludge retention time
TDS	Total dissolved solids
TKN	Total kjeldahl nitrogen
TN	Total nitrogen
TP	Total phosphorus
TSS	Total suspended solids
UASB	Up-flow anaerobic sludge blanket
UF	Ultrafiltration
VFAs	Volatile fatty acids
WO3	Tungsten oxide

1 Introduction

Textile wastewater is a typical source of industrial wastewater. The rapid growth of the global population and the development of light industry has meant that the amount of textile wastewater produced is constantly increasing [4]. The textile industry discharges a large number of hazardous chemicals during production. According to a recent report, approximately 25% of the chemicals produced globally are used in the textile industry, and 20% of the total global water pollution results from the processing of textiles [20]. Textile wastewater that is directly discharged into natural bodies of water is known to cause serious pollution and harm the ecological environment [28]. Research and development that is based on textile wastewater treatment is therefore particularly meaningful. At present, many common biological methods are available for treating textile wastewater, including activated sludge [60], granular sludge [70], membrane bioreactors (MBR) [42], the up-flow anaerobic sludge blanket (UASB) [14], and sequential batch reactors [61]. This chapter introduces the general situation surrounding the use of MBRs in the treatment of textile wastewater.

MBR is a comprehensive wastewater treatment process that combines biological wastewater treatment with membrane separation technology [79]. The technique retains all the benefits from the biological treatment process while also providing effective solid–liquid separation, meaning that the sedimentation tank required for traditional biological treatment can be omitted during MBR treatment [12]. The three basic types of MBRs (Fig. 1) are side-stream, submerged, and external submerged [65, 93], 99]. Generally, side-stream MBRs provide more direct hydrodynamic control in terms of fouling and have the advantages of easier membrane replacement, but this is at the expense of the need for frequent membrane cleaning and high energy consumption [52]. Compared to the side-stream configuration, the membrane module in submerged MBRs is placed directly into the bioreactor. The distinct advantages of submerged MBRs include lower energy consumption, milder operating conditions, and easier control of fouling because of the lower tangential velocity [58]. Development of the external submerged configuration has further improved the maintenance of the submerged MBR. However, this system uses more energy because a recycling pump is required between the membrane unit and the main reactor of the MBR [62].

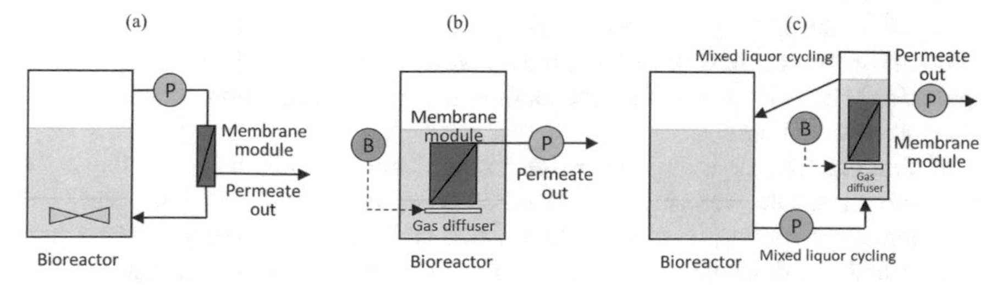

Fig. 1 Three typical configurations of MBR: **a** side-stream, **b** submerged, and **c** external submerged

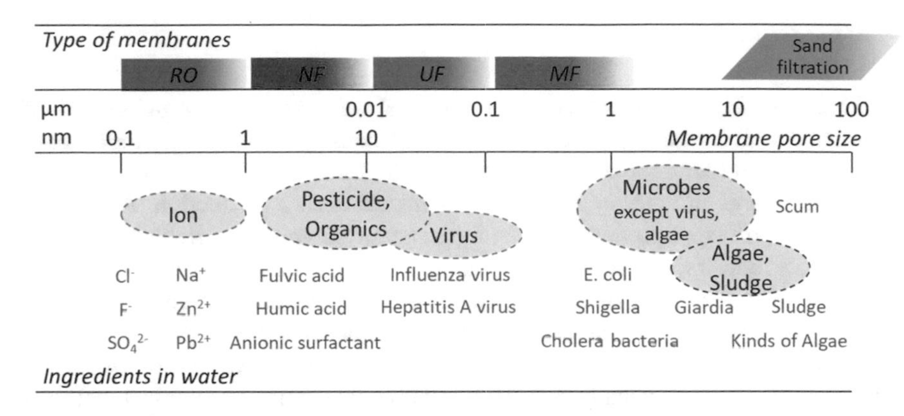

Fig. 2 Types of membranes and substances that can be filtered according to the membrane pore size

The membrane module that is used in MBRs was initially in the form of a flat sheet and is already widely applied around the world. The hollow fiber membrane that was subsequently developed has become increasingly mainstream in recent years. Distinguished by pore size, the membranes that are used for filtration are divided into microfiltration (MF, 0.1–5 μm), ultrafiltration (UF, 0.01–0.1 μm), nanofiltration (NF, 1–10 nm), and reverse osmosis (RO, 0.1–1 nm or less), which are used for removing ingredients of different sizes from wastewater [111]. The types of membranes and substances that can be filtered according to the membrane pore size are shown in Fig. 2.

At present, polypropylene (PP), polyethersulfone resin (PES), polysulfone (PSU) polyvinylidene difluoride (PVDF), polytetrafluoroethylene (PTFE), and ceramic membranes are widely used in biological wastewater treatment [3, 53, 54]. These materials have different characteristics; for example, a glass transition temperature of 225 °C has been reported for PES membranes [1], PSU membranes are pH stable and oxidation resistant, PVDF membranes exhibit thermal stability along with solvent and chemical resistance [54], PTFE exhibits high heat and chemical resistance and is characterized by soft mechanical properties, high fracture toughness, and a low friction coefficient [25], and ceramic membranes have chemical stability and a good anti-fouling ability [54]. In addition, low cost dynamic membranes have also been reported for the treatment of textile industry wastewater [83, 112], providing another option for MBRs. However, all these membranes have the common critical problem of membrane fouling.

Membrane fouling is currently the predominant limiting factor in the MBR treatment of textile wastewater (and other types of wastewater) [64]. Research into membrane fouling is an important topic, and the actual biological treatment process and the development of the material comprising the membrane are also meaningful. Low-cost operation is one of the objectives pursued via factors such as selecting the appropriate membrane pore size, the mode of operation, the online

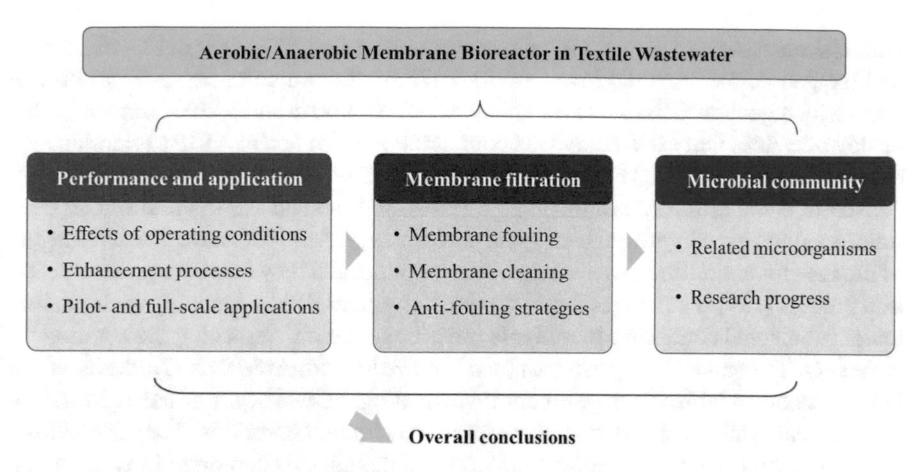

Fig. 3 The outline diagram of this chapter

backwashing cleaning method (that is used to clear foulants and pollutants), and treatment processes. Biogas such as methane, which is produced in anaerobic treatment processes, can be used to produce energy [46], therefore, the anaerobic wastewater treatment process that has been developed in recent years may bring greater application prospects and social value than the original aerobic wastewater treatment processes.

In this chapter, the performance and application of MBRs, the membrane filtration, and the microbial community was reviewed based on the previous research related to textile wastewater and MBRs. The outline of this chapter is shown in Fig. 3. In terms of performance and application, key parameters that effecting treatment performance, enhanced MBRs treatment processes, and application of MBRs in pilot-scale and full-scale was described in detail. Then, typical membrane cleaning methods were introduced to address the membrane fouling. For achieving a sustainable flux, antifouling strategies for the operation of MBR were explained. Besides, studies related to the microbial community in the biological treatment of textile wastewater were described. The status of microorganism analysis in MBR treating textile wastewater was also discussed.

2 Key Parameters Affecting Treatment Performance

2.1 Effects of Operating Conditions

MBRs can operate under both aerobic and anaerobic conditions, while aerobic MBRs (AeMBRs) considered more proportion in the application of textile wastewater treatment. Jegatheesan et al. [42] suggested that the AeMBR can treat textile wastewater with chemical oxygen demand (COD) concentrations of 500–6000 mg/L, biochemical oxygen demand (BOD) levels of 90–1375 mg/L, and color at 70–2700 Pt–Co units with a general COD removal efficiency of over 80% and a color removal efficiency of >70%. Only one study was conducted into the use of MBRs under anaerobic conditions (AnMBR) before 2013 [109]. The main feature of both AeMBRs and AnMBRs is the difference in the sludge retention time and the hydraulic retention time, resulting in an improved biological performance. AnMBRs have the advantages of biogas production and lower energy consumption, and they do not require an aeration process [109]. The biogas that is produced by AnMBRs is generally recirculated under submerged conditions to mitigate membrane fouling, replacing the aeration in AeMBRs. To compare the performance of AeMBRs with AnMBRs, Yurtsever et al. [115] operated a lab-scale experiment investigating the two types of MBR by using both to treat synthetic azo dye wastewater under similar conditions. They found that both MBRs had high COD removal capacities, although the decolorization capacities differed. This phenomenon is due to the different redox potentials of the systems. The removal of dye benefited from the low redox potential in the AnMBR as the oxidation–reduction reaction was almost complete when using this system. On the contrary, the AeMBR used oxygen rather than azo dye as the electron acceptor and only removed 30–50% of the color. However, from the perspective of membrane performance, the filterability of the AeMBR was 4–8 times higher than that of the AnMBR. Furthermore, the cake layer formed in the AeMBR had no influence on COD removal, while 20% of the supernatant COD was rejected by the gel or cake layer formed in the AnMBR [113].

A previous study has reported that MBRs can achieve stable and excellent performance in textile wastewater treatment under various influent conditions with the assistance of a membrane [110]. Several operating parameters have been discussed. Hydraulic retention time (HRT), which directly affects the volume of a reactor, is one of the most important aspects in engineering applications. Li et al. [56] operated an MBR with nanofiltration recirculation to treat real textile wastewater under different HRTs and found that the system performance was not significantly affected when the HRT was changed from 24 to 10 h. Similar results were obtained by Feng et al., who conducted a Fenton oxidation-MBR hybrid process for the treatment of wastewater in an integrated dyeing wastewater treatment plant under different HRTs (10, 18, and 25 h) [24]. Tian et al. [98] operated an anoxic/oxic-hybrid MBR system to treat dry-spun acrylic fiber wastewater under HRTs of 24–48 h and found that the removal of typical pollutants was not significantly improved at HRTs above 32 h because the organic loading was below 0.1 kg-COD/kg-MLSS/d while the biomass concentration was high. The retention of biomass is the most important feature of MBRs as it increases the biomass concentration and prolongs the sludge retention time (SRT). Khouni et al. [50] reported the effect of the dye mass loading rate on the performance of AeMBRs for textile wastewater treatment and found that the system achieved an 80–90% removal efficiency for soluble organic matter under a dye mass loading rate of 320 mg/g-MLVSS/d. The dye mass load requires more consideration when using AnMBRs because of the biogas generation. Spagni et al. [92] used reactive orange16 as a model to investigate the effect of azo dye loading on the performance of an AnMBR and found that the decolorization of the submerged AnMBR was only slightly affected, although the permeate still contained 30–40 mg/L of residual reactive orange16 with azo dye concentrations of up to 3.2 g/L. However, the performance of the anaerobic digestion was inhibited, especially with the accumulation of volatile fatty acids (VFAs), which resulted in low methane production. The influence of SRT on organic matter and color removal also appears to differ under anaerobic and aerobic conditions. Yurtsever et al. [113] reported that the decolorization efficiency of AnMBRs was independent of the SRT, while the removal of both COD and sulfate were reduced by decreasing the SRT. However, little effect was observed in the removal of COD as a result of changing the SRT in AeMBRs, and the color was observed to increase. A similar result was also reported by Yigit et al. [110], who suggested that a slightly better effluent quality was achieved under an SRT of 25 d than under an infinite SRT for a pilot-scale AeMBR that was developed for use in the treatment of wastewater from the denim textile industry. Brik et al. [13] suggested that low sludge ages were recommended to obtain color removal via adsorption by biomass.

2.2 MBR Enhanced by Combined Process for Textile Wastewater Treatment

A combined process is a process in which two or more treatment steps are conducted sequentially to achieve high effluent quality or cost- and time-saving processes [85]. Although MBRs are regarded as promising in terms of biological wastewater treatment, the application of this method is limited by the low biodegradability of textile wastewater and the aim of reusing the treated water. The process, when combined with MBR, can be divided into three types: pretreatment, which increases the biodegradability of the feed; enhancement during the MBR process; and post-treatment, which improves the quality of the permeate.

Photochemical technology is generally used as a pretreatment in textile wastewater treatment. Sathya et al. [87] conducted photocatalysis with tungsten oxide (WO_3) and WO_3/1% graphene oxide (GO) as a powdered photocatalyst with the function of dye adsorption as a pretreatment for MBR to achieve a COD removal of 78%. They reported that photocatalysis also improved the biodegradability of the textile wastewater with a BOD/COD ratio of 0.2–0.4, while the effluent from the MBR had a BOD/COD ratio of 0.5. On the other hand, the advantages of using nanomaterials have also been applied in membrane modification to improve membrane performance. Cheng et al. [17] used a GO/polysulfone composite membrane together with photosynthetic bacteria for textile wastewater treatment. Although the wastewater was synthetic, the reactor showed better performance in terms of color, ammonia–nitrogen, and COD removal than conventional MBR systems did.

Advanced oxidation is thought to be a feasible process for the decomposition of pollutants. Electron beam radiation, as a special advanced oxidation process, has been reported as an efficient pretreatment to enhance the biodegradability of textile wastewater. Sun et al. [95] used an MBR-coupled electron beam radiation to obtain good stability for COD removal and system maintenance in textile effluent that contained polyvinyl alcohol, which cannot be treated by a single MBR. Fenton oxidation can also generate highly reactive hydroxyl radicals that remove color and improve the biodegradability of dyeing wastewater [24]. Internal micro-electrolysis can also be used in the oxidization of organic contaminants and includes the additional function of releasing iron ions. Qin et al. [78] reported the development of an AeMBR for treating anthraquinone dye wastewater with internal micro-electrolysis as pretreatment and found that the release of iron ions from galvanic corrosion appeared to be a double-edged sword in the MBR system. On one hand, the iron ions resulted in the formation of bio-ferric flocs that enhanced denitrification and decolorization via adsorption and biodegradation. However, this was accompanied by severe inorganic membrane fouling.

In addition to enhancing biodegradability, the addition of coagulant has also been used as a strategy to improve the total phosphorus (TP) removal in MBR treatment. One example that included the intermittent addition of poly-aluminum chloride (PACl) to MBR influent was reported by Yan et al. [108]. The addition improved the TP removal efficiency from $28.61 \pm 10.94\%$ to $97.63 \pm 1.59\%$, which was due to the stable precipitation of $AlPO_4$ by the supplied $Al^{3+}$and the P in the wastewater. The addition of saturated granular activated carbon to create an anaerobic zone in aerobic MBRs has also been reported as an effective strategy to improve decolorization [32]. The adsorption performance of the powder-activated carbon in the MBR depended on whether the dye had been biosorbed [33].

Furthermore, because the membrane units used in MBRs for textile wastewater treatment are normally inserted MF or UF membranes through which dye molecules with low molecular weights can pass, the addition of permeate treatment is necessary to obtain better usage of the effluent [34]. NF has been reported as having a promising capacity to further improve the quality of effluent produced by MBR systems, especially in the removal of conductivity and biologically refractory colors [29]. Meanwhile, RO has been reported to completely remove mineral salts, hydrolyzed reactive dyes, chemical auxiliaries, ions, and larger species after MBR and could be another means of realizing the reuse of wastewater from the dyeing processes to produce water that reaches the specifications required for drinking [21]. The energy demanded for membrane fouling mitigation and the economic cost of system maintenance also need further investigation. The combination of a UV-unit and active carbon filter was investigated as a post-treatment for the MBR process. Rondon et al. [80] reported that this process can completely remove all dye after more than 90% of the dye, the COD, and total nitrogen has been removed by an MBR.

2.3 *Pilot- and Full-Scale Applications*

With the increasing attention on the development of new membrane biotechnology, the advantage of using MBRs has been proven on a large scale. Table 1 presents the application of MBR-related processes for textile wastewater treatment on both pilot- and full-scales. Until now, the large-scale application of MBRs has mainly been conducted under aerobic conditions, and its development is based on the fundamentals of conventional biological aerobic, anaerobic, and anoxic processes. Some applications of MBRs have involved either feeding directly with wastewater or adding influent after pH adjustment [27, 55]. However, the function of the conventional biological processes cannot be ignored. Generally, the anaerobic process can partially convert refractory organics to intermediates that are more readily degradable, while the anoxic-aerobic process achieves biological nitrogen removal via predenitrification and aerobic nitrification [94]. The addition of MBRs can allow sufficient time for slow-growing bacteria to multiply, especially those that specialize in degrading refractory compounds. Sun et al. [94] reported a combined process that takes advantage of the traditional anaerobic anoxic oxic (A^2O) system and an MBR that efficiently removed organic matter and nitrogen in an auxiliary textile wastewater treatment. Hydrolysis/acidification and denitrification occurred in the anaerobic and anoxic tanks, respectively. An aerobic tank that acted as an MBR realized the nitrification and degradation of the remaining organic matter. Jager et al. [21] studied a dual-stage MBR and subsequent RO system that also consisted of an anaerobic tank in which the azo bonds of the reactive dyes were cleaved, an anoxic tank for denitrification, and an aerobic tank for nitrification and mineralization of the aromatic amines. Yan et al. [108] used an anoxic baffled reactor (ABR) as a hydrolysis process to enhance the biodegradability of organic matter in wastewater. The enhanced strategy was also adapted to work with conventional processes. Tian et al. [98] reported a hybrid anoxic/oxic-MBR system that contained five plug-flow compartments in an oxic tank to improve the selectivity of microorganisms and sludge diversity and obtain stable hydraulic conditions. Polypropylene fiber media were hung in this region to provide a carrier for the biofilm.

Other pretreatment processes have also been worthy of notice, although few reports have been produced in this area. Sathya et al. [86] reported a pilot-scale ozonized MBR using UV-enhanced ozonation as a pretreatment for the treatment of dark dye textile wastewater and found ozonation to be a clean technology with no sludge generation and a positive influence on the system performance. Another study conducted by the same research group used a photochemically integrated MBR [87]; this study was discussed in the previous section.

In terms of the integrity of the entire processing flow, large-scale applications require more attention do than small-scale applications. Especially from the perspective of the economy, the reuse of MBR permeate is a highlighted point in the treatment procedure, and the target of reuse determines the degree to which the permeate quality is improved [110]. Li et al. [56] reported that the reuse of an NF permeate from an MBR that was reduced as cooling water for the UF unit reduced water consumption,

Table 1 Pilot-scale and full-scale application of MBR process for textile wastewater treatment

Scale	Wastewater type	Treatment system	Influent characteristics	Effluent characteristics	Membrane type	Membrane conditions	Operation conditions	References
Pilot-scale	Real textile wastewater	Bio-P/Anoxic/ Aerobic-MBR	COD 960 mg/L; TN 44 mg/L; TSS 260 mg/L; BOD5 282 mg/L; TP 5 mg/L; pH 8.8	COD 82 mg/L; TN 6.2 mg/L; BOD5 1.5 mg/L; TP 1.6 mg/L; pH 8.1	Submerged/hollow fiber membrane/PVDF	Pore size 30 nm; average flux 20 LMH; membrane aera 32 m^2; specific air demanded 1⁻3 $Nm^3/m^2/h$	Flow rate 10 m^3/d; total SRT 15 d; Temp. 28 °C;	[22]
Pilot-scale	Textile dyeing wastewater	Photocatalytic AeMBR	COD 1980 mg/L; pH 9.2 TDS 8920 mg/L; TSS 228 mg/L	COD 76% removed; color 70% removed;	Submerged/hollow fiber membrane/PES	Pore size 0.1 µm; membrane area 0.6 m^2	HRT 10 h; working volume 20 L	[87]
Full-scale	Real textile wastewater	Anoxic/oxic biofilm-AeMBR	NO_3^--N 30 mg/L; TN 40 mg/L; COD 100 mg/L; NH_4^+-N 5 mg/L	NO_3^--N 20 mg/L; TN 40 mg/L; COD 50 mg/L; NH_4^+-N 12 mg/L; NO_2^--N 0.7 mg/L	Submerged/hollow fiber/PVDF	Pore size 0.4 µm; membrane area 60 m^2	30–40 m^3/d; HRT 2 d	[118]

(continued)

Table 1 (continued)

Scale	Wastewater type	Treatment system	Influent characteristics	Effluent characteristics	Membrane type	Membrane conditions	Operation conditions	References
Pilot-scale	Real textile wastewater	Anoxic-aerobic-MBR + RO	COD 900 ± 51 mg/L; NH_4^+-N 22.4 ± 4.3 mg/L; pH 7.96 ± 0.65	COD 16.3 ± 5.5 mg/L; NH_4^+-N 2.1 mg/L	Silicon carbide Flat-sheet ceramic membrane/UF	Pore size 0.1 µm; membrane area 1.06 m^2	Feed flow rate 50 L/h; SRT 20 d; filtration 25 LMH; CFV 0.2–0.4 m/s	[11]
Pilot-scale	Real textile wastewater	Ozone/UV + MBR + photocatalytic reactor	COD 1970 mg/L; pH 8.2; Color 1500 Pt–Co; TDS 9960 mg/L; TSS 1340 mg/L	COD 140 mg/L; pH 7.9; Color 90 Pt–Co; TDS 6120 mg/L; TSS < 1 mg/L	Submerged/hollow fiber membrane/PVDF	Pore size 0.1 µm	Working volume 20 L	[86]
Pilot-scale	Effluent of an anaerobic tank for textile wastewater treatment	AeMBR-NF	COD 507.97 ± 176.22 mg/L; pH 8.57 ± 0.44; color 1378.9 ± 405.5 HUs	COD < 20 mg/L; Color 47.6 ± 15.3 HUs	Submerged/flat-sheet membrane/PVDF	Pore size 0.1 µm; membrane area 10 m^2	HRT 15 h; SRT 20d; working volume 1.5 m^3; Temp 14.5–25.1 °C	[55]
Pilot-scale	Textile auxiliaries wastewater	A^2O-MBR	COD 372–1291 mg/L; pH 7.29–9.37; NH_4 + -N 87–94 mg/L; TN 105–121 mg/L	COD 131 mg/L; NH_4 + -N 1–13 mg/L; TN 39–62 mg/L	PVDF/hollow fiber membrane + flat-sheet membrane	Pore size 0.1 µm; surface area 12.5 m^2 (H) & 10.4 m^2 (F)	Flow rate 250 L/h; temp: 20–25 °C	[94]

(continued)

Table 1 (continued)

Scale	Wastewater type	Treatment system	Influent characteristics	Effluent characteristics	Membrane type	Membrane conditions	Operation conditions	References
Pilot-scale	Real textile wastewater	AeMBR	pH 8; COD 293–618 mg/L; TSS 0.04–0.24 g/L	Color 100% removed; COD 98% removed; SS 100% removed	Submerged/flat-sheet membrane	Membrane area 0.39 m2; cutoff 150 kDa; Aeration rate 1–2 m^3/h	Working volume 60 L; Temp 24–29 °C; HRT 2.6–1 d	[27]
Pilot-scale	Dry-spun acrylic fiber wastewater treatment	A/O-MBR	COD 610.6–1094.5 mg/L; TN 178.5–310.4 mg/L; pH 7.4–7.8	COD 380.2 mg/L; NH_4^+-N 12.2 mg/L	Submerged/hollow fiber/PVDF	Pore size 0.2 μm; membrane area 6.6 m^2	HRT 24–48 h	[98]
Pilot-scale	Real textile wastewater	dual-stage UF-MBR + RO system	COD 728–1033 mg/L; pH 7.8–10.4; TN 1.2–2.3 mg/L; TSS 16.0–46.0 mg/L	COD 87 mg/L; pH 9.5; Nitrate 1.3 mg/L	Side-stream/UF	High 3 m; membrane area 5.1 m^2	Flow rate 100 L/h	[21]
Pilot-scale	Real textile wastewater	ABR + hybrid coagulation/AeMBR	COD 383.1–534.8 mg/L; pH 8.12–9.23	COD 42.3 ± 4.5 mg/L	Submerged/Hollow fiber membrane/PVDF	Pore size 0.2 μm; membrane area 25 m^2; working volume 1.67 m^3	Flow rate 1000 m^3/d	[108]

(continued)

Table 1 (continued)

Scale	Wastewater type	Treatment system	Influent characteristics	Effluent characteristics	Membrane type	Membrane conditions	Operation conditions	References
Pilot-scale	Denim producing textile wastewater	AeMBR	COD 686–2278 mg/L; TKN 2.3–7.9 mg/L; TN 19.5–81.0 mg/L; Color 286–8100 Pt Co	COD 13–60 mg/L; TKN 0.3–2.0 mg/L; TN 2.1–21.0 mg/L; Color 32–105 Pt Co	Submerged/Hollow fiber membrane	Pore size 0.04 μm; working volume 230 L	Flow rate 380–1500 L/d; HRT 14 h	[110]

resulting in a net income of 1.24 M USD/y. In most biological processes, partially recycling the permeate is a common strategy to enhance the pollutant removal efficiency. This type of recirculation has also been reported in the post-treatment of MBR via concentrate recirculation, which can also achieve a higher water recovery. However, recycling may increase the membrane fouling risk as a result of refeeding refractory compounds. A previous study reported that although the whole system maintained a stable permeate quality with the additional NF process in the MBR system, the recycling of concentrate caused refractory compounds to accumulate, including protein-like substances and humic acid substances, which did not obviously inhibit the sludge activity but contributed to membrane fouling in the NF unit [55].

3 Membrane Fouling, Membrane Cleaning, and Operational Strategies

3.1 Membrane Filtration and Fouling

The membrane separation technology used in biological wastewater treatment can achieve high-efficiency sludge-water separation via filtration [101]. Whether the wastewater treatment process uses an AeMBR or AnMBR, the membrane generally plays a significant role in separating the solid and liquid, meaning that a sedimentation tank is not required [18, 105]. Studies have also reported the direct use of membranes to remove pollutants from textile wastewater via membrane filtration. For example, sand filtration and UF have been applied as pre-treatments followed by NF and RO for the reuse of textile wastewater [63]. The efficient removal of persistent organic pollutants from textile wastewater by UF, NF, and RO has also been reported [71]. Therefore, membrane filtration can provide an additional guarantee in biological wastewater treatment with regard to overall treatment.

As the filtering during wastewater treatment progresses, pollutants and various microbial metabolites can block the surface and pores of a membrane in a process generally known as membrane fouling [7, 31]. In principle, membrane fouling occurs via continuous accumulation [105], and the filtration resistance increases as membrane fouling progress [69, 91]. When a certain level of membrane fouling occurs, filtration becomes very difficult owing to the high resistance, and the wastewater processing cannot be completed [6, 31]. In this situation, the membranes need to be replaced or cleaned (maintenance). Membrane fouling is generally caused by adsorption, cake layer, gel layer, and pore plugging [66]. These types of fouling lead to the resistance that hinders the membrane filtration process (Fig. 4).

The foulants that occur during the treatment of textile wastewater by MBRs are mainly due to the presence of soluble microbial products (SMP) and extracellular

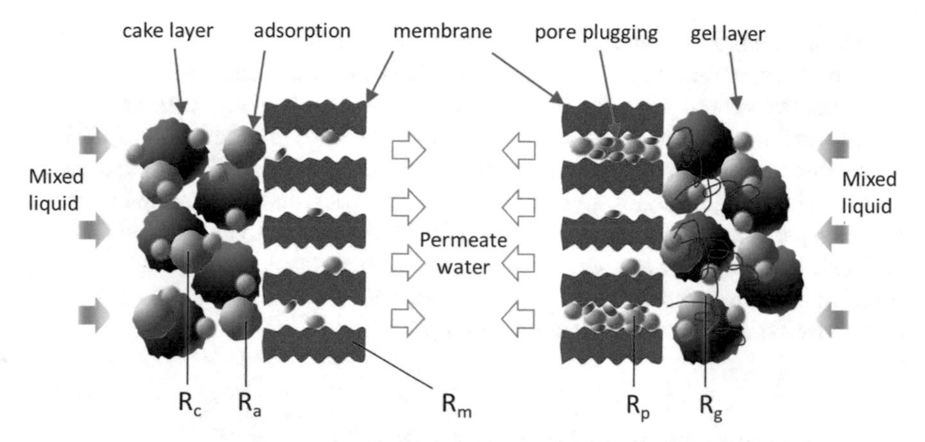

Fig. 4 Types of membrane fouling and the resulting resistances (R_a, resistance to adsorption; R_c, resistance of cake layer; R_g, resistance of gel layer; R_m, resistance of membrane; and R_p, resistance of pore plugging)

polymeric substances (EPS) in the cake layer, which have been observed during analysis as carbohydrate and protein compounds [116]. Organic matter of high molecular weight was observed in a previous study using gel permeation chromatography (GPC) measurements of foulant samples on the surfaces of membranes [74]. Soluble organic compounds with high molecular weights can be derived from cell debris, as well as SMP and EPS [114]. Accompanied by membrane fouling, more pollutants were retained in the reactor. Thus, organic matter with a molecular weight lower than 15 kDa remained in the cake layer even during the MF process, which was further performed as UF due to the presence of the cake layer [116].

3.2 Conventional Membrane Cleaning Methods

Membrane cleaning is generally carried out either physically or chemically [52], 103]. Physical cleaning includes the use of sponges or other soft materials to wipe and clean the surface of a membrane [51]. Backwashing with clean water under pressure is also often used [47]. High-temperature incineration is an efficient and inexpensive physical cleaning method for use with ceramic membranes. A previous study investigated the regeneration of ceramic membranes by calcining at 400 °C for 30 min used in a dye separation process [84]. Chemical cleaning generally involves the immersion of membranes in different chemicals according to the material composition of the foulants [104]. Sodium hypochlorite and citric acid solutions can be used to remove organic and inorganic foulants, respectively, to mitigate against membrane fouling [67, 107]. Sulfuric acid with a pH of 2 has also reported as a chemical cleaning solution for anaerobic and aerobic membrane bioreactors that treat textile wastewater [113]. Noel Dow et al. reported the implementation of a 1.5-wt% NaOH solution at

55 °C for membrane cleaning [23]. Depending on the fouling situation, the soaking time could range from several hours to several days. In the biological wastewater treatment process, various products such as EPSs and SMPs, which may cause membrane fouling, are produced as a result of biochemical reactions. The removal of EPS and SMP often requires chemical cleaning [39]. Therefore, a combination of physical and chemical cleaning is usually applied to clean membranes and recover the membrane filtration performance to a maximum level [103]. According to the cleaning method adopted, foulants have been reported as removable (removable by physical cleaning), irremovable (removable by chemical cleaning), and irreversible (foulants that cannot be removed through any systematic cleaning) [66].

3.3 *Operation Strategy for Anti-Fouling*

Because replacing membranes is expensive, increasing the service life of a membrane is a very important issue [9]. Reasonable operation methods or appropriate device improvements can not only reduce the filtration resistance (Fig. 5), increase the membrane life, and lower the cost, but can also ensure the smooth progress of the wastewater treatment process [6].

One of the most common ways to reduce membrane filtration resistance is to install a diffuser at the bottom of the membrane module [88]. The shear force provided by gas flowing alongside the filtration can to some extent keep contaminants away from the surface of a membrane [37]. The aeration that is associated with aerobic treatment provides a decent shear force at the surface of the membrane [26]. However, the circulation of biogas is required in the anaerobic process, following a benefit that mechanical stirring can be dispensed [57]. Another commonly used operation strategy is intermittent filtration during the treatment process [44]. As mentioned above, membrane fouling is an accumulation process that occurs continually during filtration; therefore, the resistance may suddenly result in the filtration process being unable to proceed normally after fouling and reach a critical level. The use of intermittent filtration can alleviate this problem to some extent as compared to continuous

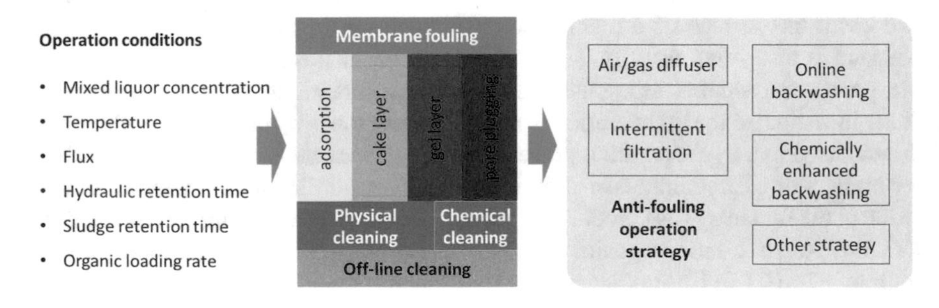

Fig. 5 Membrane fouling and anti-fouling operation strategies

filtration [5, 15]. The specific operation mode (time period of filtration and relaxation) needs to be determined according to the actual conditions, which are mainly related to the filtration flux, sludge concentration, and membrane performance [106]. An advanced treatment of dyeing wastewater using an AeMBR employed a filtration process using an intermittent filtration cycle of 8 min filtration and 2 min relaxation [24]. An operation mode of a 5 min filtration followed by 1 min relaxation was also reported for dealing with both synthetic and real textile wastewaters [115, 116].

Specific strategies have also been implemented for the treatment of textile wastewater. For example, hybrid electrocoagulation was combined with the NF process for treating real textile wastewater with efficient dye/salt fractionation and low membrane fouling [97]. Another study found that backwashing with a saturated CO_2 solution was more effective than general backwashing methods in which Milli-Q water and chemicals are used [2]. Changing the character of the mixed liquid in the reactor is also an effective approach. One of the common and low-cost methods is the use of flocculants to coagulate suspended solids into larger particles. PACl and aluminum sulfate have been used as coagulants to decrease membrane fouling in a process based on hollow fiber membranes and coagulation during the treatment of textile wastewater [3]. COD removal of up to 66% was obtained, and the use of coagulant led to a significant decrease in TMP during filtration with three kinds of membranes (PP, PES, and PVDF).

Regular backwashing using Milli-Q can remove the accumulated foulants to a certain extent. Another operation strategy uses a method called on-site chemically enhanced backwashing [38]. In previous research, backwashing by chemicals was found to be efficient and to further extend the membrane life to ensure the progress of the wastewater treatment process [73]. Especially for the anaerobic treatment process, opening an airtight reactor for maintenance is costly, and there is a time cost in restarting the operation [77]. Because of the relationship between the type of membrane and the structure of membrane fouling, a low-pressure backwash should be used when chemical cleaning is the mainstay [59], and higher pressure should be used when physical cleaning is the mainstay [43]. For applications, different plans (chemical agents, cleaning time, and cleaning frequency) are therefore required according to the type of foulant and wastewater treatment process used [42]. Because the main pollutants in textile wastewater contain both organic and inorganic substances, sodium hypochlorite and citric acid are generally used for online backwashing [40, 116]. In one study, a biomimetic dynamic membrane was developed and applied to dye removal [16]. After backwash cleaning, a thin layer of dye remained on the membrane. According to the results, the backwashing process could effectively remove the inactive bionic layer but was unable to regenerate the pollutant removal ability of the membrane [16]. However, because different from traditional membranes, the familiar dynamic membrane uses the trapped pollutants to self-form a new bio-membrane layer with smaller pore sizes to further block the pollutants, online backwashing is not applicable, which limits the practical application of the dynamic membrane [96].

4 Microbial Community in the MBR Treating Textile Wastewater

4.1 Research Status of the Microorganisms Related to the Removal of Dyes with Biological Treatment Methods

Textile wastewater has always been a major problem in the field of biodegradation because of its complex ingredients and toxicity. As a key element, the microorganisms involved have received increasing attention in many studies. Because textile wastewater contains organic matter that is difficult to degrade, it is often treated via the simultaneous combination of aerobic and anaerobic methods. This method involves the degradation and decolorization of hardly degradable dyes under anaerobic conditions followed by treatment of the obtained low-COD wastewater under aerobic conditions. Regardless of whether a treatment process is aerobic or anaerobic, some specific microorganisms are required to treat a particular dye. For example, most microorganisms in these systems are able to break down the azo bonds (-N = N-) that are found in most dyes [35], while some microorganisms can decompose anthraquinone dyes (such as Reactive Blue 19, Reactive Blue 4, and Disperse Blue 2BLN) [102]. To be able to clearly ascertain the mechanism of various microorganisms in the treatment of dyes, many studies have focused on the degradation effects that single culture strains have on dyes. Table 2 lists the known isolated bacteria that can degrade azo bond dyes. In addition, previous studies have summarized the effects of using mixed cultures, fungi, yeast, and algae to remove dyes [35, 100].

Regrettably, although studies using MBRs for textile wastewater treatment are not uncommon, very few articles have provided an in-depth elaboration of the microorganisms involved. The reason may be that the composition and structure of the microbial community in the sludge are more connected to the types of substrates and the reaction conditions (aerobic, anaerobic, temperature, etc.). Hence, based on previous experience, the microbial diversity and dominant microorganisms will not change much if only the reactor type is changed. However, it is necessary to investigate the microorganisms involved in the removal of dyes by MBRs. Similar to the case of dye removal by the UASB reactor [14] the growth and structure of various microorganisms in the granular sludge differ, and the mechanism of the dye decomposition process can be explored. Because of the existence of the membrane, the microbial composition in the reactor and the membrane surface will inevitably differ after long-term stable operation of the MBR. The different structures of the specific dye-removing microorganisms and other microorganisms in the cake layer on the membrane surface have important scientific value for understanding the transfer of substances onto the membrane surface and the analysis of the textile wastewater treatment mechanism. This is also useful for understanding and improving the treatment performance, and further research in this field is expected in the future.

Table 2 Isolated bacteria species that can degrade azo bonds dyes

Dye name	Taxa	References
Reactive orange 16	*Nocardiopsis* sp.	[19]
	Pseudomonas aeruginosa SVM16	[68]
Reactive orange 13	*Alcaligenes faecalis* PMS-1	[89]
Reactive red 21	*Pseudomonas aeruginosa* SVM16	[69]
Reactive red 2	*Pseudomonas sp.* SUK1	[48]
Reactive red 184	*Halomonas sp.* strain A55	[30]
Reactive red 120	*Bacillus cohnii* RAPT1	[76]
Reactive yellow	*Pseudomonas mendocina*	[81]
Reactive yellow 2	*Serratia sp.* RN34	[72]
Reactive blue 160	*Pseudomonas sp.* BDN4	[8]
Reactive Blue 13	*Proteus mirabilis* LAG	[75]
Reactive Blue 19	*Enterobacter sp.* F NCIM 5545	[36]
Reactive Black 5	*Psychrobacter alimentarius* KS23	[82]
	Pseudomonas entomophila BS1	[49]
Direct Blue 14	*Bacillus fermus*	[41]
Methylene Blue	*Alcaligenes faecalis*	[10]
Methyl Red	*Aeromonas jandaei* strain SCS5	[90]

4.2 Current Status of Research on Microorganisms in MBRs

As mentioned in the previous article, there are relatively few studies in this area, but there are some meaningful related studies. In addition to simply using MBRs for biological treatment, the addition of activated carbon, flocculants, and other reagents is often used to assist in the removal and decolorization of dyes during pretreatment. Therefore, the diversity and functions of microorganisms that are associated with the biological treatment of pre-treated textile wastewater with high salt concentrations need to be clearly identified. Zhou et al. reported the reduction of assimilated and dissimilated sulfates and the diversity of biological bacteria in the biofilm-membrane bioreactor during the treatment of textile wastewater [117]. The authors used metagenomic sequencing to distinguish assimilatory and dissimilatory sulfate reduction (ASR and DSR) in an MBR and revealed that there is a two-stage sulfur reduction process, explaining the pathways of ASR and DSR in the process. In another study, Sun et al. treated textile wastewater containing polyvinyl alcohol with electron beam radiation and the MBR coupling method to obtain a COD removal rate of 31 $\pm$ 7%

during the stable period [95]. Filamentous bacteria were observed by scanning electron microscopy (SEM) and were dominant in the activated sludge. The radiolysis intermediates of polyvinyl alcohol and other organic pollutants in textile wastewater are beneficial to the growth of diatoms, leading to the production of O_2 and providing a suitable environment for aerobic microorganisms that are further conducive to the removal of pollutants from textile wastewater.

SEM was also applied to observe the microbial morphology of the membrane surface in textile wastewater by Sathya et al., who reported that the pores of the contaminated membrane surface were occupied by microorganisms and other substances [86]. The study used 16S sequencing to explore and compare the difference between the microbial diversity in the MBR and on the membrane surface.

Research on the microorganisms that are used to treat textile wastewater in MBRs has gradually changed from exploring the impact of simple reaction conditions (such as SRT) to exploring the microbial diversity [113] to qualitative separation [102] and the application of functional bacteria to remove specific pollutants (such as malachite green) [45]. Various advanced exploration and analysis methods, such as 16S analysis, metagenomic analysis, bacterial isolation and culture, and microscopic observation, have been applied to investigate the microbial diversity, structure, and metabolism in the incremental treatment of textile wastewater. Such methods are helpful in deepening the understanding of how microbial metabolism affects dyes and other organic or inorganic substances in the MBR and in further improving the treatment method and efficiency. It is worth noting that there are still significant deficiencies in the related research in terms of quantity and content. More systematic and in-depth research is still required in order to expand the knowledge in this field, which will be the responsibility of future researchers.

5 Conclusions

MBRs in both aerobic and anaerobic processes are efficient for the removal of dye, COD, nitrogen, phosphorus, and other specific contaminants produced by the textile industry. During long-term operation, multiple approaches to the anti-fouling strategy have been found useful for extending the life of a membrane to achieve low-cost and sustainable filtration. In addition, although microbial community analysis related to the treatment of textile wastewater are abundant, studies related to the MBR treatment of textile wastewater are still very insufficient and incomplete. More diverse research on systematic microbial analysis that is based on the microorganisms and pollutants in the process of textile wastewater treatment via MBRs is therefore required.

References

1. Ahmad AL, Abdulkarim AA, Ooi BS, Ismail S (2013) Recent development in additives modifications of polyethersulfone membrane for flux enhancement. Chem Eng J 223:246–267. https://doi.org/10.1016/j.cej.2013.02.130
2. Al-Ghamdi MA, Alhadidi A, Ghaffour N (2019) Membrane backwash cleaning using CO_2 nucleation. Water Res 165, 114985.https://doi.org/10.1016/j.watres.2019.114985
3. Alibeigi-Beni S, Habibi Zare M, Pourafshari Chenar M, Sadeghi M, Shirazian S (2021) Design and optimization of a hybrid process based on hollow-fiber membrane/coagulation for wastewater treatment. Environ Sci Pollut Res 28:8235–8245. https://doi.org/10.1007/s11356-020-11037-y
4. Alinsafi A, Evenou F, Abdulkarim EM, Pons M-N, Zahraa O, Benhammou A, Yaacoubi A, Nejmeddine A (2007) Treatment of textile industry wastewater by supported photocatalysis. Dye Pigment 74:439–445
5. An Y, Wu B, Wong FS, Yang F (2010) Post-treatment of upflow anaerobic sludge blanket effluent by combining the membrane filtration process: fouling control by intermittent permeation and air sparging. Water Environ J 24:32–38. https://doi.org/10.1111/j.1747-6593.2008.00152.x
6. Aslam M, Charfi A, Lesage G, Heran M, Kim J (2017) Membrane bioreactors for wastewater treatment: a review of mechanical cleaning by scouring agents to control membrane fouling. Chem Eng J 307:897–913. https://doi.org/10.1016/j.cej.2016.08.144
7. Bai R, Leow HF (2002) Microfiltration of activated sludge wastewater—the effect of system operation parameters. Sep Purif Technol 29:189–198. https://doi.org/10.1016/S1383-5866(02)00075-8
8. Balapure KH, Jain K, Chattaraj S, Bhatt NS, Madamwar D (2014) Co-metabolic degradation of diazo dye-Reactive blue 160 by enriched mixed cultures BDN. J Hazard Mater 279:85–95. https://doi.org/10.1016/j.jhazmat.2014.06.057
9. Bennett A (2005) Membranes in industry: facilitating reuse of wastewater. Filtr Sep 42:28–30. https://doi.org/10.1016/S0015-1882(05)70658-3
10. Bharti V, Vikrant K, Goswami M, Tiwari H, Sonwani RK, Lee J, Tsang DCW, Kim KH, Saeed M, Kumar S, Rai BN, Giri BS, Singh RS (2019) Biodegradation of methylene blue dye in a batch and continuous mode using biochar as packing media. Environ Res 171:356–364. https://doi.org/10.1016/j.envres.2019.01.051
11. Bilici Z, Unal BO, Ozay Y, Keskinler B, Karagunduz A, Orhon D, Dizge N (2020) Effluent reuse potential of a dual-stage ceramic mbr coupled wiro treatment for textile wastewater. Desalin Water Treat 177:374–383. https://doi.org/10.5004/dwt.2020.25006
12. Brepols C, Dorgeloh E, Frechen F-B, Fuchs W, Haider S, Joss A, de Korte K, Ruiken C, Schier W, van der Roest H (2008) Upgrading and retrofitting of municipal wastewater treatment plants by means of membrane bioreactor (MBR) technology. Desalination 231:20–26
13. Brik M, Schoeberl P, Chamam B, Braun R, Fuchs W (2006) Advanced treatment of textile wastewater towards reuse using a membrane bioreactor. Process Biochem 41:1751–1757. https://doi.org/10.1016/j.procbio.2006.03.019
14. Carvalho JRS, Amaral FM, Florencio L, Kato MT, Delforno TP, Gavazza S (2020) Microaerated UASB reactor treating textile wastewater: the core microbiome and removal of azo dye Direct Black 22. Chemosphere 242:125157.https://doi.org/10.1016/j.chemosphere.2019.125157
15. Cerón-Vivas A, Morgan-Sagastume JM, Noyola A (2012) Intermittent filtration and gas bubbling for fouling reduction in anaerobic membrane bioreactors. J. Memb. Sci. 423–424:136–142. https://doi.org/10.1016/j.memsci.2012.08.008
16. Chen W, Mo J, Du X, Zhang Z, Zhang W (2019) Biomimetic dynamic membrane for aquatic dye removal. Water Res 151:243–251. https://doi.org/10.1016/j.watres.2018.11.078

17. Cheng J, Wu X, Jin B, Zhang C, Zheng R, Qin L (2021) Coupling of immobilized photosynthetic bacteria with a graphene oxides/PSF composite membrane for textile wastewater treatment: biodegradation performance and membrane anti-fouling behavior. Membranes (Basel) 11.https://doi.org/10.3390/membranes11030226
18. Ching KF (2010) Design and operation of MBR type sewage treatment plant at Lo Wu Correctional Institution , Hong Kong
19. Chittal V, Gracias M, Anu A, Saha P, Bhaskara Rao KV (2019) Biodecolorization and biodegradation of azo dye reactive orange-16 by marine nocardiopsis sp. Iran. J Biotechnol 17:18–26. https://doi.org/10.29252/ijb.1551
20. Choudhury AKR (2017) 10—Sustainable chemical technologies for textile production. In: Muthu SS (Ed), Sustainable fibres and textiles, the textile institute book series. Woodhead Publishing, pp 267–322. https://doi.org/10.1016/B978-0-08-102041-8.00010-X
21. De Jager D, Sheldon MS, Edwards W (2014) Colour removal from textile wastewater using a pilot-scale dual-stage MBR and subsequent RO system. Sep Purif Technol 135:135–144. https://doi.org/10.1016/j.seppur.2014.08.008
22. Doğruel S, Altun A, Çokgör EU, Insel G, Keskinler B, Orhon D (2021) Anatomy of the organic carbon in an industrial wastewater: implications of particle size distribution, respirometry and process modelling. Process Saf Environ Prot 146:257–266. https://doi.org/10.1016/j.psep.2020.09.002
23. Dow N, Villalobos García J, Niadoo L, Milne N, Zhang J, Gray S, Duke M (2017) Demonstration of membrane distillation on textile waste water assessment of long term performance, membrane cleaning and waste heat integration. Environ Sci Water Res Technol 3:433–449. https://doi.org/10.1039/c6ew00290k
24. Feng F, Xu Z, Li X, You W, Zhen Y (2010) Advanced treatment of dyeing wastewater towards reuse by the combined Fenton oxidation and membrane bioreactor process. J Environ Sci 22:1657–1665. https://doi.org/10.1016/S1001-0742(09)60303-X
25. Feng S, Zhong Z, Wang Y, Xing W, Drioli E (2018) Progress and perspectives in PTFE membrane: preparation, modification, and applications. J. Memb. Sci. 549:332–349. https://doi.org/10.1016/j.memsci.2017.12.032
26. Field RW, Pearce GK (2011) Critical, sustainable and threshold fluxes for membrane filtration with water industry applications. Adv Colloid Interface Sci 164:38–44. https://doi.org/10.1016/j.cis.2010.12.008
27. Friha I, Bradai M, Johnson D, Hilal N, Loukil S, Ben Amor F, Feki F, Han J, Isoda H, Sayadi S (2015) Treatment of textile wastewater by submerged membrane bioreactor: In vitro bioassays for the assessment of stress response elicited by raw and reclaimed wastewater. J Environ Manage 160:184–192. https://doi.org/10.1016/j.jenvman.2015.06.008
28. Gita S, Hussan A, Choudhury TG (2017) Impact of textile dyes waste on aquatic environments and its treatment. Environ Ecol 35:2349–2353
29. Grilli S, Piscitelli D, Mattioli D, Casu S, Spagni A (2011) Textile wastewater treatment in a bench-scale anaerobic-biofilm anoxic-aerobic membrane bioreactor combined with nanofiltration. J Environ Sci Heal Part A Toxic/Hazardous Subst. Environ Eng 46:1512–1518. https://doi.org/10.1080/10978526.2011.609078
30. Guadie A, Gessesse A, Xia S (2018) Halomonas sp. strain A55, a novel dye decolorizing bacterium from dye-uncontaminated Rift Valley Soda lake. Chemosphere 206:59–69. https://doi.org/10.1016/j.chemosphere.2018.04.134
31. Guo W, Ngo HH, Li J (2012) A mini-review on membrane fouling. Bioresour Technol 122:27–34. https://doi.org/10.1016/j.biortech.2012.04.089
32. Hai FI, Yamamoto K, Nakajima F, Fukushi K (2011) Bioaugmented membrane bioreactor (MBR) with a GAC-packed zone for high rate textile wastewater treatment. Water Res 45:2199–2206. https://doi.org/10.1016/j.watres.2011.01.013
33. Hai FI, Yamamoto K, Nakajima F, Fukushi K (2008) Removal of structurally different dyes in submerged membrane fungi reactor-Biosorption/PAC-adsorption, membrane retention and biodegradation. J Memb Sci 325:395–403. https://doi.org/10.1016/j.memsci.2008.08.006

34. Hoinkis J, Deowan SA, Panten V, Figoli A, Huang RR, Drioli E (2012) Membrane bioreactor (MBR) technology—a promising approach for industrial water reuse. Procedia Eng 33:234–241. https://doi.org/10.1016/j.proeng.2012.01.1199
35. Holkar CR, Jadhav AJ, Pinjari DV, Mahamuni NM, Pandit AB (2016) A critical review on textile wastewater treatments: Possible approaches. J Environ Manage 182:351–366. https://doi.org/10.1016/j.jenvman.2016.07.090
36. Holkar CR, Pandit AB, Pinjari DV (2014) Kinetics of biological decolorisation of anthraquinone based Reactive Blue 19 using an isolated strain of Enterobacter sp.F NCIM 5545. Bioresour Technol 173:342–351. https://doi.org/10.1016/j.biortech.2014.09.108
37. Hong H, Zhang M, He Y, Chen J, Lin H (2014) Fouling mechanisms of gel layer in a submerged membrane bioreactor. Bioresour Technol 166:295–302. https://doi.org/10.1016/j.biortech.2014.05.063
38. Hu Y, Cheng H, Ji J, Li Y-Y (2020) A review of anaerobic membrane bioreactors for municipal wastewater treatment with a focus on multicomponent biogas and membrane fouling control. Environ Sci Water Res Technol 6:2641–2663. https://doi.org/10.1039/D0EW00528B
39. Iorhemen OT, Hamza RA, Tay JH (2016) Membrane bioreactor (MBR) technology for wastewater treatment and reclamation: Membrane fouling. Membranes (Basel). 6:13–16. https://doi.org/10.3390/membranes6020033
40. Isik Z, Arikan EB, Bouras HD, Dizge N (2019) Bioactive ultrafiltration membrane manufactured from Aspergillus carbonarius M333 filamentous fungi for treatment of real textile wastewater. Bioresour Technol Reports 5:212–219. https://doi.org/10.1016/j.biteb.2019.01.020
41. J., N.N., Sandesh, K., K., G.K., Chidananda, B., Ujwal, P., (2019) Optimization of Direct Blue-14 dye degradation by Bacillus fermus (Kx898362) an alkaliphilic plant endophyte and assessment of degraded metabolite toxicity. J Hazard Mater 364:742–751. https://doi.org/10.1016/j.jhazmat.2018.10.074
42. Jegatheesan V, Pramanik BK, Chen J, Navaratna D, Chang CY, Shu L (2016) Treatment of textile wastewater with membrane bioreactor: a critical review. Bioresour Technol 204:202–212. https://doi.org/10.1016/j.biortech.2016.01.006
43. Jepsen KL, Bram MV, Hansen L, Yang Z, Lauridsen SMØ (2019) Online backwash optimization of membrane filtration for produced water treatment. Membranes (Basel). 9:1–18. https://doi.org/10.3390/membranes9060068
44. Ji J, Chen Y, Hu Y, Ohtsu A, Ni J, Li E, Sakuma S, Hojo T, Chen R, Li Y-Y, (2021a) One-year operation of a 20-L submerged anaerobic membrane bioreactor for real domestic wastewater treatment at room temperature: pursuing the optimal HRT and sustainable flux. Sci Total Environ 775:145799.https://doi.org/10.1016/j.scitotenv.2021.145799
45. Ji J, Kulshreshtha S, Kakade A, Majeed S, Li X, Liu P (2020) Bioaugmentation of membrane bioreactor with Aeromonas hydrophila LZ-MG14 for enhanced malachite green and hexavalent chromium removal in textile wastewater. Int Biodeterior Biodegrad 150:104939.https://doi.org/10.1016/j.ibiod.2020.104939
46. Ji J, Ni J, Ohtsu A, Isozumi N, Hu Y, Du R, Chen Y, Qin Y, Kubota K, Li Y-Y (2021b) Important effects of temperature on treating real municipal wastewater by a submerged anaerobic membrane bioreactor: removal efficiency, biogas, and microbial community. Bioresour Technol 336:125306. https://doi.org/10.1016/j.biortech.2021.125306
47. Ji J, Sakuma S, Ni J, Chen Y, Hu Y, Ohtsu A, Chen R, Cheng H, Qin Y, Hojo T, Kubota K, Li YY (2020) Application of two anaerobic membrane bioreactors with different pore size membranes for municipal wastewater treatment. Sci Total Environ 745:140903.https://doi.org/10.1016/j.scitotenv.2020.140903
48. Kalyani DC, Telke AA, Dhanve RS, Jadhav JP (2009) Ecofriendly biodegradation and detoxification of Reactive Red 2 textile dye by newly isolated Pseudomonas sp. SUK1. J Hazard Mater 163:735–742. https://doi.org/10.1016/j.jhazmat.2008.07.020
49. Khan S, Malik A (2016) Degradation of Reactive Black 5 dye by a newly isolated bacterium Pseudomonas entomophila BS1. Can J Microbiol 62:220–232. https://doi.org/10.1139/cjm-2015-0552

50. Khouni I, Louhichi G, Ghrabi A (2020) Assessing the performances of an aerobic membrane bioreactor for textile wastewater treatment: Influence of dye mass loading rate and biomass concentration. Process Saf Environ Prot 135:364–382. https://doi.org/10.1016/j.psep.2020.01.011
51. Kimura K, Uchida H (2019) Intensive membrane cleaning for MBRs equipped with flat-sheet ceramic membranes: Controlling negative effects of chemical reagents used for membrane cleaning. Water Res 150:21–28. https://doi.org/10.1016/j.watres.2018.11.030
52. Le-Clech P, Chen V, Fane TAG (2006) Fouling in membrane bioreactors used in wastewater treatment. J Memb Sci 284:17–53. https://doi.org/10.1016/j.memsci.2006.08.019
53. Le NL, Nunes SP (2016) Materials and membrane technologies for water and energy sustainability. Sustain Mater Technol 7:1–28. https://doi.org/10.1016/j.susmat.2016.02.001
54. Lee A, Elam JW, Darling SB (2016) Membrane materials for water purification: design, development, and application. Environ Sci Water Res Technol 2:17–42. https://doi.org/10.1039/c5ew00159e
55. Li K, Jiang C, Wang J, Wei Y (2016) The color removal and fate of organic pollutants in a pilotscale MBR-NF combined process treating textile wastewater with high water recovery. Water Sci Technol 73:1426–1433. https://doi.org/10.2166/wst.2015.623
56. Li K, Liu Q, Fang F, Wu X, Xin J, Sun S, Wei Y, Ruan R, Chen P, Wang Y, Addy M (2020) Influence of nanofiltration concentrate recirculation on performance and economic feasibility of a pilot-scale membrane bioreactor-nanofiltration hybrid process for textile wastewater treatment with high water recovery. J Clean Prod 261:121067.https://doi.org/10.1016/j.jclepro.2020.121067
57. Liao B-Q, Kraemer JT, Bagley DM (2006) Anaerobic membrane bioreactors: applications and research directions. Crit Rev Environ Sci Technol 36:489–530
58. Lin H, Peng W, Zhang M, Chen J, Hong H, Zhang Y (2013) A review on anaerobic membrane bioreactors: applications, membrane fouling and future perspectives. Desalination 314:169–188. https://doi.org/10.1016/j.desal.2013.01.019
59. Liu C, Caothien S, Hayes J, Caothuy T, Otoyo T, Ogawa T (2001) Membrane chemical cleaning : from art to science. Pall Corp Port Washington, NY, 11050
60. Lotito AM, De Sanctis M, Di Iaconi C, Bergna G (2014) Textile wastewater treatment: aerobic granular sludge vs activated sludge systems. Water Res 54:337–346
61. Lourenco ND, Novais JM, Pinheiro HM (2001) Effect of some operational parameters on textile dye biodegradation in a sequential batch reactor. J Biotechnol 89:163–174
62. Mahboubi A, Ylitervo P, Doyen W, De Wever H, Taherzadeh MJ (2016) Reverse membrane bioreactor: Introduction to a new technology for biofuel production. Biotechnol Adv 34:954–975. https://doi.org/10.1016/j.biotechadv.2016.05.009
63. Marcucci M, Nosenzo G, Capannelli G, Ciabatti I, Corrieri D, Ciardelli G (2001) Treatment and reuse of textile effluents based on new ultrafiltration and other membrane technologies. Desalination 138:75–82. https://doi.org/10.1016/S0011-9164(01)00247-8
64. Marrot B, Barrios-Martinez A, Moulin P, Roche N (2004) Industrial wastewater treatment in a membrane bioreactor: a review. Environ Prog 23:59–68. https://doi.org/10.1002/ep.10001
65. Martinez R, Ruiz MO, Ramos C, Camara JM, Diez V (2020). Comparison of external and submerged membranes used in anaerobic membrane bioreactors: fouling related issues and biological activity. Biochem Eng J 159:107558
66. Meng F, Chae SR, Drews A, Kraume M, Shin HS, Yang F (2009) Recent advances in membrane bioreactors (MBRs): Membrane fouling and membrane material. Water Res 43:1489–1512. https://doi.org/10.1016/j.watres.2008.12.044
67. Metzger U, Le-Clech P, Stuetz RM, Frimmel FH, Chen V (2007) Characterisation of polymeric fouling in membrane bioreactors and the effect of different filtration modes. J Memb Sci 301:180–189. https://doi.org/10.1016/j.memsci.2007.06.016
68. Mishra S, Mohanty P, Maiti A (2019) Bacterial mediated bio-decolourization of wastewater containing mixed reactive dyes using jack-fruit seed as co-substrate: process optimization. J Clean Prod 235:21–33. https://doi.org/10.1016/j.jclepro.2019.06.328

69. Mitra S, Daltrophe NC, Gilron J (2016) A novel eductor-based MBR for the treatment of domestic wastewater. Water Res 100:65–79. https://doi.org/10.1016/j.watres.2016.04.057
70. Muda K, Aris A, Salim MR, Ibrahim Z, Yahya A, van Loosdrecht MCM, Ahmad A, Nawahwi MZ (2010) Development of granular sludge for textile wastewater treatment. Water Res 44:4341–4350
71. Mustereţ CP, Teodosiu C (2007) Removal of persistent organic pollutants from textile wastewater by membrane processes. Environ Eng Manag J 6:175–187. https://doi.org/10.30638/eemj.2007.022
72. Najme R, Hussain S, Maqbool Z, Imran M, Mahmood F, Manzoor H, Yasmeen T, Shahzad T (2015) Biodecolorization of Reactive Yellow-2 by <I>Serratia</I> sp. RN34 Isolated from Textile Wastewater. Water Environ Res 87:2065–2075. https://doi.org/10.2175/106143015x14362865226031
73. Nilusha RT, Wang T, Wang H, Yu D, Zhang J, Wei Y (2020) Optimization of in situ backwashing frequency for stable operation of anaerobic ceramic membrane bioreactor. Processes 8.https://doi.org/10.3390/PR8050545
74. Okamura D, Mori Y, Hashimoto T, Hori K (2009) Identification of biofoulant of membrane bioreactors in soluble microbial products. Water Res 43:4356–4362. https://doi.org/10.1016/j.watres.2009.06.042
75. Olukanni OD, Osuntoki AA, Kalyani DC, Gbenle GO, Govindwar SP (2010) Decolorization and biodegradation of Reactive Blue 13 by Proteus mirabilis LAG. J Hazard Mater 184:290–298. https://doi.org/10.1016/j.jhazmat.2010.08.035
76. Padmanaban VC, Geed SRR, Achary A, Singh RS (2016) Kinetic studies on degradation of Reactive Red 120 dye in immobilized packed bed reactor by Bacillus cohnii RAPT1. Bioresour Technol 213:39–43. https://doi.org/10.1016/j.biortech.2016.02.126
77. Pretel R, Robles A, Ruano MV, Seco A, Ferrer J (2014) The operating cost of an anaerobic membrane bioreactor (AnMBR) treating sulphate-rich urban wastewater. Sep Purif Technol 126:30–38. https://doi.org/10.1016/j.seppur.2014.02.013
78. Qin L, Zhang G, Meng Q, Xu L, Lv B (2012) Enhanced MBR by internal micro-electrolysis for degradation of anthraquinone dye wastewater. Chem Eng J 210:575–584. https://doi.org/10.1016/j.cej.2012.09.006
79. Radjenović J, Matošić M, Mijatović I, Petrović M, Barceló D (2008) membrane bioreactor (MBR) as an advanced wastewater treatment technology. In: Barceló D, Petrovic M (Eds), Emerging contaminants from industrial and municipal waste: removal technologies. Springer Berlin Heidelberg, Berlin, Heidelberg, pp 37–101. https://doi.org/10.1007/698_5_093
80. Rondon H, El-Cheikh W, Boluarte IAR, Chang CY, Bagshaw S, Farago L, Jegatheesan V, Shu L (2015) Application of enhanced membrane bioreactor (eMBR) to treat dye wastewater. Bioresour Technol 183:78–85. https://doi.org/10.1016/j.biortech.2015.01.110
81. Roy U, Sengupta S, Banerjee P, Das P, Bhowal A, Datta S (2018) Assessment on the decolourization of textile dye (Reactive Yellow) using Pseudomonas sp. immobilized on fly ash: Response surface methodology optimization and toxicity evaluation. J Environ Manage 223:185–195. https://doi.org/10.1016/j.jenvman.2018.06.026
82. Saba B, Khalid A, Nazir A, Kanwal H, Mahmood T (2013) Reactive black-5 azo dye treatment in suspended and attach growth sequencing batch bioreactor using different co-substrates. Int Biodeterior Biodegrad 85:556–562. https://doi.org/10.1016/j.ibiod.2013.05.005
83. Sahinkaya E, Yurtsever A, Çınar Ö (2017) Treatment of textile industry wastewater using dynamic membrane bioreactor: Impact of intermittent aeration on process performance. Sep Purif Technol 174:445–454. https://doi.org/10.1016/j.seppur.2016.10.049
84. Saikia J, Sarmah S, Bora JJ, Das B, Goswamee RL (2019) Preparation and characterization of low cost flat ceramic membranes from easily available potters' clay for dye separation. Bull Mater Sci 42:1–13. https://doi.org/10.1007/s12034-019-1767-7
85. Samsami S, Mohamadi M, Sarrafzadeh MH, Rene ER, Firoozbahr M (2020) Recent advances in the treatment of dye-containing wastewater from textile industries: Overview and perspectives. Process Saf Environ Prot 143:138–163. https://doi.org/10.1016/j.psep.2020.05.034

86. Sathya U, Keerthi N, M., Balasubramanian, N., (2019) Evaluation of advanced oxidation processes (AOPs) integrated membrane bioreactor (MBR) for the real textile wastewater treatment. J Environ Manage 246:768–775. https://doi.org/10.1016/j.jenvman.2019.06.039
87. Sathya U, Keerthi P, Nithya M, Balasubramanian N (2021) Development of photochemical integrated submerged membrane bioreactor for textile dyeing wastewater treatment. Environ Geochem Health 43:885–896. https://doi.org/10.1007/s10653-020-00570-x
88. Seib MD, Berg KJ, Zitomer DH (2016) Low energy anaerobic membrane bioreactor for municipal wastewater treatment. J Memb Sci 514:450–457. https://doi.org/10.1016/j.memsci.2016.05.007
89. Shah PD, Dave SR, Rao MS (2012) Enzymatic degradation of textile dye Reactive Orange 13 by newly isolated bacterial strain Alcaligenes faecalis PMS-1. Int Biodeterior Biodegrad 69:41–50. https://doi.org/10.1016/j.ibiod.2012.01.002
90. Sharma SCD, Sun Q, Li J, Wang Y, Suanon F, Yang J, Yu CP (2016) Decolorization of azo dye methyl red by suspended and co-immobilized bacterial cells with mediators anthraquinone-2,6-disulfonate and Fe3O4 nanoparticles. Int Biodeterior Biodegrad 112:88–97. https://doi.org/10.1016/j.ibiod.2016.04.035
91. Smithers LG, Markrides M, Gibson RA (2010) Human milk fatty acids from lactating mothers of preterm infants: a study revealing wide intra- and inter-individual variation. Prostaglandins Leukot. Essent. Fat. Acids 83:9–13. https://doi.org/10.1016/j.plefa.2010.02.034
92. Spagni A, Casu S, Grilli S (2012) Decolourisation of textile wastewater in a submerged anaerobic membrane bioreactor. Bioresour Technol 117:180–185. https://doi.org/10.1016/j.biortech.2012.04.074
93. Stricot M, Filali A, Lesage N, Spérandio M, Cabassud C (2010) Side-stream membrane bioreactors: Influence of stress generated by hydrodynamics on floc structure, supernatant quality and fouling propensity. Water Res 44:2113–2124
94. Sun F, Sun B, Hu J, He Y, Wu W (2015) Organics and nitrogen removal from textile auxiliaries wastewater with A2O-MBR in a pilot-scale. J Hazard Mater 286:416–424. https://doi.org/10.1016/j.jhazmat.2015.01.031
95. Sun W, Chen J, Chen L, Wang J, Zhang Y (2016) Coupled electron beam radiation and MBR treatment of textile wastewater containing polyvinyl alcohol. Chemosphere 155:57–61. https://doi.org/10.1016/j.chemosphere.2016.04.030
96. Sun XY, Chu HQ, Zhang YL, Zhou XF (2012) Review on dynamic membrane reactor (DMBR) for municipal and industrial wastewater treatment. Adv Mater Res 455–456:1278–1284. https://doi.org/10.4028/www.scientific.net/AMR.455-456.1278
97. Tavangar T, Jalali K, Alaei Shahmirzadi MA, Karimi M (2019) Toward real textile wastewater treatment: Membrane fouling control and effective fractionation of dyes/inorganic salts using a hybrid electrocoagulation—Nanofiltration process. Sep Purif Technol 216:115–125. https://doi.org/10.1016/j.seppur.2019.01.070
98. Tian Z, Xin W, Song Y, Li F (2015) Simultaneous organic carbon and nitrogen removal from refractory petrochemical dry-spun acrylic fiber wastewater by hybrid A/O-MBR process. Environ Earth Sci 73:4903–4910. https://doi.org/10.1007/s12665-015-4210-4
99. van't Oever, R., (2005) MBR focus: is submerged best? Filtr Sep 42:24–27
100. Varjani S, Rakholiya P, Ng HY, You S, Teixeira JA (2020) Microbial degradation of dyes: an overview. Bioresour Technol 314.https://doi.org/10.1016/j.biortech.2020.123728
101. Visvanathan C (2009) Treatment of Industrial Wastewater by Membrane Bioreactors. EOLSS Publications, Water and Wastewateer Treatment Technologies-Volume II
102. Wang Y, Wang H, Wang X, Xiao Y, Zhou Y, Su X, Cai J, Sun F (2020) Resuscitation, isolation and immobilization of bacterial species for efficient textile wastewater treatment: a critical review and update. Sci Total Environ 730:139034.https://doi.org/10.1016/j.scitotenv.2020.139034
103. Wang Z, Ma J, Tang CY, Kimura K, Wang Q, Han X (2014) Membrane cleaning in membrane bioreactors: a review. J. Memb. Sci. 468:276–307. https://doi.org/10.1016/j.memsci.2014.05.060

104. Wang Z, Ma J, Tang CY, Kimura K, Wang Q, Han X (2014) Membrane cleaning in membrane bioreactors : a review. J Memb Sci 468:276–307
105. Watanabe R, Qiao W, Norton M, Wakahara S, Li Y-Y (2014) Recent developments in municipal wastewater treatment using anaerobic membrane bioreactor: a review. J Water Sustain 2:101–122. https://doi.org/10.11912/jws.4.2.101-122
106. Wen C, Huang X, Qian Y (1999) Domestic wastewater treatment using an anaerobic bioreactor coupled with membrane filtration. Process Biochem 35:335–340. https://doi.org/10.1016/S0032-9592(99)00076-X
107. Wu J, Le-Clech P, Stuetz RM, Fane AG, Chen V (2008) Effects of relaxation and backwashing conditions on fouling in membrane bioreactor. J Memb Sci 324:26–32. https://doi.org/10.1016/j.memsci.2008.06.057
108. Yan Z, Wang S, Kang X, Ma Y (2009) Pilot-scale hybrid coagulation/membrane bioreactor (HCMBR) for textile dyeing wastewater advanced treatment. 3rd Int. Conf Bioinforma Biomed Eng iCBBE 2009:1–4. https://doi.org/10.1109/ICBBE.2009.5163198
109. Yang X, Crespi M, López-Grimau V (2018) A review on the present situation of wastewater treatment in textile industry with membrane bioreactor and moving bed biofilm reactor. Desalin Water Treat 103:315–322. https://doi.org/10.5004/dwt.2018.21962
110. Yigit NO, Uzal N, Koseoglu H, Harman I, Yukseler H, Yetis U, Civelekoglu G, Kitis M (2009) Treatment of a denim producing textile industry wastewater using pilot-scale membrane bioreactor. Desalination 240:143–150. https://doi.org/10.1016/j.desal.2007.11.071
111. Yoon S-H (2015) Membrane bioreactor processes: principles and applications. CRC Press
112. Yurtsever A, Basaran E, Ucar D (2020a) Process optimization and filtration performance of an anaerobic dynamic membrane bioreactor treating textile wastewaters. J Environ Manage 273:111114.https://doi.org/10.1016/j.jenvman.2020.111114
113. Yurtsever A, Calimlioglu B, Sahinkaya E (2017) Impact of SRT on the efficiency and microbial community of sequential anaerobic and aerobic membrane bioreactors for the treatment of textile industry wastewater. Chem Eng J 314:378–387. https://doi.org/10.1016/j.cej.2016.11.156
114. Yurtsever A, Çinar Ö, Sahinkaya E (2016) Treatment of textile wastewater using sequential sulfate-reducing anaerobic and sulfide-oxidizing aerobic membrane bioreactors. J Memb Sci 511:228–237. https://doi.org/10.1016/j.memsci.2016.03.044
115. Yurtsever A, Sahinkaya E, Aktaş Ö, Uçar D, Çinar Ö, Wang Z (2015) Performances of anaerobic and aerobic membrane bioreactors for the treatment of synthetic textile wastewater. Bioresour Technol 192:564–573. https://doi.org/10.1016/j.biortech.2015.06.024
116. Yurtsever A, Sahinkaya E, Çınar Ö (2020b) Performance and foulant characteristics of an anaerobic membrane bioreactor treating real textile wastewater. J Water Process Eng 33.https://doi.org/10.1016/j.jwpe.2019.101088
117. Zhou L, Ou P, Zhao B, Zhang W, Yu K, Xie K, Zhuang WQ (2021a) Assimilatory and dissimilatory sulfate reduction in the bacterial diversity of biofoulant from a full-scale biofilm-membrane bioreactor for textile wastewater treatment. Sci Total Environ 772.https://doi.org/10.1016/j.scitotenv.2021.145464
118. Zhou L, Zhao B, Ou P, Zhang W, Li H, Yi S, Zhuang WQ (2021b) Core nitrogen cycle of biofoulant in full-scale anoxic & oxic biofilm-membrane bioreactors treating textile wastewater. Bioresour Technol 325:124667. https://doi.org/10.1016/j.biortech.2021.124667

Constructed Wetlands in Dye Removal

Chandra Wahyu Purnomo and Fitri Ramdani

Abstract Wastewater from textile industry contains many harmful substances in the form of organic and inorganic moieties. The characteristic of dye molecule itself is difficult to be degraded by conventional wastewater treatments. It is common that the color is persistent in the treated effluent even after sequential chemical, physical, and biological processes. Vegetations as living organism offer several advantages by their ability in sequestering the pollution into internal tissue and decomposing the complex aromatic dye molecule into less toxic metabolites by enzymatic mechanism. The main application of phytoremediation is by constructing artificial wetland. In a wetland system, not only vegetation is important but also other components such as bed media, wastewater flow pattern and also wetland configuration. Even though this natural process tends to be slow, constructed wetland (CW) can be adjusted and enhanced to be able to optimally treat challenging dye wastewater. Furthermore, CW is considered to be a robust system that can withstand with various parameter fluctuations of wastewater. In this chapter, CW will be comprehensively discussed from phytoremediation principles to wetland application to treat various dye using different configurations and vegetation types.

Keywords Phytoremediation · Constructed wetland · Dye removal · Vegetation · Enzymes · Decolorization · Phytodegradation

C. W. Purnomo (✉) · F. Ramdani
Chemical Engineering Department, Universitas Gadjah Mada, Jl. Grafika no 2, Bulaksumur, Mlati, Sleman, Yogyakarta, Indonesia 55281
e-mail: chandra.purnomo@ugm.ac.id

C. W. Purnomo
Agrotechnology Innovation Center UGM (PIAT UGM), Tanjungtirto, Kalitirto, Berbah, Sleman, Yogyakarta, Indonesia

A. Khadir et al. (eds.), *Biological Approaches in Dye-Containing Wastewater*, Sustainable Textiles: Production, Processing, Manufacturing & Chemistry,
https://doi.org/10.1007/978-981-19-0545-2_10

1 Introduction

Recently, more than 10,000 dyes products are used in the coloring and printing industry worldwide which many of them are toxic and carcinogenic [1]. About ten to fifteen percent of the dyes used in the textile dyeing process are lost and released into effluent stream [2]. Dyes can be included into recalcitrant pollutants which transformed or get dispersed and persist in the receiving environments. They are synthesized to resist deterioration by sunlight exposure, oxidizing agents, surfactants (soaps) and cannot be easily degraded by common wastewater treatment processes [3]. Most of the dye molecules have complex aromatic structures which makes them highly resistant to degradation [4]. One of the most used dyes is Azo dye that characterized by strong —N = N— with structural stability against biological and chemical degradation.

Dyes concentration in the textile industry wastewater is in the range from 10 to 200 mg/L [5]. The common characteristic of textile wastewater is high BOD and COD, high level of Total Dissolved Solids (TDSs) and Total Suspended Solids (TSSs), highly acidic or alkaline and intense in color. Textile wastewater also contains some dangerous heavy metals. Some dyes can contain copper or other metals as integrated part of its chromophore molecular structure for stronger attachment to the fabric [6].

Currently, phytoremediation is heavily explored as an alternative of textile wastewater treatment. Lower cost and a much greener approach of this technology open a wide possibility to be implemented in various remediation of various contaminations. Compare to a single chemical or physical agents, plants have a combination mechanism of absorption, accumulation, and enzymatic machinery for degrading complex molecular structures in different plant's part [4]. It is expected that the three removal mechanisms of the plant can be utilized to degrade recalcitrant dye molecules effectively. In this chapter, a short description of phytoremediation is provided in the first section, followed by the mechanism and factors affecting the constructed wetland system, and then the application of wetland to treat various dye wastewater.

2 Phytoremediation

Knowing the quantity and properties of textile wastewater above, it is necessary to adopt an environmentally friendly and sustainable treatment method. Phytoremediation, a pollutant removal method using plants, is among the low cost and sustainable approaches for waste treatment [7, 8]. Phytoremediation is a low energy process that powered by natural solar energy. Plants possess biodegradation ability by enzymatic mechanism and uptake processes that can break down organic pollutants or accumulate the toxic substances inside the plant tissue [4].

Detoxification and decolorization of dye molecules using vegetation may provide ecofriendly and cost-effective approach. The advantage of this process is solar driven nature, aesthetic side, durability, and flexibility of implementation on site. Direct application of this method at the polluted sites makes it more interesting and relatively cost-effective strategy over other conventional treatment methods [9–11].

Up until now, various plants have been applied to remediate dyes from textile wastewater. Each plants species have specific pollutants removal potential and stress tolerance to particular contaminants [12]. Rane et al. [13] and Kabra et al. [4] has listed many macrophyte plants that being used for phytoremediation of various dyes which are *Phragmites australis, Typha angustifolia, Thymus vulgaris, Blumea malcolmii, Rosmarinus officinalis, Portulaca grandifora, Typhonium fagelliforme, Ipomoea hederifolia, Aster amellus, Glandularia pulchella, Petunia grandifora, Zinnia angustifolia and Tagetes patula, B. juncea, and Sesuvium portulacastrum.* Table 1. shows the performance of selected plant species for phytodegradation of certain dyes.

From Table 1, it is shown that the treatment time is vary largely from 6 h to 30 d with low to medium dye initial concentration (20–70 mg/L). This removal performance implies that the phytodegradation of dye molecules is relatively slow processes. It requires a series of stages for dye molecule to be metabolized from dye uptake by the root system, then enzyme degradation inside the plant membrane and metabolites released back to the solution [4]. The vegetation selection for a specific dye removal also appears to be an important issue.

Plant enzymes have capacity to metabolize several toxic organic compounds to produce nontoxic simpler molecules. At least nine different enzymes have been identified taking a role in dye metabolism which are lignin peroxidase (LiP), laccase, dichlorophenolindophenol (DCIP) reductase, tyrosinase, azo reductase, veratryl alcohol oxidase (VAO), riboflavin reductase, catalase, and superoxide dismutase (SOD) [13]. Plants' oxidoreductases are the main biodegrading enzymes which break the complex dyes molecules by asymmetric and oxidative cleavage [21].

One of the real scale applications of phytoremediation to treat real industrial wastewater is using constructed wetland system (CW). In CW, vegetation plays an important role in the remediation of various pollutants. Indeed, several other factors are also important such as bed media, wastewater load and flow pattern and additional treatment such as aeration, electric discharge, and wastewater re-circulation.

3 Artificial Wetland System (Constructed Wetlands)

Constructed Wetlands are an artificial ecosystem to be used for sewage treatment mimicking of the water purification process that occurs in swamps (natural wetlands). Aquatic plants (hydrophyte) that grow in the area plays an important role in the process of recovery of the water quality. Wetlands are defined as a treatment system by areas that are saturated by water and support the life of aquatic plants. Various

Table 1 Plant species application for phytoremediation of dyes

Plant species	Dye (conc.)	Decolorization (contact time)	References
Sunflower *(Helianthus annuus L.)*	Evans Blue, Bismark Browny, and Orange G (100 mg/L)	62.64% (4 d)	Huicheng et al. [14]
Narrow-leaved cattail *(Typha angustifolia L.)*	Reactive Red 141 (20 mg/L)	60% (14 d)	Nilratnisakorn et al. [15]
duckweed *(L. minor)*	C.I Acid Blue 92 (20 mg/L)	75% (6 d)	Khataee et al. [16]
Tagetes patula L	Reactive Blue 160 (20 mg/L)	90% (4 d)	Patil and Jadhav[10]
Aster amellus L	Remazol Red (RR); a mixture of dyes (MD) and a textile effluent (TE) (20 mg/L)	RR 96% (60 h); MD 47% and TE 62% (60 h)	Khandare et al. [17]
Alternanthera philoxeroides Griseb	Remazol Red (RR) (70 mg/L)	100% (72 h)	Rane et al. [13]
Glandularia pulchella (Sweet) Tronc	Scarlet RR	97% (72 h)	Rane et al. [13]
Glandularia pulchella (Sweet) Tronc	Scarlet RR (SRR), Rubine GFL (RGFL), Brilliant Blue R (BBR), Navy Blue 2R (NB2R) and Red HE3B (RHE3B) (50 mg/L)	RGFL and SRR 100% (72 h) BBR 85%; RHE3B 55%; NB2R 60% (96 h)	Kabra et al. [4]
Duckweed (Lemna minor)	methylene blue (MB) (50 mg/L)	80,56% (24 h)	Imron et al. [18]
Petunia grandiflora	Mix of Brilliant Blue G, Direct Blue GLL, Rubin GFL, Scarlet RR, and Brown3 REL (50 mg/L)	76% (36 h)	Watharkar and Jadhav [11]
Gaillardia grandiflora	As above	62% (36 h)	Watharkar and Jadhav [11]
Portulaca grandiflora Hook	Navy Blue HE2R (NBHE2R) (20 mg/L)	98% (40 h)	Khandare et al. [17]
Pennywort *(Hydrocotyle vulgaris)*	Acid Blue 92 (AB92) (20 mg/L)	30% (6 h)	Vafaei et al. [19]
T. patula, A. amellus, P. grandiflora and G. grandiflora	Textile wastewater/effluent treatment plant	59, 50, 46 and 73%, respectively (30 d)	Chandanshive et al. [20]

processes may incur in the wetland which are sedimentation, filtration, gas transfer, adsorption, chemical, and biological processing.

Differ from natural swamps, artificial wetland system can be adjusted to meet several design parameters such as discharge rate, organic load, depth of media, type of plant for ensuring the effluent quality of the system. Furthermore, based on both technical and cost-effectiveness approaches, the system was preferred for the following reasons:

- The construction of wetlands systems is often cheaper than other alternative waste treatment systems.
- Low operational and maintenance costs and periodic operating time, no need to be continuous.
- This wetlands system has a high tolerance for fluctuations in wastewater discharge.
- Able to treat wastewater with different types of pollutants and concentrations.
- Allows for the implementation of reuse and recycling of water.

In general, there are two main types of constructed wetland, namely, the Surface Flow Constructed Wetland (SFCW) and the Sub-Surface Flow Constructed Wetland system or SSFCW. The major difference between the two systems is the surface water level. SFCW has water level above the ground level so that there is always free water flow on the surface, while SSFCW the water table is under the ground level. In most cases, SSFCW is preferred than SFCW, since SSFCW with no water above the ground can suppress odor and also mosquito breeding.

According to [22], wastewater treatment with the SSF system is recommended for several reasons: potential in processing many type of wastewater from domestic, agricultural, and some industrial wastes including heavy metals, high processing efficiency, and ease of planning, operation, and maintenance without high skills workers requirement.

3.1 Wastewater Purification Mechanism

According to Rana and Maiti [23] the mechanism for absorbing pollutants in artificial wetlands is generally through abiotic (physical and chemical) or biotic (microbial and plant) processes and a combination of the two processes. Abiotic mechanism is through settling and sedimentation, adsorption and absorption, oxidation and reduction, photodegradation/oxidation, and volatilization. Meanwhile, the biotic processes carried out by microbes and plants in wetlands include the following:

- Phytoextraction, the process of uptake of pollutants by roots system of the plant from soil or water and accumulation of the substance into plant biomass.
- Rhizofiltration, the process of absorbing, precipitating, and concentrating toxic substances from water or soil by plant roots system.

- Phytostabilization, the immobilization of pollutants in the substrate to contain the harmful substance on site preventing from being transported to the air or groundwater.
- Phytodegradation, the process of pollutant uptake, metabolization, and degradation to produce non-toxic metabolites with the help of enzymes generated by plants and its rhizospheric microorganisms.
- Phytovolatilization, the release of harmful substances into the air via plant surfaces (leaves) after taking up the pollutants.

3.2 Factors Affecting CW Performance

There are four main factors that affect the performance of the system [24], as described below.

1. Media

The main roles of the bed media in constructed wetlands are a place to grow for plants, media for the breeding of microorganisms, help the sedimentation process, assisting the absorption (adsorption) of odors from the gases of biodegradation, as well as a place for chemical transformation processes to occur, and storage of nutrients needed by plants.

The media commonly used in processing CW consist of soil, sand, and gravel. Variations in the use of media were also developed to support microbial-plant synergetic systems such as peats, biochar, husks, bagasse, zeolites, and others [25–27]. The combination of using media can also improve the performance of the constructed wetland. The adsorption capacity using a combination of porous media is higher than using only one medium. The level of permeability and hydraulic conductivity of the media greatly affects the residence time of wastewater, where sufficient contact time will provide an opportunity for better wastewater biodegradation, as well as oxygen released by plant roots.

2. Vegetations

Plants are the most important component that functions as a recycler of pollutants in wastewater to become biomass which has an economic impact and supplies oxygen to the bottom of the water or into the substrate with anaerobic conditions.

Broadly speaking, aquatic plants are divided into three groups emergent (rooted at the bottom while leaves and part of the stem sticking out of the water surface), submerged (floating below the water surface or drowning in the water column), and floating (free floating or rooted in the water). Aquatic plants rooted at the bottom can act as stabilizers for the bottom of the waters. Aquatic plants can also act as traps for organic matter in eutrophic waters. Certain aquatic plants have luxury uptake properties, which are active or able to absorb certain substances or nutrients in excess of their needs.

The emergent type of water plants is known as the most productive of all plants in the world. According to [22], when compared to other types of aquatic plants, the sticking type with tall, straight leaves requires less land, so that it can reduce land use and is more economical and ecologically efficient, which means that the sticking type of water plant is good when used. as a nutrient absorber. Aquatic plants which have hyper-tolerant properties against heavy metals are the key to the characteristics indicating their hyper-accumulator properties. Plants can be called hyper-accumulators if they have the following characteristics:

1. Plants have a rate of absorption of certain elements from their environment more than other types of plants.
2. Plants can tolerate a high level of an element in the root and canopy tissue.
3. Plants have a high translocation rate of heavy metals from root to canopy so that their accumulation in the canopy is higher than that of the roots.

The types of plants that are often used for subsurface artificial wetlands are water plants or plants that can survive in stagnant water (submerged plants or amphibious plants). Plants as phytoremediators in the constructed wetland processing system are the most important components capable of transforming nutrients through physical, chemical, and microbiological (biological) processes. Plants transfer oxygen through the roots and rhizome system to the bottom of the media and provide an underwater medium for microorganism's attachment. Plants are very influential on flow velocity, increase detention time and facilitate the deposition of suspended particles. The characteristics of plants used in the phytoremediation process should be able to grow rapidly, have high biomass, withstand to pollutants, and have a large nutrient uptake capacity.

3. Microorganisms

The microorganisms that are expected to grow and develop in wetland media are aerobic heterotrophic types, because processing takes place faster than anaerobic microorganisms. The activity of microorganisms in the constructed wetland processing system affects the performance of chemical and biological processes. In this system, microorganisms transform large amounts of organic and inorganic materials into innocuous or easily biodegradable materials. In addition, microorganisms also act as a medium for reducing and oxidizing (redox) reaction in changing the substrate content and affecting the wetland's ability, as well as a medium for recycling pollutants into nutrients.

4. Temperature

The temperature/temperature of wastewater will affect the activity of microorganisms and plants, so that it will affect the performance of wastewater treatment that enters the CW cell that will be used. Temperature will be able to influence the reaction, where every 10 °C temperature increase will increase the reaction 2–3 times faster. In addition, temperature is also a limiting factor for microorganism life. Although the limit of microorganism mortality is in a fairly wide temperature area (0–90 °C), the optimal life for each species has a certain range. Based on this, there are 3 (three) groups of microorganisms, namely,

- Psychrophilic Microorganisms (Optimal growth at 15 °C).
- Mesophyll microorganisms (optimal growth at 25–37 °C).
- Thermophile microorganisms (optimal growth at a temperature of 55–60 °C).

Considering climate conditions in tropical regions in which most textile industry located generally have a relatively small range of daily temperature differences (amplitude). In this region temperature is no longer a limiting factor, so that microbial life can be optimal throughout the year. Thus, phytoremediation in tropical countries using wetland is a promising method.

4 Constructed Wetland for Dye Wastewater Treatment

Actually, CW is not a new method in waste removal. Indeed, the application of CW to treat dye wastewater is a growing area. Several types of CW have been used to treat dye wastewater from the conventional horizontal flow CW to a combination of horizontal and vertical flow. The summary of CW usages for dye wastewater treatment is presented in Table 2. Types of CW that used for the dye wastewater treatment will be further discussed.

4.1 Free Water Surface Constructed Wetland (FWSCW)

Free water surface constructed wetland wastewater flows over the surface (Fig. 1). Since this type of CW has growth media in the bottom and water surface on the top, many options of plants can be used which are emergent plants (cattail, *P. australis*), submerged plants as well as floating plants (water hyacinth, lotus). However, as mentioned before this FWS type of CW is prone to odor emission and insect breeding that hinder the vast application especially closer to residential area. This type of CW is hardly reported to be used for dye wastewater. [28] reported the development of a real scale facility of FWS wetland in Gadoon Industrial Estate, Pakistan. The Gadoon FWSCW occupies more than 6000 m^2 of land area. The ponds are segregated by horizontal baffles that direct the wastewater flow into 11 separated ponds with meander pattern. The first pond is used for sedimentation pond in which 455 m^3 dye wastewater entering every day. The second pond is the deepest pond and acting as anaerobic degradation pond, while the third shallow pond is used for aerobic pond. There is no vegetation in the first three ponds. In ponds 4 and 8, *Typha latifolia* is dominant species while the rest are fully covered by *E. crassipes*. This unique pond configuration are able to retain heavy metals and COD up to 87% from high strength dye wastewater from more than 100 industries' effluents.

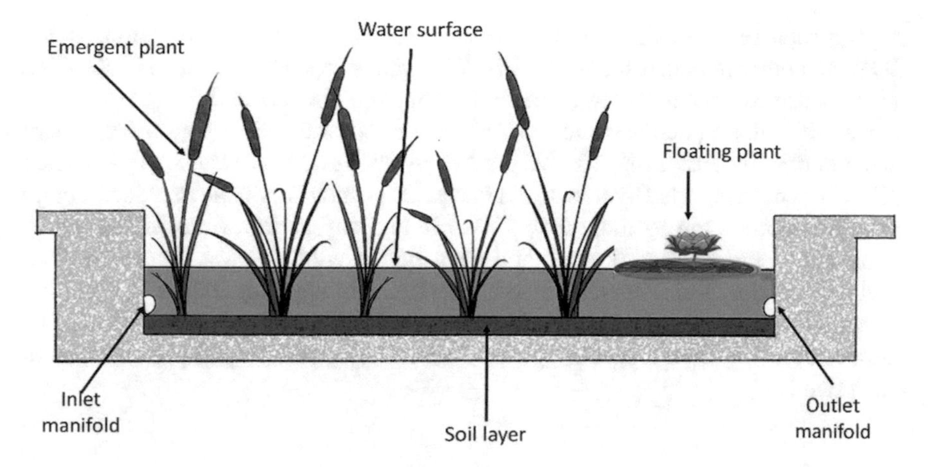

Fig. 1 Free water surface constructed wetland [23]

4.2 *Horizontal Flow Constructed Wetland (HFCW)*

The horizontal flow of CW is a simple treatment method with an ease of construction and operation. Most of this CW can be operated naturally utilizing gravity flow of wastewater through the bed media. The wastewater flows through in the media of the constructed wetland. The wastewater undergoes aerobic degradation at the rhizosphere (root zone) and the organic matter is digested anaerobically at lower zone. In this CW type only emergent plant can be cultivated such as Cattail and *Phragmites australis*. In general, HFCW requires larger land area than vertical flow (VF) CW since it has a low removal efficiency of pollutant, the detail will be discussed in the next section. Thus, to achieve high removal rate a vast area of CW with horizontal flow is needed.

A pilot-scale study of HFCW using real textile wastewater effluent has been reported by Hussain et al. [29]. Besides treatment using HFCW vegetated with *L. fusca.*, they also did bacteria inoculation to improve the performance. HFSW alone showed potential to treat the textile wastewater (80% for COD, 76% for BOD, 29% for TSS, and 76% for color), while bacterial incorporation further improved the pollutant removal, i.e., 86% for COD, 78% for BOD, 35% for TDS, and 90% for color.

4.3 *Vertical Flow Constructed Wetland (VFCW)*

Vertical flow constructed wetland controls wastewater flows vertically through the bed. Mostly the flow in VFCW is downward but it also can be upward. The removal efficiency of most pollution using VFCW is relatively larger than HFCW due to larger aerobic regime inside the media, therefore, achieved higher nitrification, BOD, COD, and other pollutant removals. VFCW requires less land area than HFCW but maintenance cost might be larger due to higher clogging potential.

Davies et al. [30] demonstrated a significant removal of acid orange 7 (AO7) dye concentration by 70% using VFCW planted by *Phragmites australis*. A pilot scale VFCW has been reported by Bedah and Faisal [31] which is suitable to reduce Congo red dye concentration by more than 90% within 5 days. Meanwhile, Up flow CW planted with *Phragmites australis* and Manchurian wild rice was chosen by Ong et al. [32] to treat Acid Orange 7 dye wastewater. By applying 3 days of retention time in a 70 cm length of tube with diameter of 18 cm, the dye concentration can be reduced up to 95%. To maintain upward flow of wastewater, a pump is supplied to the system.

4.4 Hybrid Constructed Wetland

Hybrid constructed wetland comprises multistage treatment by a combination of VFCW and HFCW using the separate pond. This treatment combination is pursued to optimize the removal efficiency of the both systems. For treating dye wastewater, usually VFCW is used in the first stage then followed by HFCW. This configuration is optimum for decolorizing the wastewater by 90% or above [33].

Another type of flow combination in CW system can be done within a pond by adding vertical baffles (Fig. 2). Tee et al. [27] modify HFCW by baffles which separated the pond into 5 segments with a sequential up flow and down flow pattern. By this modification high color removal more than 90% of Acid Orange 7 can be achieved by a retention time of 5 days (Table 2).

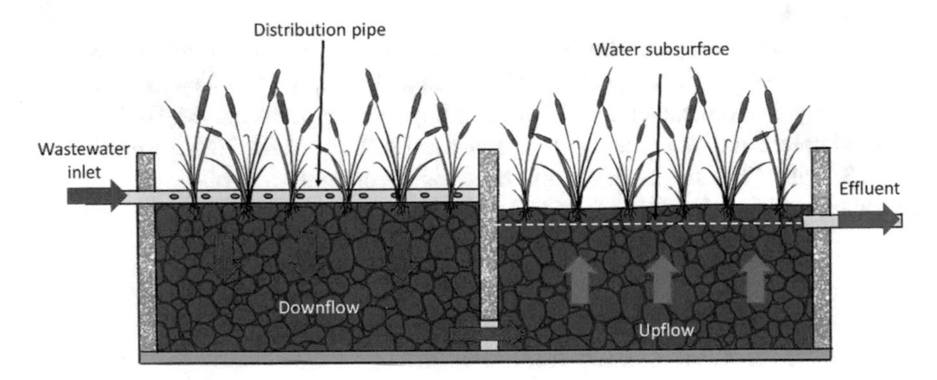

Fig.2 HFCW with baffle (Baffled Subsurface Flow CW) [27]

Table 2 Various CW configuration for dye wastewater treatment

Type of CW	Type of dye (strength)	plant	Pollutants (%removal)	References
FWS with baffles	Mixture dye (COD 2842 mg/L)	Cattail (*Typha latifolia*); Water hyacinth (*E. crassipes*)	COD (53%) Sulfate (88%) Sulfide (63%)	Nawab et al. [28]
HFCW	Textile wastewater (COD 483 mg/L)	*Leptochloa fusca*	COD 80% BOD 76% TSS 29% Color 76%	Hussain et al. [29]
HFCW	Textile dye waste (COD 167 mg/L)	*Prescaria barbata*	COD 43.11% Color 78.74%	Saba et al. [25]
HFCW with re-circulation	Amaranth azo dye (25 mg/L)	*Typha domingensis*	Color 92% COD 56% NO3 92% NH4 + 97%	Haddaji et al. [34]
HFCW with baffles	Acid Orange 7 (300 mg/L)	Cattails (*Typha latifolia*)	100% (5 day)	Tee et al. [27]
VFCW	Acid Orange 7	*P. australis*	Discoloration (71%)	Davies et al. [30]
VFCW	Acid Blue 113 (AB113) and Basic Red 46 (BR46))	*Phragmites australis*	Color 94% COD 82%,	Hussein and Scholz [35]
VFCW	Congo red dye (10 mg/L)	*Phragmites australis Typha domingensis*	dye 99.34 COD 86.0%	Bedah and Faisal [31]
Up flow constructed wetlands (UFCW)	Acid Orange 7 (50 mg/L)	Manchurian wild rice *Phragmites australis*	AO7 95% COD 78–82% NH4-N 41–48%	Ong et al. [32]
VF + HFCW (hybrid)	dye-rich textile wastewater	*P. australis*	COD (84%) BOD5 (66%) TOC (89%) Ntotal (52%) Norganic (87%) SO4 2 – (88%) TSS (93%) color (90%)	Bulc and Ojstršek [33]
VF + HFCW (hybrid)	Textile wastewater	VF section *P. australis, D. Sanderina, Asplenium Platyneuron* HF section *P. australis D. Sanderina*	BOD 79.2 COD 62.5, NH4-N 66.4, Turbidity 67.6 BOD 76.1 COD 69.7 NH4-N 42.1 Turbidity 68.1	Saeed and Sun [26]

5 Conclusion

CW method offers many advantages in dye wastewater treatment in terms of low cost and one-pot complete removal mechanism (biological, chemical, and physical). However, the real application should be carefully designed to meet any expected treatment performances since this natural biodegradation requires quite long retention time. The large option of flow pattern, type of vegetation, bed media, and additional treatment (aeration, bacterial incorporation) should be properly selected to achieve a high dye removal using CW system for a specific textile industrial effluent.

References

1. Javaid R, Qazi UY (2019) Catalytic oxidation process for the degradation of synthetic dyes: an overview. Int J Environ Res Public Health 16(11):1–27
2. Sponza DT, Işik M (2005) Toxicity and intermediates of C.I. Direct Red 28 dye through sequential anaerobic/aerobic treatment. Process Biochem 40(8):2735–2744
3. Seshadri S, Bishop PL, Agha AM (1994) Anaerobic/Aerobic treatment of selected azo dyes in wastewater. Waste Manag 14(2):127–137
4. Shertate RS, Thorat P (2014) Biotransformation of textile dyes: a bioremedial aspect of marine environment. Am J Environ Sci 10(5):489–499
5. Zille A (2005) Laccase reactions for textile applications Andrea Zille Laccase reactions for textile applications. Uminho 1–168
6. Kabra AN, Khandare RV, Waghmode TR, Govindwar SP (2012) Phytoremediation of textile effluent and mixture of structurally different dyes by Glandularia Pulchella (Sweet) tronc. Chemosphere 87(3):265–272
7. Ashraf S, Ali Q, Zahir ZA, Ashraf S, Asghar HN (2019) Phytoremediation: environmentally sustainable way for reclamation of heavy metal polluted soils. Ecotoxicol Environ Saf 174(November 2018):714–27
8. Anning AK, Akoto R (2018) Assisted phytoremediation of heavy metal contaminated soil from a mined site with Typha Latifolia and Chrysopogon Zizanioides. Ecotoxicol Environ Saf 148(October 2017):97–104
9. Khandare RV, Kabra AN, Tamboli DP, Govindwar SP (2011) The role of Aster amellus Linn. in the degradation of a sulfonated azo dye Remazol Red: a phytoremediation strategy. Chemosphere 82(8):1147–1154
10. Patil AV, Jadhav JP (2013) Evaluation of phytoremediation potential of Tagetes Patula L. for the Degradation of textile dye Reactive Blue 160 and assessment of the toxicity of degraded metabolites by cytogenotoxicity. Chemosphere 92(2):225–232
11. Watharkar AD, Jadhav JP (2014) Detoxification and decolorization of a simulated textile dye mixture by phytoremediation using petunia grandiflora and gailardia grandiflora: a plant-plant consortial strategy. Ecotoxicol Environ Saf 103(1):1–8
12. El-Esawi MA, Elkelish A, Soliman M, Elansary HO, Zaid A, Wani SH (2020) Serratia marcescens BM1 enhances cadmium stress tolerance and phytoremediation potential of soybean through modulation of osmolytes, leaf gas exchange, antioxidant machinery, and stress-responsive genes expression. Antioxidants 9(1):1–17
13. Huicheng X, Chongrong L, Jihong L, Li W (2012) Phytoremediatgion of wastewater containing azo dye by sunflowers and their photosynthetic response. Acta Ecol Sin 32(5):240–243
14. Nilratnisakorn S, Thiravetyan P, Nakbanpote W (2007) Synthetic reactive dye wastewater treatment by Narrow-Leaved Cattails (Typha Angustifolia Linn.): effects of dye, salinity and metals. Sci Total Environ 384(1–3):67–76

15. Khataee AR, Movafeghi A, Torbati S, Salehi Lisar SY, Zarei M (2012) Phytoremediation potential of duckweed (Lemna Minor L.) in degradation of C.I. Acid Blue 92: Artificial neural network modeling. Ecotoxicol Environ Saf 80:291–298
16. Rane NR, Chandanshive VV, Watharkar AD, Khandare RV, Patil TS, Pawar PK, Govindwar SP (2015) Phytoremediation of sulfonated Remazol Red dye and textile effluents by Alternanthera philoxeroides: an anatomical, enzymatic and pilot scale study. Water Res 83:271–281
17. Imron MF, Kurniawan SB, Soegianto A, Wahyudianto FE (2019) Phytoremediation of methylene blue using duckweed (Lemna Minor). Heliyon 5(8):e02206
18. Khandare RV, Kabra AN, Kurade MB, Govindwar SP (2011) Phytoremediation potential of portulaca grandiflora hook. (Moss-Rose) in degrading a sulfonated diazo reactive dye navy blue HE2R (Reactive Blue 172). Biores Technol 102(12):6774–6777
19. Vafaei F, Movafeghi A, Khataee A (2013) Evaluation of antioxidant enzymes activities and identification of intermediate products during phytoremediation of an anionic dye (C.I. Acid Blue 92) by Pennywort (Hydrocotyle Vulgaris). J Environ Sci (China) 25(11):2214–2222
20. Chandanshive VV, Kadam SK, Khandare RV, Kurade MB, Jeon BH, Jadhav JP, Govindwar SP (2018) In situ phytoremediation of dyes from textile wastewater using garden ornamental plants, effect on soil quality and plant growth. Chemosphere 210:968–976
21. Khandare RV, Govindwar SP (2015) Phytoremediation of textile dyes and effluents: current scenario and future prospects. Biotechnol Adv 33(8):1697–1714
22. Wijaya DS, Hidayat T, Dadan Sumiarsa TB, Kurnaeni B, Kurniadi D (2016) A review on subsurface flow constructed wetlands in tropical and sub-tropical countries. Open Sci J 1(2):1–11
23. Rana V, Maiti SK (2020) Municipal and industrial wastewater treatment using constructed wetlands. In: Smmaefsky BR (ed). Springer, Cham, pp 329–367
24. Crites R, Tchobanoglous G (1998) Small & decentralized wastewater management. First edit. McGraw-Hill, Singapore
25. Saba B, Jabeen M, Mahmood T, Aziz I (2014) Treatment comparison efficiency of microbial amended agro-waste biochar constructed wetlands for reactive black textile dye. In: International proceedings of chemical, biological and environmental engineering, vol 65, pp 12–16
26. Saeed T, Sun G (2013) A lab-scale study of constructed wetlands with sugarcane bagasse and sand media for the treatment of textile wastewater. Biores Technol 128:438–447
27. Tee HC, Lim PE, Seng CE, Nawi MAM, Adnan R (2015) Enhancement of azo dye Acid Orange 7 removal in newly developed horizontal subsurface-flow constructed wetland. J Environ Manag 147:349–355
28. Nawab B, Esser KB, Jenssen PD, Nyborg ILP, Baig SA (2018) Technical viability of constructed wetland for treatment of dye wastewater in Gadoon Industrial Estate, Khyber Pakhtunkhwa, Pakistan. Wetlands 38(6):1097–1105
29. Hussain Z, Arslan M, Malik MH, Mohsin M, Iqbal S, Afzal M (2018) Integrated perspectives on the use of bacterial endophytes in horizontal flow constructed wetlands for the treatment of liquid textile effluent: phytoremediation advances in the field. J Environ Manag 224(May):387–395
30. Davies LC, Carias CC, Novais JM, Martins-Dias S (2005) Phytoremediation of textile effluents containing azo dye by using phragmites australis in a vertical flow intermittent feeding constructed wetland. Ecol Eng 25(5):594–605
31. Bedah BJ, Faisal AAH (2020) Use of vertical subsurface flow constructed wetland for reclamation of wastewater contaminated with congo red dye. Plant Arch 20(2):8784–8792
32. Ong SA, Uchiyama K, Inadama D, Yamagiwa K (2009) Simultaneous removal of color, organic compounds and nutrients in azo dye-containing wastewater using up-flow constructed wetland. J Hazard Mater 165(1–3):696–703

33. Bulc TG, Ojstršek A (2008) The use of constructed wetland for dye-rich textile wastewater treatment. J Hazard Mater 155(1–2):76–82
34. Haddaji D, Ghrabi-Gammar Z, Hamed KB, Bousselmi L (2019) A re-circulating horizontal flow constructed wetland for the treatment of synthetic azo dye at high concentrations. Environ Sci Pollut Res 26(13):13489–13501
35. Hussein A, Scholz M (2017) Dye wastewater treatment by vertical-flow constructed wetlands. Ecol Eng 101:28–38

Printed in the United States
by Baker & Taylor Publisher Services